Lecture Notes in Physics

Founding Editors

Wolf Beiglböck

Jürgen Ehlers

Klaus Hepp

Hans-Arwed Weidenmüller

Volume 1050

Series Editors

Roberta Citro, Salerno, Italy

Peter Hänggi, Augsburg, Germany

Betti Hartmann, London, UK

Morten Hjorth-Jensen, Oslo, Norway

Maciej Lewenstein, Barcelona, Spain

Satya N. Majumdar, Orsay, France

Luciano Rezzolla, Frankfurt am Main, Germany

Angel Rubio, Hamburg, Germany

Stefan Theisen, Potsdam, Germany

James D. Wells, Ann Arbor, MI, USA

Gary P. Zank, Huntsville, AL, USA

The series Lecture Notes in Physics (LNP), founded in 1969, reports new developments in physics research and teaching - quickly and informally, but with a high quality and the explicit aim to summarize and communicate current knowledge in an accessible way. Books published in this series are conceived as bridging material between advanced graduate textbooks and the forefront of research and to serve three purposes:

- to be a compact and modern up-to-date source of reference on a well-defined topic;
- to serve as an accessible introduction to the field to postgraduate students and non-specialist researchers from related areas;
- to be a source of advanced teaching material for specialized seminars, courses and schools.

Both monographs and multi-author volumes will be considered for publication. Edited volumes should however consist of a very limited number of contributions only. Proceedings will not be considered for LNP.

Volumes published in LNP are disseminated both in print and in electronic formats, the electronic archive being available at springerlink.com. The series content is indexed, abstracted and referenced by many abstracting and information services, bibliographic networks, subscription agencies, library networks, and consortia.

Proposals should be sent to a member of the Editorial Board, or directly to the responsible editor at Springer:

Dr Lisa Scalone
lisa.scalone@springernature.com

Carlo F. Barenghi · Thomas Bland ·
Nick G. Parker

Quantum Fluids, Solitons, and Vortices

A Primer with Worked Problems

Second Edition

Carlo F. Barenghi
School of Mathematics, Statistics
and Physics
Newcastle University
Newcastle upon Tyne, UK

Nick G. Parker
School of Mathematics, Statistics
and Physics
Newcastle University
Newcastle upon Tyne, UK

Thomas Bland
Department of Physics
Lund University
Lund, Sweden

ISSN 0075-8450
ISSN 1616-6361   (electronic)
Lecture Notes in Physics
ISBN 978-3-032-20170-6
ISBN 978-3-032-20171-3   (eBook)
https://doi.org/10.1007/978-3-032-20171-3

Preface

In the decade which has elapsed since our *A Primer on Quantum Fluids* was first drafted, the field of quantum fluids has expanded considerably outside the original contexts of atomic Bose Einstein condensates and superfluid helium and the two communities of atomic physics and low temperature physics. For example, there have been remarkable applications of quantum fluids to quantum simulations of many-body systems in optical lattices and to astrophysics (from neutron stars to fuzzy dark matter to analogue gravity). These developments, and the success of the first edition of the "Primer" as a practical tool to help beginners to quickly start their own research projects, has led us to write a second upgraded and expanded edition.

The focus and the core material of this second edition are still the same: providing the necessary theoretical background information in order to quickly move on to one's research project. But some topics are now presented with more depth to facilitate the reader's journey into the expanded range of applications. Besides our own arbitrary tastes, the choice of topics has been guided more by the need to keep the material simple than by the aim of being exhaustive at all costs. We have thus greatly expanded the sections dealing with solitons and vortices, including quantum turbulence, and now also describe dipolar gases, droplets and two-component condensates.

The readers of the "Primer" liked the problems at the end of the chapters. Ten years ago no collection of problems in quantum fluids suitable to beginners was easily available. Since this is still the case, and since problem-solving is an essential part of learning, we have increased the number of problems, integrated them better with the text, and present the solutions.

Newcastle upon Tyne, UK

Carlo F. Barenghi
carlo.barenghi@newcastle.ac.uk

Lund, Sweden
Newcastle upon Tyne, UK
December 2025

Thomas Bland
Nick G. Parker
nick.parker@newcastle.ac.uk

Competing Interests The authors have no competing interests to declare that are relevant to the content of this manuscript.

Contents

Natural Constants

Bohr's radius $a_B = 5.29 \times 10^{-11}$ m
Bohr's magneton $\mu_B = 9.274 \times 10^{-24}$ J T^{-1}
Boltzmann's constant $k_B = 1.381 \times 10^{-23}$ m^2 Kg s^{-2} K^{-1}
Planck's constant $h = 6.626 \times 10^{-34}$ J s
Reduced Planck's constant $\hbar = h/(2\pi) = 1.055 \times 10^{-34}$ J s
Vacuum magnetic permeability $\mu_0 = 1.257 \times 10^{-6}$ Kg m s^{-2} A^{-2}
^{4}He mass $m = 6.644 \times 10^{-27}$ kg
^{4}He quantum of circulation $\kappa = h/m = 0.997 \times 10^{-7}$ m^{2}s^{-1}
^{4}He vortex core radius $a_0 \approx 10^{-10}$ m

Introduction

1

Abstract

Quantum fluids have emerged from scientific efforts to cool matter to colder and colder temperatures, representing staging posts towards absolute zero (Fig. 1.1). As a macroscopic manifestation of quantum mechanics they have contributed to our understanding of the quantum world, captivating and intriguing scientists with their counter-intuitive non-classical properties, and are increasingly finding applications in quantum technologies. Here we summarize the background of the two main quantum fluids to date, superfluid helium and atomic Bose-Einstein condensates.

1.1 Towards Absolute Zero

The nature of cold has intrigued humankind. Its explanation as a primordial substance, *primum frigidum*, prevailed from the ancient Greeks until Robert Boyle pioneered the scientific study of the cold in the mid 1600 s. Decrying the "almost totally neglect" of the nature of cold, he set about hundreds of experiments which systematically disproved the ancient myths and seeded our modern understanding. While working on an air-based thermometer in 1703, French physicist Guillaume Amontons observed that air pressure was proportional to temperature; extrapolating towards zero pressure led him to predict an "absolute zero" of approximately -240 °C in today's units, not far from the modern value of -273.15 °C (or 0 K). The implication was profound: the realm of the cold was much vaster than anyone had dared believe. An entertaining account of low temperature exploration is given by Ref. [1].

The liquefaction of the natural gases became the staging posts as low temperature physicists, with increasingly complex apparatus, raced to explore the undiscovered territories of the "map of frigor". Chlorine was liquefied at 239 K in 1823, and oxygen and nitrogen at $T = 90$ K and 77 K, respectively, in 1877. In 1898 the English physicist James Dewar liquefied what was believed to be the only remaining

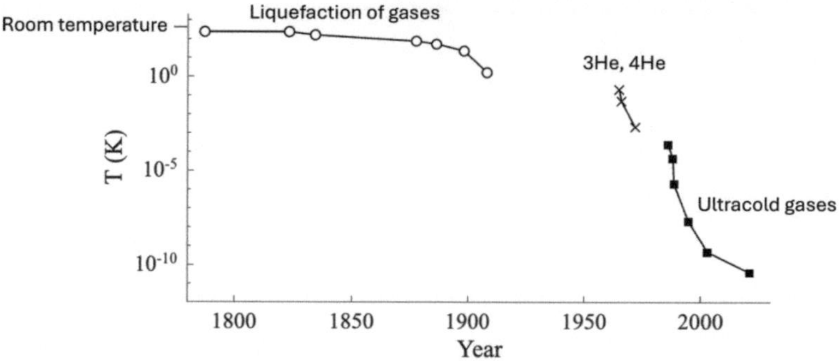

Fig. 1.1 Timeline of some of the coldest temperatures achieved in gases and liquids

elementary gas, hydrogen, at 23 K, helped by his invention of the vacuum flask. Concurrently, however, chemists discovered helium on Earth. Although helium is the second most common element in the Universe and known to exist in the Sun, its presence on Earth is tiny. With helium's even lower boiling point, a new race was on. A dramatic series of lab explosions and a lack of helium supplies meant that Dewar's main competitor, Heike Kamerlingh Onnes, pipped him to the post, liquefying helium at 4 K in 1908. This momentous achievement led to Onnes being awarded the 1913 Nobel Prize in Physics.

1.1.1   Discovery of Superconductivity and Superfluidity

These advances enabled scientists to probe the fundamental behaviour of materials at the depths of cold. Electricity was widely expected to grind to a halt as in this limit. Using liquid helium to cool mercury, Onnes instead observed its resistance to simply vanish below 4 K. *Superconductivity*, the flow of electrical current without resistance, has since been observed in many materials, at up to 130 K, and has found applications in medical MRI scanners, particle accelerators and levitating "maglev" trains.

Onnes and his co-workers also observed unusual behaviour in liquid helium itself. At around 2.2 K its heat capacity undergoes a discontinuous change, termed the "lambda" transition due to the shape of the curve. Since such behaviour is characteristic of a phase change, the idea developed that liquid helium existed in two phases: helium I for $T > T_\lambda$ and helium II for $T < T_\lambda$, where T_λ is the critical temperature. Later experiments revealed helium II to have unusual properties, such as it remaining a liquid even as absolute zero is approached, the ability to move through extremely tiny pores and the reluctance to boil. These two liquid phases, and the fact that helium remains liquid down to $T \to 0$ (at atmospheric pressure), mean that the phase diagram of helium (Fig. 1.2) is very different to a conventional liquid (inset). In 1938, landmark experiments by Allen and Misener and by Kapitza revealed the most striking property of helium II: its ability to flow without viscosity. The amazing internal mobility of the fluid, analogous to superconductors, led Kapitza to coin the

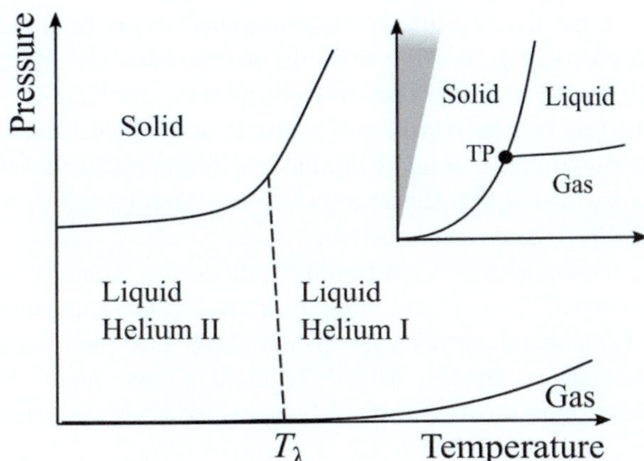

Fig. 1.2 Phase diagram of helium. For a conventional substance (inset), there exists a *triple point* (TP), where solid, liquid and gas coexist. Helium lacks such a point. The shaded region illustrates where Bose-Einstein condensation is predicted to occur for an ideal gas

term "superfluid". Other strange observations followed, including "fluid creep" (the ability of helium to creep up the walls of a vessel and over the edge) and the "fountain effect" (generation of a persistent fountain when heat was applied to the liquid).

1.1.2 Bose-Einstein Condensation

Superfluidity and superconductivity were at odds with classical physics and required a new way of thinking. In 1938 London resurrected an obscure 1925 prediction of Einstein to explain superfluidity. Considering an ideal gas of quantum particles, Einstein (having developed the ideas put forward by Bose for photons) had predicted the effect of *Bose-Einstein condensation*, that at low temperatures a large proportion of the particles would condense into the same quantum state—the *condensate* - and the remainder of the particles would behave conventionally. This idea stalled, however, since the conditions for this gaseous phenomena lay in the solid region of the pressure-temperature diagram (shaded region in Fig. 1.2(inset)), making it inaccessible. We will follow Einstein's derivation in Chap. 2. Einstein's model predicts a discontinuity in the heat capacity, suggestively similar to that observed in helium. This, in turn, led to the development of the successful two-fluid model by Tisza and Landau, in which helium-II is regarded as a combination of a viscosity-free superfluid and a viscous "normal fluid".

Bose-Einstein condensation applies to bosons (particles with integer spin, such as photons and ^{4}He atoms), but not to fermions (particles with half-integer spin, such as protons, neutrons and electrons). The Pauli exclusion principle prevents more than one identical fermion occupying the same quantum state. How then could Bose-Einstein condensation be responsible for the flow of electrons in superconductivity? The answer, put forward in 1957 by Bardeen, Cooper and Schrieffer was for the electrons to form *Cooper pairs*; these composite bosons could then undergo Bose-Einstein condensation. The observation of superfluidity in the fermionic helium isotope ^{3}He in 1972 (at around 2 mK) further cemented this pairing mechanism. More information on superconductivity can be found in Ref. [2].

Superfluid helium and superconductors are both manifestations of Bose-Einstein condensation. Arising from the macroscopic quantum state that is the condensate, they represent fluids governed by quantum mechanics, i.e. quantum fluids (superconductors can be considered as fluids of charged Cooper pairs). However, the strong particle interactions in liquids and solids mean that these systems are much more complicated that Einstein's ideal-gas paradigm, and it took until the 1990s for such an ideal state to be created.

Hallmarks of superfluidity include the capacity to flow without viscosity, the presence of a critical velocity above which superflow breaks down, the presence of quantised vortices, persistent flow, and macroscopic tunneling in the form of Josephson currents. We will detail all of these superfluid phenomena throughout this book, with the exception of Josephson currents which can be studied elsewhere [2].

1.2   Ultracold Quantum Gases

1.2.1   Laser Cooling and Magnetic Trapping

Liquids have since been cooled down to milliKelvin using cryogenic refrigeration techniques [3]. Meanwhile, the cooling of gases was advanced greatly by *laser cooling*, developed in the 1980s [4]. Atoms and molecules in a gas are in constant random motion with an average speed related to temperature, for example, around 300 m/s in room temperature air. For a laser beam incident upon a gas of atoms (in a vacuum chamber), and under certain conditions, the photons in the beam can be made to impart, on average, momentum to atoms travelling towards the beam, thus slowing them down in that direction; applying laser beams in multiple directions then allows three-dimensional (3D) cooling. In 1985 this "optical molasses" produced a gas at $240\,\mu$K, with average atom speeds of ~ 0.5 m/s. A few years later, $2\,\mu$K was achieved (~ 1 cm/s). These vapours were extremely dilute, with typical number densities of $n \sim 10^{20}$ m^{-3} (c.f. $n \sim 10^{28}$ m^{-3} for room temperature air); this made the transition from a gas to a solid, the natural process at such cold temperatures (inset of Fig. 1.2), so slow as to be insignificant on the experimental timescales. In addition, magnetic fields allowed the creation of traps, bowl-like potentials to confine the atoms and keep them away from hot surfaces; with experimental advances, it is now possible to create such ultracold gases in a variety of configurations, from toruses to periodic potentials, and manipulate them in time. The development of laser cooling and magnetic trapping techniques was recognised with the 1997 Nobel Prize in Physics [5]; further details of these techniques can be found elsewhere [4,6].

1.2.2   Bose-Einstein Condensate à la Einstein

The achievement of ultracold gases put Einstein's gaseous condensate within sight and a new race was on. Einstein's model predicted the condensate to form below a critical temperature $T_{\rm c} \sim 10^{-19} n^{2/3}$, but the low gas densities employed predicted

$T_c \sim 1\,\mu\mathrm{K}$, colder than achievable by laser cooling alone. To cool even further, a stage of *evaporative cooling* was employed whereby the hottest atoms were selectively removed, just like how evaporation cools a cup of coffee.

In 1995 Cornell and Wieman cooled a gas of rubidium atoms down to 200 nanoKelvin (200 billionths of a degree above absolute zero) to realize the first gaseous Bose-Einstein condensate (BEC) [7]. Figure 1.3 shows the famous experimental signature of this new state of matter. These images were obtained by releasing the trap which confines the gas, thus letting the atoms to fly away and the gas to expand. Above T_c (left plot), the gas was an energetic "thermal" gas of atoms characterised by a wide distribution of speed; upon opening the trap, atoms with large speeds moved far away, hence the broad picture in the left plot. As the temperature was cooled through T_c, a narrow distribution emerged from the thermal gas (middle and right plots), characteristic of accumulation of atoms into a state of almost zero energy and speed; these atoms are the Bose-Einstein condensate. We derive these thermal and condensate profiles in Chap. 2. A few months later, Ketterle independently formed a BEC of sodium atoms [8]. Seventy years on, Einstein's prediction had been realized at the depths of absolute zero. Cornell, Wieman and Ketterle shared the 2001 Nobel Prize for this landmark achievement [9].

There are now over 100 BEC experiments worldwide. These gases are typically 10–100 micro-meters across (about the width of a human hair), exist in the temperature range 1 to 100 nK, contain $10^3 - 10^9$ atoms, and are many times more dilute than room temperature air. At the time of going to press, the coldest temperature ever engineered is a 38 picoKelvin BEC [10]. BECs are most commonly formed with rubidium (^{87}Rb) and sodium (^{23}Na) atoms, but many other atomic species, and a growing number of molecular species, have been condensed, as shown in the periodic table in Fig. 1.4. It is also possible to create multi-component condensates, where two or more condensates co-exist. These gases constitute the purest and simplest

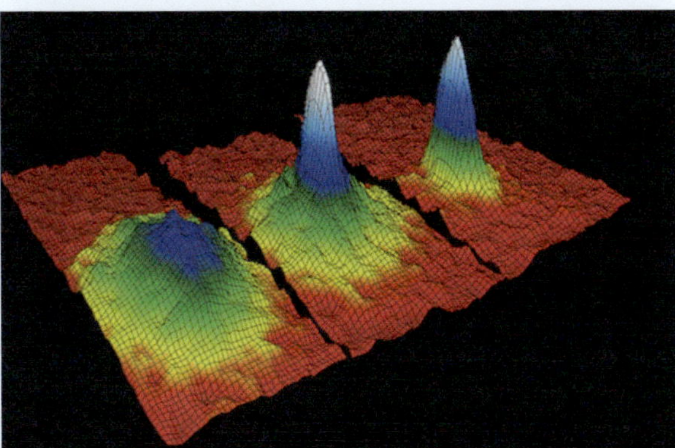

Fig. 1.3 The first observation of a gas Bose-Einstein condensate [7], showing the momentum distribution of a dilute gas of ^{87}Rb atoms, confined in a harmonic trap. As the temperature was reduced, the gas changed from a broad, energetic thermal gas (left) to a narrower distribution (right), characteristic of the condensate. Image reproduced from the NIST Image Gallery (Reference NIST/JILA/CU-Boulder)

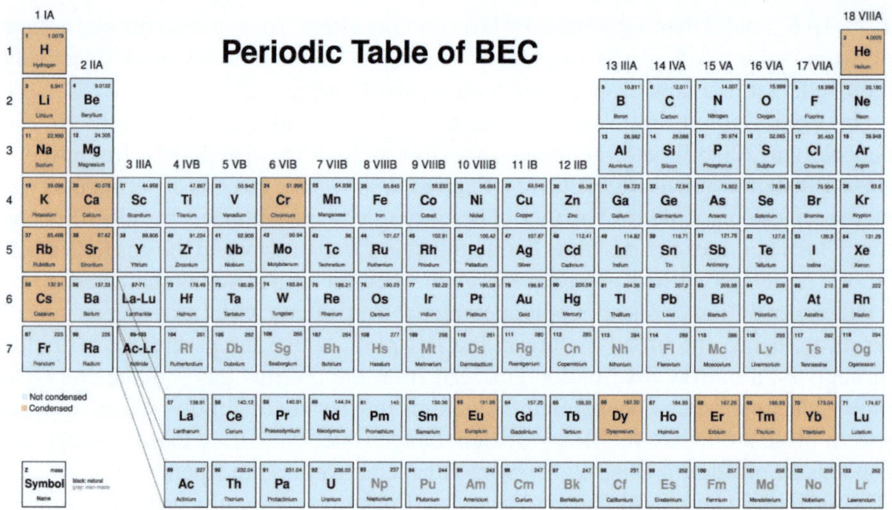

Fig. 1.4 Elements that have been condensed into gaseous Bose-Einstein condensates (orange) [7, 8, 11–25]

quantum fluids available, with typically 99% of the atoms lying in the condensed state. The last property makes condensates amenable to first-principles modelling; the work-horse model is provided by the Gross-Pitaevskii equation, which will be introduced and analysed in Chap. 3. Gaseous condensates have remarkable properties, such as superfluidity. Unlike superfluid helium (see Chap. 6, the interaction between the atoms is very weak, which makes them very close to Einstein's original concept of an ideal gas.

1.2.3 Degenerate Fermi Gases

For a fermionic gas, cooled towards absolute zero, the particles (in the absence of Cooper pairing) are forbidden to enter the same quantum state by the Pauli exclusion principle. Instead, they are expected fill up the quantum states, from the ground state upwards, each with unit occupancy. This effect was observed in 1999 when a degenerate Fermi gas was formed by cooling potassium (^{40}K) atoms to below 300 nK [26]. In this limit, the gas was seen to saturate towards a relatively wide distribution, indicating the higher average energy of the system, relative to a BEC. The Pauli exclusion principle exerts a very strong "pressure" against further contraction, an effect which is believed to stabilize neutron stars against collapse. A striking experimental comparison between bosonic and fermionic gases as the temperature is reduced is shown in Fig. 1.5: the distribution of the fermionic system cannot contract as the bosonic one. Experiments with Fermi gases have been able to precisely probe Cooper pairing, the essential process that underpins fermionic superfluidity and superconductivity, including direct observation of the pairs [27,28].

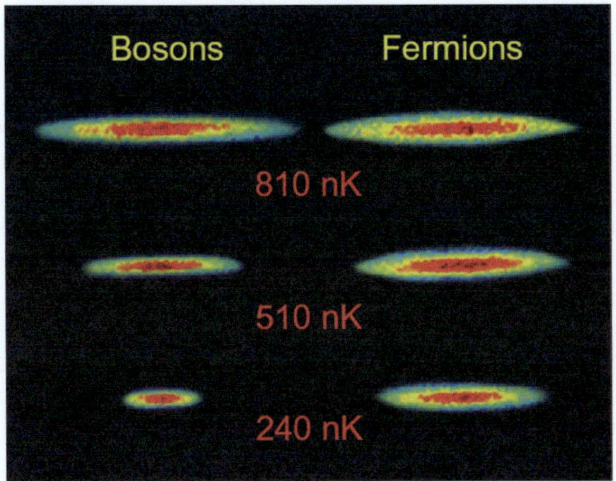

Fig. 1.5 Change in density profile as a ^{7}Li bosonic gas and a ^{6}Li fermionic gas are cooled towards absolute zero. The bosonic gas reduces to a narrow distribution corresponding to the low-energy condensate, while the fermionic gas saturates to a larger distribution due to the outwards Pauli pressure imposed by the fermions. Reproduced from www.apod.nas.gov with permission from A. G. Truscott and R. G. Hulet, and corresponding to the experiment of Ref. [29]

1.3   Quantum Fluids Today

We have briefly told the story of the discoveries of superfluid helium and atomic condensates, but what about the wider implications of these discoveries and the current status of the field? Here we list some examples.

Many-body quantum systems: Quantum fluids embody quantum behaviour on a macroscopic scale of many many particles; it is this property that gives rise to their remarkable properties. As such, quantum fluids provide fundamental insight into quantum many-body physics [30]. Moreover, for the case of condensates, the experimental capacity to engineer the system, e.g. its interactions, dimensionality, and the presence of disorder and periodicity, allows the controlled investigation of diverse many-body scenarios and emulation of complex condensed matter systems such as superconductors.

Nonlinear systems: Quantum fluids represent a prototype fluid, free from viscosity (as we see in Chap. 3) and whose vorticity is constrained to take the form of discrete, uniformly-sized mini-tornadoes. Turbulence, a notorious difficult problem of nonlinear statistical physics, assumes new aspects in quantum fluids: we discuss this *quantum turbulence* [31,32] in Chap. 10. Condensates also provide an idealized system to study nonlinear phenomena, such as solitons, which we shall meet in Chap. 5. The atomic interactions in a condensate give rise to a well-defined nonlinearity, and experimental tricks allow this nonlinearity to be controlled in size and nature (e.g. local versus non-local nonlinearity). Nonlinear effects such as solitons and four-wave mixing have been experimentally studied in quantum fluids.

Quantum fluids with enriched particle interactions: It is possible to create quantum fluids within which the particle-particle interactions are multiplex and give rise to new phenomena and ways to control the fluids. For example, by combining different quantum fluid components, their mutual interaction leads to regimes of miscibility and immiscibility, supporting new soliton and vortex states, and driving atypical instabilities [33], as will be discussed in Chap. 11. A further example are quantum fluids composed of particles with strong dipolar interactions, such as with atoms with strong magnetic dipole moments [34] discussed in Chap. 12 or with dipolar molecules; this adds an additional anisotropic interaction which can be controlled via external fields and opens the door to studying long-range dipolar effects in strongly correlated liquids.

Quantum liquids and droplets: Conventional atomic condensates are dominated by the averaged, mean-field behaviour of the particles, masking the underlying quantum fluctuations. However, by careful tuning of the interactions, it is possible to suppress mean-field effects and elevate the quantum fluctuations such that they dominate the system. This gives rise to an unusual liquid-like state which is highly dilute yet is highly incompressible due to the strong repulsive effects of quantum fluctuations [36]. This regime is of much fundamental interest and will be explored in Chap. 13—it can support intriguing phenomena such as self-bound droplets and supersolid states [35], and provides a lens to probe and understand the nature of quantum fluctuations.

Extra-terrestrial phenomena: Condensates are analogous to curved space-time, and have been used to simulate quantum field theory [37] and effects such as analog black holes [38,39], Hawking radiation [40], superradiance[41]. Both condensates and helium provide analogs of the quantum vacuum believed to permeate the universe and be responsible for its development from the Big Bang. Moreover, the capture and release of superfluid vortices from pinning sites can form avalanches emulating neutron star glitches [42]. Finally, condensates are currently being used to model dark matter galactic halos [43]. All these phenomena are not experimentally accessible on Earth but, using condensates, can be mimicked and explored in controlled, laboratory-based experiments.

Cooling: The excellent thermal transport property of helium II lends to its use as a coolant; helium is therefore present in superconducting systems, from MRI machines in hospitals to the Large Hadron Collider at CERN [44].

Sensors: Condensates are easily affected by external forces, and experiments have demonstrated extreme sensitivity to magnetic fields, gravity and rotational forces. Considerable efforts are currently underway to develop these ideas into next-generation sensors, for applications such as testing fundamental laws of physics, geological mapping and navigation [45].

Since 2000, Bose-Einstein condensation has also been achieved in several new systems: magnons (magnetic quasi-particle) in magnetic insulators, polaritons (coupled light-matter quasi-particles) in semiconductor microcavities, and photons in optical microcavities. In particular, the latter two systems have realized quantum fluids of light, with superfluid properties.

Problems

1.1 The energy per nucleon in the particle beam accelerated by CERN's Large Hadron Collider is 6.5TeV. Express this energy in Kelvin degrees.

References

1. T. Shachtman, *Absolute Zero and the Conquest of Cold* (Houghton-Mifflin, Boston, 2001)
2. J.F. Annett, *Superconductivity* (Superfluids and Condensates Oxford University Press, Oxford, 2004)
3. F. Pobell, *Matter and Methods at Low Temperatures*, 3rd edn. (Springer, Berlin, 2007)
4. H.J. Metcalf, P. van der Straten, *Laser Cooling and Trapping*, Graduate Texts in Contemporary Physics (Springer, Berlin, 2001)
5. The Nobel Prize in Physics (1997). www.nobelprize.org/nobel_prizes/physics/laureates/1997/
6. C.J. Pethick, H. Smith, *Bose-Einstein Condensation in Dilute Gases* (Cambridge University Press, Cambridge, 2008)
7. M.H. Anderson et al., Science **269**, 198 (1995)
8. K.B. Davis et al., Phys. Rev. Lett. **75**, 3969 (1995)
9. The Nobel Prize in Physics (2001). http://www.nobelprize.org/nobel_prizes/physics/laureates/2001/
10. C. Deppner et al., Phys. Rev. Lett. **127**, 100401 (2021)
11. C.C. Bradley, C.A. Sackett, J.J. Tollett, R.G. Hulet, Phys. Rev. Lett. **75**, 1687 (1995)
12. C.C. Bradley, C.A. Sackett, R.G. Hulet, Phys. Rev. Lett. **78**, 985 (1997)
13. D.G. Fried, T.C. Killian, L. Willmann, D. Landhuis, S.C. Moss, D. Kleppner, T.J. Greytak, Phys. Rev. Lett. **81**, 3811 (1998)
14. A. Robert, O. Sirjean, A. Browaeys, J. Poupard, S. Nowak, D. Boiron, C.I. Westbrook, A. Aspect, Science **292**, 461 (2001)
15. F. Pereira Dos Santos, J. Léonard, J. Wang, C.J. Barrelet, F. Perales, E. Rasel, C.S. Unnikrishnan, M. Leduc, C. Cohen-Tannoudji, Phys. Rev. Lett. **86**, 3459 (2001)
16. G. Modugno, G. Ferrari, G. Roati, R.J. Brecha, A. Simoni, M. Inguscio, Science **294**, 1320 (2001)
17. T. Weber, J. Herbig, M. Mark, H.-C. Nägerl, R. Grimm, Science **299**, 232 (2003)
18. Y. Takasu, K. Maki, K. Komori, T. Takano, K. Honda, M. Kumakura, T. Yabuzaki, Y. Takahashi, Phys. Rev. Lett. **91**, 040404 (2003)
19. A. Griesmaier, J. Werner, S. Hensler, J. Stuhler, T. Pfau, Phys. Rev. Lett. **94**, 160401 (2005)
20. S. Kraft, F. Vogt, O. Appel, F. Riehle, U. Sterr, Phys. Rev. Lett. **103**, 130401 (2009)
21. S. Stellmer, M.K. Tey, B. Huang, R. Grimm, F. Schreck, Phys. Rev. Lett. **103**, 200401 (2009)
22. M. Lu, N.Q. Burdick, S.H. Youn, B.L. Lev, Phys. Rev. Lett. **107**, 190401 (2011)
23. K. Aikawa, A. Frisch, M. Mark, S. Baier, A. Rietzler, R. Grimm, F. Ferlaino, Phys. Rev. Lett. **108**, 210401 (2012)

24. E.T. Davletov, I.N. Gurtovoi, A.V. Dolgikh, A.V. Bel'kov, D.A. Tayurskii, A.I. Safonov, Phys. Rev. A **102**, 011302(R) (2020)
25. Y. Miyazawa, R. Inoue, H. Matsui, G. Nomura, M. Kozuma, Phys. Rev. Lett. **129**, 223401 (2022)
26. B. DeMarco, D.S. Jin, Science **285**, 1703 (1999)
27. K. Levin, R.G. Hulet, *The Fermi Gases and Superfluids: Experiment and Theory* in *Ultracold Bosonic and Fermionic Gases*, ed. by K. Levin, A.L. Fetter, D.M. Stamper-Kurn (Elsevier, Oxford, 2012)
28. M. Holten et al., Nature **606**, 287 (2022)
29. A.G. Truscott, K.E. Strecker, W.I. McAlexander, G.B. Partridge, R.G. Hulet, Science **291**, 2570 (2001)
30. I. Bloch, J. Dalibard, W. Zwergefr, Rev. Mod. Phys. **88**, 885 (2008)
31. C.F. Barenghi, L. Skrbek, K. Sreenivasan, *Quantum Turbulence* (Cambridge University Press, 2023)
32. M. Tsubota, K. Kasamatsu, *Quantum Hydrodynamics and Turbulence* (Oxford University Press, 2025)
33. C. Baroni, G. Lamporesi, M. Zaccanti, Nat. Rev. Phys. **6**, 736 (2024)
34. T. Lahaye, C. Menotti, L. Santos, M. Lewenstein, T. Pfau, Rep. Prog. Phys. **72**, 126401 (2009)
35. A. Recati, S. Stringari, Nat. Rev. Phys. **5**, 735 (2023)
36. F. Bottcher et al., Rep. Prog. Phys. **84**, 012403 (2021)
37. K. Falque, A. Delhom, Q. Glorieux, E. Giacobino, A. Bramati, M.J. Jacquet, Phys. Rev. Lett. **135**, 023401 (2025)
38. H.S. Nguyen, D. Gerace, I. Carusotto, D. Sanvitto, E. Galopin, A. Lemaître, I. Sagnes, J. Bloch, A. Amo, Phys. Rev. Lett. **114**, 036402 (2015)
39. P. Svancara, P. Smaniotto, L. Solidoro, J.F. MacDonald, S. Patrick, R. Gregory, C.F. Barenghi, S. Weinfurtner, Nature **628**, 66 (2024)
40. J.R. Munoz de Nova, K. Golubkov, V.I. Kolobov, J. Steinhauer, Nature **569**, 688 (2019)
41. S. Inouye, A.P. Chikkatur, D.M. Stamper-Kurn, J. Stenger, D.E. Pritchard, W. Ketterle, Science **285**, 571 (1999)
42. I.-K. Liu, A.W. Baggaley, C.F. Barenghi, T.S. Wood, Ap. J. **984**, 83 (2025)
43. I.-K. Liu, N.P. Proukakis, G. Rigopoulos, MNRAS **521**, 3625 (2023)
44. S.W. Van Sciver *Helium Cryogenics*, 2nd ed. (Springer, Berlin, 2015)
45. C.L. Degen, F. Reinhard, P. Cappellaro, Rev. Mod. Phys. **89**, 035002 (2017)

Classical and Quantum Ideal Gases

Abstract

Bose and Einstein's prediction of Bose-Einstein condensation came out of their theory for how quantum particles in a gas behaved, and was built on the pioneering statistical approach of Boltzmann for classical particles. Here we follow Boltzmann, Bose and Einstein's footsteps, leading to the derivation of Bose-Einstein condensation for an ideal gas and its key properties.

2.1 Introduction

Consider the air in the room around you. We ascribe properties such as temperature and pressure to characterise it, motivated by our human sensitivity to these properties. However, the gas itself has a much finer level of detail, being composed of specks of dust, molecules and atoms, all in random motion. How can we explain the macroscopic, coarse-grained appearance in terms of the fine scale behaviour? An exact classical approach would proceed by solving Newton's equation of motion for each particle, based on the forces it experiences. For a typical room (volume $\sim 50\,\mathrm{m}^3$, air particle density $\sim 2 \times 10^{25}\,\mathrm{m}^{-3}$ at room temperature and pressure) this would require solving around 10^{28} coupled ordinary differential equations, an utterly intractable task. Since the macroscopic properties we experience are *averaged* over many particles, a particle-by-particle description is unnecessarily complex. Instead it is possible to describe the fine-scale behaviour *statistically* through the methodology of statistical mechanics. By specifying rules about how the particles behave and any physical constraints (boundaries, energy, etc.), the most likely macroscopic state of the system can be deduced.

We develop these ideas for an ideal gas of N identical and non-interacting particles, with temperature T and confined to a box of volume $\mathcal{V}$. The system is isolated, with no energy or particles entering or leaving the system[1] Our aim is to predict

---

[1] In the formalism of statistical mechanics, this is termed the *microcanonical ensemble*.

C. F. Barenghi et al., *Quantum Fluids, Solitons, and Vortices*, Lecture Notes in Physics 1050, https://doi.org/10.1007/978-3-032-20171-3_2

the equilibrium state of the gas. After performing this for classical (point-like) particles, we extend it to quantum (blurry) particles. This leads directly to the prediction of Bose-Einstein condensation of an ideal gas. In doing so, we follow the seminal works of Boltzmann, Bose and Einstein. Further information can be found in an introductory statistical physics textbook, e.g., [1] or [2].

2.2   Classical Particles

The state of a classical particle is specified by its position $\mathbf{r}$ and momentum $\mathbf{p}$. In the 3D Cartesian world, this requires six coordinates (x, y, z, p_x, p_y, p_z). Picturing the world as an abstract six dimensional *phase space*, the instantaneous state of the particle is a point in this space, and traces out a trajectory as it evolves. Accordingly, an N-particle gas is specified by N points/trajectories in this phase space. The accessible range of phase space is determined by the box (which provides a spatial constraint) and the energy of the gas (which determines the maximum possible momentum). Figure 2.1 (left) illustrates two particle trajectories in 1D phase space (x, p_x).

Classically, a particle's state (its position and momentum) can be determined to arbitrary precision. As such, classical phase space is continuous and contains an infinite number of accessible states. This also implies that each particle can be independently tracked, that is, that they are *distinguishable* from each other.

2.3   Ideal Classical Gas

We develop an understanding of the macroscopic behaviour of the gas from these microscopic rules (particle distinguishability, continuum of accessible states) following the pioneering work of Boltzmann in the late 1800s on the kinetic theory of gases. Boltzmann's work caused great controversy, as its particle and statistical basis was at odds with the accepted view of matter as being continuous and deterministic.

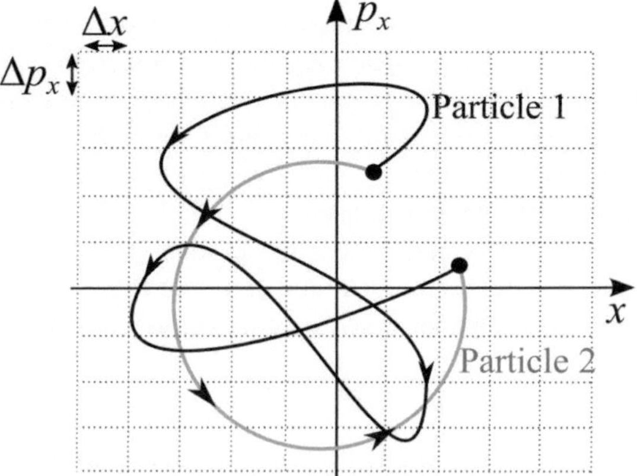

Fig. 2.1 Two different classical particle trajectories through 1D phase space (x, p_x), with the same initial and final states. While classical phase space is a continuum of states, it is convenient to imagine phase space to be discretised into finite-sized cells, here with size Δp_x and Δx

To overcome the practicalities of dealing with the infinity of accessible states, we imagine phase space to be discretised into cells of finite (but otherwise arbitrary) size, as shown in Fig. 2.1, and our N particles to be distributed across them randomly. Let there be I accessible cells, each characterised by its average momentum and position. The number of particles in the ith cell - its *occupancy number* - is denoted as N_i. The number configuration across the whole system is specified by the full set of occupancy numbers $\{N_1, N_2, \ldots, N_I\}$. We previously assumed that the total particle number is conserved, that is,

$$N = \sum_{i=1}^{i=I} N_i.$$

Conservation of energy provides a further constraint; for now, however, we ignore energetic considerations.

2.3.1 Macrostates, Microstates and the Most Likely State of the System

The macroscopic, equilibrium state of the gas is revealed by considering the ways in the particles can be distributed across the cells. In the absence of energetic constraints, each cell is equally likely to be occupied. Consider two classical particles, A and B (the distinguishability of the particles is equivalent to saying we can label them), and three such cells. The nine possible configurations, shown in Fig. 2.2, are termed *microstates*. Six distinct sets of occupancy numbers are possible, $\{N_1, N_2, N_3\} = \{2, 0, 0\}, \{0, 2, 0\}, \{0, 0, 2\}, \{1, 1, 0\}, \{1, 0, 1\}$ and $\{0, 1, 1\}$; these are termed *macrostates*. Each macrostate may be achieved by one or more microstates.

The particles are constantly moving and interacting/colliding with each other in a random manner, such that, after a sufficiently long time, they will have visited all available microstates, a process termed *ergodicity*. It follows that each microstate is equally likely (the assumption of "equal a priori probabilities"). Thus the most probable macrostate of the system is the one with the most microstates. In our example, the macrostates $\{1, 1, 0\}, \{1, 0, 1\}$ and $\{0, 1, 1\}$ are most probable (having 2 microstates each). In a physical gas, each macrostate corresponds to a particular macroscopic

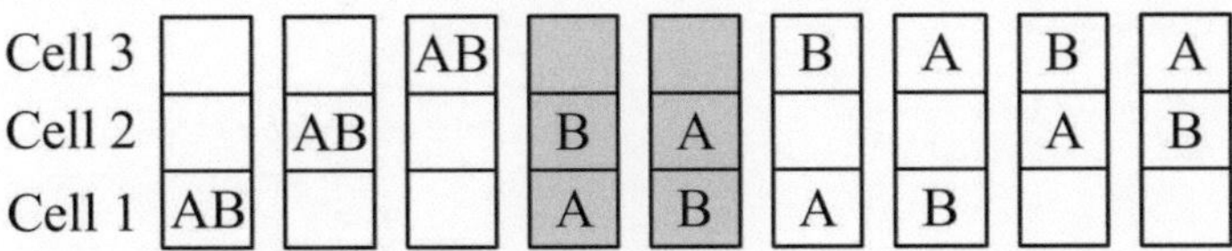

Fig. 2.2 Possible configurations of two classical particles, A and B, across three equally-accessible cells. If we treat the energies of cells 1–3 as 0, 1 and 2, respectively, and require that the total system energy is 1 (in arbitrary units), then only the shaded configurations are possible

appearance, e.g. a certain temperature, pressure, etc. Hence, these abstract probabilistic notions become linked to the most likely macroscopic appearances of the gas.

For a more general macrostate $\{N_1, N_2, N_3, \ldots, N_I\}$, the number of microstates is,

$$W = \frac{N!}{\prod_i N_i!} \tag{2.1}$$

Invoking the principle of equal a priori probabilities, the probability of being in the jth macrostate is,

$$\Pr(j) = \frac{W_j}{\sum_j W_j}. \tag{2.2}$$

W_j, and hence Pr(j), is maximised for the most even distribution of particles across the cells. This is true when each cell is equally accessible; as we discuss next, energy considerations modify the most preferred distribution across cells.

2.3.2   The Boltzmann Distribution

In the ideal-gas-in-a-box, each particle carries only *kinetic energy* $p^2/2m = (p_x^2 + p_y^2 + p_z^2)/2m$. Having discretised phase space, particle energy also becomes discretised, forming the notion of energy levels (familiar from quantum mechanics). This is illustrated in Fig. 2.3 for (x, p_x) phase space. Three energy levels, $E_1 = 0$, $E_2 = p_1^2/2m$ and $E_3 = p_2^2/2m$, are formed from the five momentum values $(p = 0, \pm p_1, \pm p_2)$. In two- and three-spatial dimensions, cells of energy E_i fall on circles and spherical surfaces which satisfy $p_x^2 + p_y^2 = 2mE_i$ and $p_x^2 + p_y^2 + p_z^2 = 2mE_i$, respectively. The lowest energy state E_1 is the *ground state*; the higher energy states are *excited states*.

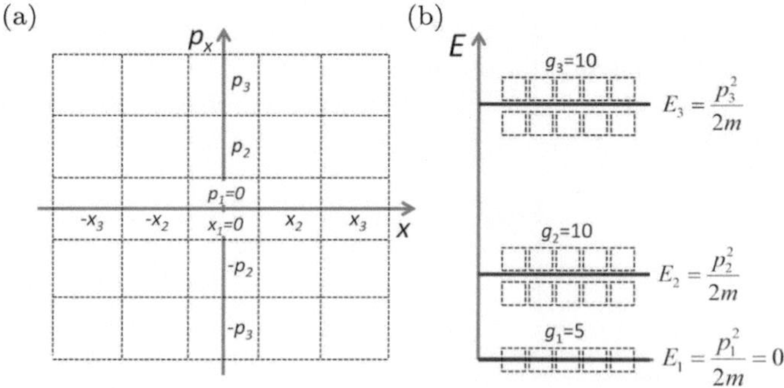

Fig. 2.3 For the phase space (x, p_x) shown in (**a**), the discretisation of phase space, coupled with the energy-momentum relation $E = p^2/2m$, leads to the formation of (**b**) energy levels. The degeneracy g of the levels is shown

The total energy of the gas U is,

$$U = \sum_i N_i E_i,$$

where E_i is the energy of cell i. Taking U to be conserved has important consequences for the microstates and macrostates. For example, imposing some arbitrary energy values in Fig. 2.2 restricts the allowed configurations. Particle occupation at high energy is suppressed, skewing the distribution towards low energy.

For a system at thermal equilibrium with a large number of particles, one macrostate (or a very narrow range of macrostates) will be greatly favoured. The preferred macrostate can be analytically predicted by maximising the number of microstates W with respect to the set of occupancy numbers $\{N_1, N_2, N_3, \ldots N_I\}$; details can be found in, e.g. [1,2]. The result is,

$$N_i = f_B(E_i), \tag{2.3}$$

where $f_B(E)$ is the famous Boltzmann distribution,

$$f_B(E) = \frac{1}{e^{(E-\mu)/k_B T}}. \tag{2.4}$$

The Boltzmann distribution tells us the most probable spread of particle occupancy across states in an ideal gas, as a function of energy. This is associated with the thermodynamic equilibrium state. Here k_B is Boltzmann's constant (1.38×10^{-23} m^2 kg s^{-2} K^{-1}) and T is temperature (in Kelvin degrees, K). On average, each particle carries kinetic energy $\frac{3}{2}k_B T$ ($\frac{1}{2}k_B T$ in each direction of motion); this property is referred to as the *equipartition theorem*.

The Boltzmann distribution function f_B is normalised to the number of particles, N, as accommodated by the chemical potential μ. Writing $A = e^{\mu/k_B T}$ gives $f_B = A/e^{E/k_B T}$, evidencing that A, and thereby μ, controls the amplitude of the distribution function.

The Boltzmann distribution function $f_B(E)$ is plotted in Fig. 2.4. Low energy states (cells) are highly occupied, with diminishing occupancy of higher energy states. As the temperature and hence the thermal energy increases, the distribution broadens as particles can access, on average, higher energy states. Remember, however, that this is the most *probable* distribution. Boltzmann's theory allows for the possibility, for example, that the whole gas of molecules of air in a room concentrates into a corner of the room. Due to the strong statistical bias towards an even distribution of energy, momenta and position, such an occurrence has incredibly low probability, but it is nonetheless possible, a fact which caused great discomfort with the scientific community at the time.

It is often convenient to work in terms of the occupancy of *energy levels* rather than *states* (phase space cells). To relate the Boltzmann result to energy levels, we must take into account the number of states in a given energy level, termed the *degeneracy*

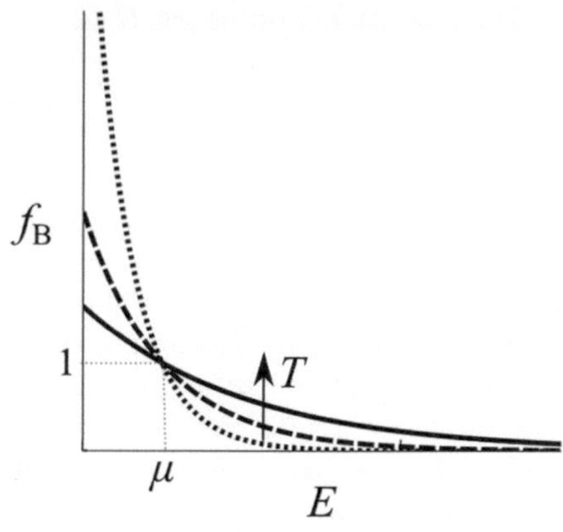

Fig. 2.4 The Boltzmann distribution function $f_B(E)$ for 3 different temperatures (the direction of increasing temperature is indicated)

and denoted g_j (we reserve i as the labelling of states). The occupation of the jth energy level is then,

$$N_j = g_j f_B(E_j). \tag{2.5}$$

2.4   Quantum Particles

Having introduced classical particles, their statistics and the equilibrium properties of the ideal gas, now we turn to the quantum case. The statistics of quantum particles, developed in the 1920s, was pivotal to the development of quantum mechanics, predating the well-known Schrödinger equation and uncertainty principle.

2.4.1   A Chance Discovery

Quantum physics arose from the failure of classical physics to describe the emission of radiation from a black body in the ultraviolet range (the "ultraviolet catastrophe"). In 1900, Max Planck discovered a formula which empirically fit the data for all wavelengths and led him to propose that energy is emitted in discrete quanta of units hf (h being Planck's constant and f the radiation frequency). Einstein extended this idea with his 1905 prediction that the light itself was quantised.

The notion of quantum particles was discovered by accident. Around 1920, the Indian physicist Satyendra Bose was giving a lecture on the failure of the classical theory of light using statistical arguments; a subtle mistake led to him prove the opposite. Indeed, he was able to derive Planck's empirical formula from first principles, based on the assumptions that a) the radiation particles are indistinguishable and b) phase space was discretised into cells of size h^3. Bose struggled at first to get these results published and sought support from Nobel Laureate Einstein; Bose's paper "Planck's law and the light quantum hypothesis" was then published in 1924 [3]. Soon after Einstein extended the idea to particles with mass in the paper "Quantum theory of the monoatomic ideal gas" [4].

The division of phase space was mysterious. Bose wrote "Concerning the kind of subdivision of this type, nothing definitive can be said", while Einstein confided in a colleague that Bose's "derivation is elegant but the essence remains obscure". It is now established as a fundamental property of particles, consistent with de Broglie's notion of wave-particle duality (that particles are smeared out, over a lengthscale given by the de Broglie wavelength $\lambda_{dB} = h/p$) and with Heisenberg's uncertainty principle (that the position and momentum of a particle have an inherent uncertainty $\Delta x \, \Delta y \, \Delta z \, \Delta p_x \, \Delta p_y \, \Delta p_z = h^3$). Each cell represents a distinct quantum state. The indistinguishability of particles follows since it becomes impossible to distinguish two blurry particles in close proximity in phase space.

2.4.2 Bosons and Fermions

Quantum particles come in two varieties—*bosons* and *fermions*:

Fermions   Soon after Bose and Einstein's work, Fermi and Dirac developed *Fermi-Dirac statistics* for fermions. Fermions possess half-integer spin, and include electrons, protons and neutrons. Fermions obey the Pauli exclusion principle (Pauli, 1925), which states that two identical fermions cannot occupy the same quantum state simultaneously.

Bosons   Bosons obey Bose-Einstein statistics, as developed by Bose and Einstein (above), and include photons and the Higgs boson. Bosons have integer spin, and since spin is additive, composite bosons may be formed from equal numbers of fermions, e.g. ^{4}He, ^{87}Rb and ^{23}Na. Unlike fermions, any number of bosons can occupy the same quantum state simultaneously.

The indistinguishability of quantum particles, and the different occupancy rules for bosons and fermions, affect their statistical behaviour. Consider 2 quantum particles across 3 cells, as shown in Fig. 2.5. Since the particles are indistinguishable, we can no longer label them. For bosons there are six microstates; for fermions there are only three (compared to nine for classical particles, Fig. 2.2). The relative probability of paired states to unpaired states is $\frac{1}{3}$, $\frac{1}{2}$ and 0 for classical particles, bosons and fermions, respectively. Bosons are the most gregarious, having the greatest tendency to bunch up, while fermions are the most anti-social of all and completely avoid each other.

2.4.3 The Bose-Einstein and Fermi-Dirac Distributions

Boltzmann's mathematical trick of discretising classical phase space becomes physical reality in the quantum world, and the same methodology can be applied to find the distribution functions for bosons and fermions (accounting for their indistinguisha-

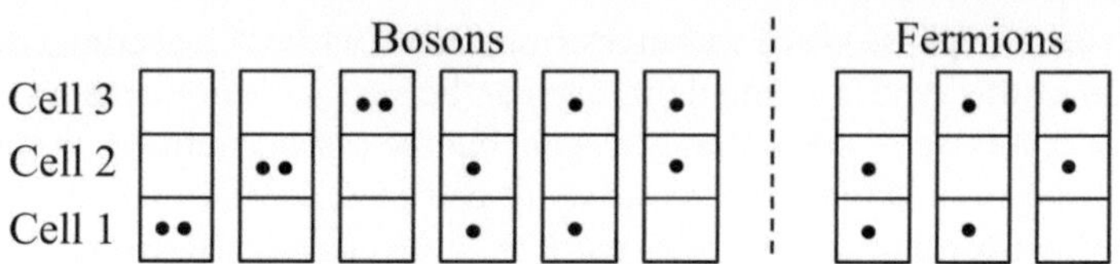

Fig. 2.5 Possible configurations of two bosons (left) and two fermions (right) across three equally-accessible cells. The classical case was shown in Fig. 2.2

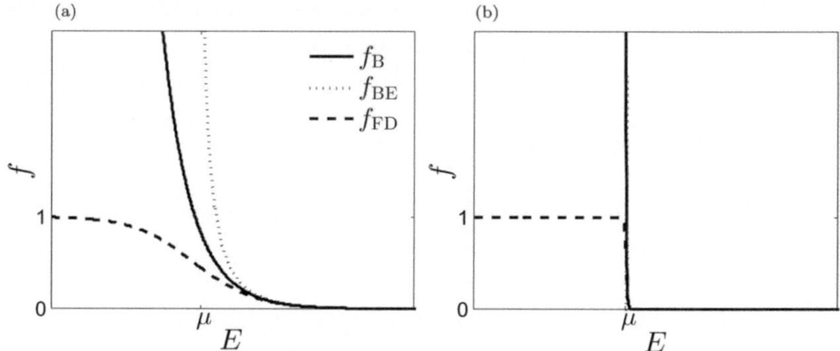

Fig. 2.6 The Boltzmann, Bose-Einstein and Fermi-Dirac distribution functions for **a** $T \gg 0$ and **b** $T \approx 0$

bility and occupancy rules). The Bose-Einstein and Fermi-Dirac particle distribution functions, which describe the mean distribution of bosons and fermions over energy E in an ideal gas, are,

$$f_{BE}(E) = \frac{1}{e^{(E-\mu)/k_B T} - 1}, \tag{2.6}$$

$$f_{FD}(E) = \frac{1}{e^{(E-\mu)/k_B T} + 1}. \tag{2.7}$$

The rather insignificant looking $-1/+1$ terms in the denominators have profound consequences. Figure 2.6 compares the Boltzmann, Bose-Einstein and Fermi-Dirac distributions.

We make the following observations of the distributions functions:

- To be physical, the distribution functions must satisfy $f \geq 0$ (for all E). This implies that $\mu \leq 0$ for the Bose-Einstein distribution. For the Fermi-Dirac and Boltzmann distributions, μ can take any value and sign.
- For $(E - \mu)/k_B T \gg 1$, the Bose-Einstein and Fermi-Dirac distributions approach the Boltzmann distribution. Here, the average state occupancy is much less than unity, such that the effects of particle indistinguishability become negligible. Note that the classical limit condition $(E - \mu)/k_B T \gg 1$ should not be interpreted too directly, as it seems to predict, counter-intuitively, that low temperatures favour

classical behaviour; this is because μ itself has a non-trivial temperature dependence.

- As $E \to \mu$ from above, the Bose-Einstein distribution diverges, i.e. particles accumulate in the lowest energy states.
- For $E \ll \mu$, the Fermi-Dirac distribution saturates to one particle per state, as required by the Pauli exclusion principle.
- For decreasing temperature, the distributions develop a sharper transition about $E = \mu$, approaching step-like forms for $T \to 0$.

2.5   The Ideal Bose Gas

A year after Einstein and Bose set forth their new particle statistics for a gas of bosons, Einstein published "Quantum theory of the monoatomic ideal gas: a second treatise" [5], elaborating on this topic. Here he predicted Bose-Einstein condensation. We now follow Einstein's derivation of this phenomena and predict some key properties of the gas.

2.5.1   Continuum Approximation and Density of States

We consider an ideal (non-interacting) gas of bosons confined to a box, with energy level occupation according to the Bose-Einstein distribution (2.6). For mathematical convenience we approximate the discrete energy levels by a continuum, valid providing there are a large number of accessible energy levels. Replacing the level variables with continuous quantities ($E_j \mapsto E$, $g_j \mapsto g(E)$ and $N_j \mapsto N(E)$), the number of particles at energy E is written,

$$N(E) = f_{\mathrm{BE}}(E)\, g(E) = \frac{g(E)}{e^{(E-\mu)/k_{\mathrm{B}}T} - 1}, \tag{2.8}$$

where $g(E)$ is the *density of states*. The total number of particles and total energy follow as the integrals,

$$N = \int N(E)\, \mathrm{d}E, \tag{2.9}$$

$$U = \int E\, N(E)\, \mathrm{d}E. \tag{2.10}$$

These are integrated in energy upwards from the $E = 0$ ($j = 1$) ground state.

The density of states $g(E)$ is defined such that the total number of possible states in phase space $\mathcal{N}_{\mathrm{ps}}$ is,

$$\mathcal{N}_{\mathrm{ps}} = \int g(E)\mathrm{d}E = \int g(p)\mathrm{d}p, \tag{2.11}$$

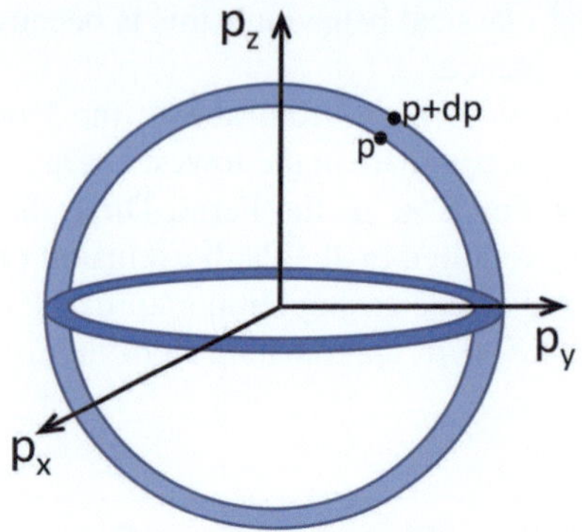

Fig. 2.7 The volume of momentum space from p to $p + dp$ is a spherical shell in 3D momentum space

where we have also provided the corresponding expression in terms of momentum p, which is more convenient to work with. The quantity $g(p)dp$ represents the number of states lying between momenta p and $p + dp$. These states occupy a (6D) volume in phase space which is the product of their (3D) volume in position space and their (3D) volume in momentum space. The former is the box volume, $\mathcal{V}$. For the latter, the range p to $p + dp$ represents a spherical shell in momentum space of inner radius p and thickness dp, as illustrated in Fig. 2.7, with momentum-space volume $4\pi p^2 dp$. Hence the phase space volume is $4\pi p^2 \mathcal{V} dp$. Now recall that each quantum state takes up a volume h^3 in phase space. Thus the number of states between p and $p + dp$,

$$g(p)dp = \frac{4\pi p^2 \mathcal{V}}{h^3} dp. \tag{2.12}$$

Using the momentum-energy relation $p^2 = 2mE$, its differential form $dp = \sqrt{m/2E}\, dE$), and the relation $g(E)\, dE = g(p)\, dp$, Eq. (2.12) leads to,

$$g(E) = \frac{2\pi (2m)^{\frac{3}{2}} \mathcal{V}}{h^3} E^{\frac{1}{2}}. \tag{2.13}$$

This is the density of states for an ideal gas confined to a box of volume $\mathcal{V}$. There are a diminishing amount of states in the limit of zero energy, and an increasing amount with larger energy.

While the *occupancy of a state* goes like $1/(e^{(E-\mu)/k_\mathrm{B}T} - 1)$ and diverges as $E \to \mu$, the *occupancy of an energy level* goes like $E^{\frac{1}{2}}/(e^{(E-\mu)/k_\mathrm{B}T} - 1)$ and diminishes as $E \to 0$ (due to the decreasing amount of available states in this limit). These two distributions are compared in Fig. 2.8a.

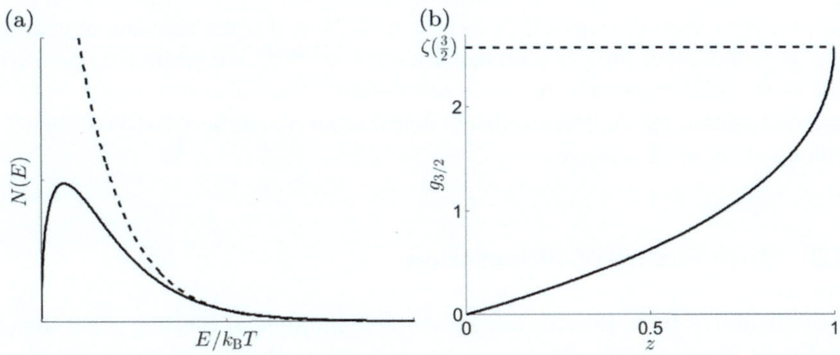

Fig. 2.8 a The occupancy of energy levels $N(E)$ (solid line), compared to the Bose-Einstein distribution f_{BE} (dashed line). The former vanishes as $E \to 0$ due to the diminishing density of states in this limit. **b** The function $g_{\frac{3}{2}}(z) = \sum_{p=1}^{\infty} z^p / p^{\frac{3}{2}}$ over the relevant range $0 < z \le 1$

2.5.2   Integrating the Bose-Einstein Distribution

Using Eqs. (2.8, 2.13) we can write the number of particles (2.9) as,

$$N = \frac{2\pi (2\,m)^{\frac{3}{2}} V}{h^3} \int_0^{\infty} \frac{E^{\frac{1}{2}}}{e^{(E-\mu)/k_B T} - 1} \mathrm{d}E. \tag{2.14}$$

We seek to evaluate this integral. To assist us, we quote the general integral,[2]

$$\int_0^{\infty} \frac{x^{\alpha}}{e^x/z - 1} \mathrm{d}x = \Gamma(\alpha + 1) g_{\alpha+1}(z), \tag{2.15}$$

where $\Gamma(x) = \int_0^{\infty} t^{x-1} e^{-t} \mathrm{d}t$ is the *Gamma function*.[3] We have also defined a new function, $g_\beta(z) = \sum_{p=1}^{\infty} \frac{z^p}{p^\beta}$; an important case is when $z = 1$ for which it reduces to the *Riemann zeta function*,[4] $\zeta(\beta) = \sum_{p=1}^{\infty} \frac{1}{p^\beta}$.

Taking $\alpha = \frac{1}{2}$, $x = E/k_B T$ and $z = e^{\mu/k_B T}$ in the general result (2.15), we evaluate Eq. (2.14) as,

$$N = \frac{(2\pi m k_B T)^{\frac{3}{2}} V}{h^3} g_{\frac{3}{2}}(z), \tag{2.16}$$

---

[2] This result can be derived by introducing new variables $z = e^{\mu/k_B T}$ and $x = E/k_B T$ to rewrite part of integrand in the form $ze^{-x}/(1 - ze^{-x})$, and then writing as a power series expansion.
[3] Relevant values for us are $\Gamma(3/2) = \sqrt{\pi}/2$ and $\Gamma(5/2) = 3\sqrt{\pi}/4$.
[4] Relevant values for us are $\zeta(3/2) = 2.612$ and $\zeta(5/2) = 1.341$.

where we have used the result $\Gamma(3/2) = \sqrt{\pi}/2$. Note that the relevant range of z is $0 < z \leq 1$: the lower limit is required since $z = e^{\mu/k_B T} > 0$ while the upper limit $z \leq 1$ is required to prevent negative populations. Note also that $\mu \leq 0$ over this range, as required for the Bose-Einstein distribution (recall Sect. 2.4.3). In Fig. 2.8b we plot $g_{\frac{3}{2}}(z)$ over this range.

2.5.3   Bose-Einstein Condensation

The prediction of Bose-Einstein condensation in the style of Einstein arises directly from Eq. (2.16). Consider adding particles to the box, while at constant temperature. An increase in N is accommodated by an increase in the function $g_{\frac{3}{2}}(z)$. However, $g_{\frac{3}{2}}(z)$ is finite, reaching a maximum value of $g_{\frac{3}{2}} = \zeta(\frac{3}{2}) = 2.612$ at $z = 1$. In other words, the system becomes *saturated* with particles. This critical number of particles, denoted N_c, follows as,

$$N_c = \frac{(2\pi m k_B T)^{\frac{3}{2}} V}{h^3} \zeta\left(\frac{3}{2}\right). \tag{2.17}$$

Our derivation predicts a limit to how many particles the Bose-Einstein distribution can hold, but common sense tells us that it should always be possible to add more particles to the box. In fact, we made a subtle mistake. In calculating N we replaced the summation over discrete energy levels (from the $i = 1$ ground state upwards) by an integral over a continuum of energies (from $E = 0$ upwards). However, this continuum approximation does not properly account for the population of the ground state, since the density of states, $g(E) \propto E^{\frac{1}{2}}$, incorrectly predicts zero population in the ground state. What we have predicted is the *saturation of the excited states*; any additional particles added to the system enter the ground state (which comes at no energetic cost). For $N \gg N_c$, the ground state acquires an anomalously large population.

As Einstein put it [5], "a number of atoms which always grows with total density makes a transition to the ground quantum state, whereas the remaining atoms distribute themselves... A separation occurs; a part condenses, the rest remains a saturated ideal gas." This effect is *Bose-Einstein condensation*, and the collection of particles in the ground state is the *Bose-Einstein condensate*. The effect is a condensation in momentum space, referring to the occupation of the zero momentum state. In practice, when the system is confined by a potential, a condensation in real space also takes place, towards the region of lowest potential. Bose-Einstein condensation is a *phase transition*, but whereas conventional phase transitions (e.g. transformation from gas to liquid or liquid to solid) are driven by particle interactions, Bose-Einstein condensation is driven by the particle statistics.

Based on the above hindsight, we note that the total atom number N appearing in Eqs. (2.9), (2.14) and (2.16) should be replaced by the number in excited states, N_{ex}.

2.5.4   Critical Temperature for Condensation

If, instead, the particle number and volume are fixed, then there exists a critical temperature T_c below which condensation occurs. The population of excited particles at a given temperature is given by Eq. (2.16. For $T > T_c$, this is sufficient to accommodate all of the particles, and the gas is in the normal phase. As temperature is lowered, however, the excited state capacity also decreases. At the point where the excited states no longer accommodate all the particles, Bose-Einstein condensation occurs. The critical temperature is obtained by setting $z = 1$ in Eq. (2.16) and rearranging for T,

$$T_c = \frac{h^2}{2\pi m k_B} \left(\frac{N}{\zeta(\frac{3}{2})\mathcal{V}} \right)^{\frac{2}{3}}. \tag{2.18}$$

For further decreases in temperature, N_{ex} decreases and so more and more particles must enter the ground state. In the limit $T \to 0$, excited states can carry no particles and all particles enter the condensate.

2.5.5   Condensate Fraction

A useful quantity for characterising the gas is the *condensate fraction*, that is, the proportion of particles which reside in the condensate, N_0/N. Let us consider its variation with temperature. Writing $N = N_0 + N_{ex}$ leads to,

$$\frac{N_0}{N} = 1 - \frac{N_{ex}}{N}. \tag{2.19}$$

For $T \leq T_c$, the excited population N_{ex} is given by Eq. (2.16) with $z = 1$, and the total population is given by Eq. (2.14) with $z = 1$ and $T = T_c$. Substituting both into the above gives,

$$\frac{N_0}{N} = 1 - \left(\frac{T}{T_c} \right)^{3/2}. \tag{2.20}$$

For $T > T_c$, we expect $N_0/N \approx 0$. This behaviour is shown in Fig. 2.9.

2.5.6   Particle-Wave Overlap

Bose-Einstein condensation occurs when $N > N_c$, with N_c given by Eq. (2.17). It is equivalent to write this criterion in terms of the number density of particles, $n = N/\mathcal{V}$, as,

$$n > \zeta\left(\frac{3}{2} \right) \left(\frac{2\pi m k_B T}{h^3} \right)^{\frac{3}{2}}. \tag{2.21}$$

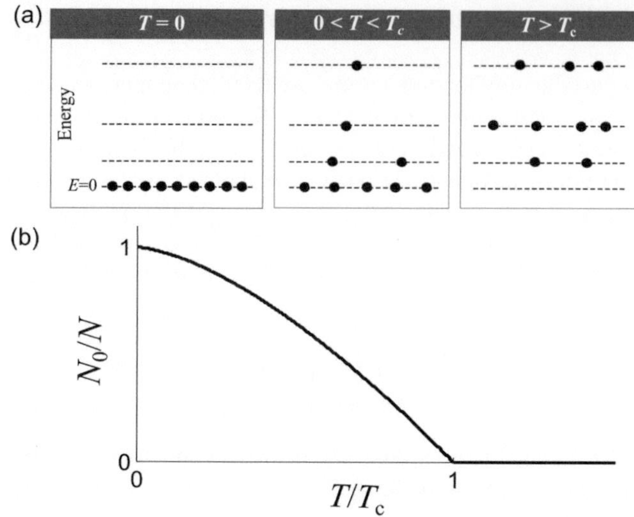

Fig. 2.9 a Illustration of energy level occupations in the boxed ideal Bose gas. At $T = 0$ all particles lie in the ground state. For $0 < T < T_c$, some particles are in excited levels but there is still macroscopic occupation of the ground state. For $T > T_c$, there is negligible occupation of the ground state. b Variation of condensate fraction, N_0/N, with temperature, as per Eq. (2.20)

According to de Broglie, particles behave like waves, with a wavelength $\lambda_{dB} = h/p$. For a thermally-excited gas, the particle wavelength is $\lambda_{dB} = \dfrac{h}{\sqrt{2\pi m k_B T}}$. Employing this, the above criterion becomes,

$$n\lambda_{dB}^3 > \zeta\left(\frac{3}{2}\right). \tag{2.22}$$

Upon noting that the average inter-particle distance $d = n^{-\frac{1}{3}}$ and $\zeta(\frac{3}{2})^{\frac{1}{3}} \sim 1$ we arrive at,

$$\lambda_{dB} \gtrsim d \tag{2.23}$$

Thus, Bose-Einstein condensation coincides with when the particle waves overlap with each other, as depicted in Fig. 2.10. The individual particles become smeared out into one giant wave of matter, the condensate.

2.5.7 Internal Energy

The internal energy of the gas U is determined by the excited states only, since the ground state possesses zero energy; therefore we can express U by integrating across the excited state particles as,

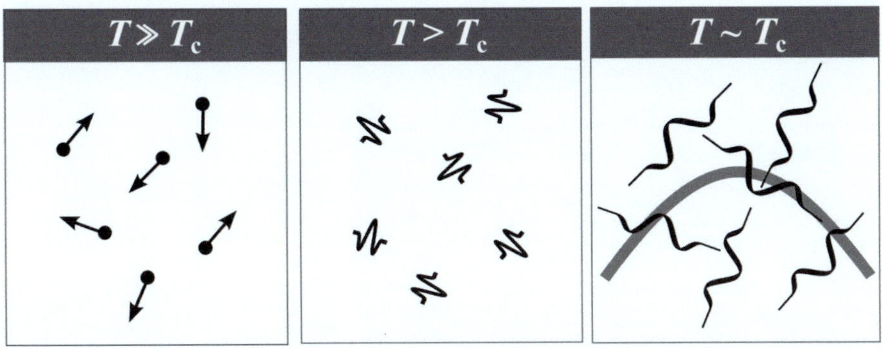

Fig. 2.10 Schematic of the transition between a classical gas and a Bose-Einstein condensate. At high temperatures ($T \gg T_c$) the gas is a thermal gas of point-like particles. At low temperatures (but still exceeding T_c) the de Broglie wavelength λ_{dB} becomes significant, yet smaller than the average spacing d. At T_c, the matter waves overlap ($\lambda_{dB} \sim d$), marking the onset of Bose-Einstein condensation

$$U = \int_0^\infty E \, N_{ex}(E) \, dE. \tag{2.24}$$

Upon evaluating this integral below and above T_c we find,

$$U = \begin{cases} \dfrac{3}{2} \dfrac{\zeta(5/2)}{\zeta(3/2)} N k_B T \left(\dfrac{T}{T_c}\right)^{3/2} & \text{for } T < T_c, \\[2ex] \dfrac{3}{2} N k_B T & \text{for } T \gg T_c. \end{cases} \tag{2.25}$$

The $T \gg T_c$ result is consistent with the classical equipartition theorem for an ideal gas, which states that each particle has on average $\frac{1}{2} k_B T$ of kinetic energy per direction of motion. The different behavior for $T < T_c$ confirms the presence of a distinct state of matter.

2.5.8   Pressure

The pressure of an ideal gas is $P = 2U/3\mathcal{V}$. From Eq. (2.25), then for $T \gg T_c$ we recover the standard result for a classical ideal gas that $P \propto T/\mathcal{V}$. For $T < T_c$, and recalling that $T_c \propto 1/\mathcal{V}^{2/3}$, we find that $P \propto T^{5/2}$. The pressure of the condensate is zero at absolute zero and does not depend on the volume of the box! A consequence of this is that the condensate has infinite compressibility, as explored in Problem 2.6.

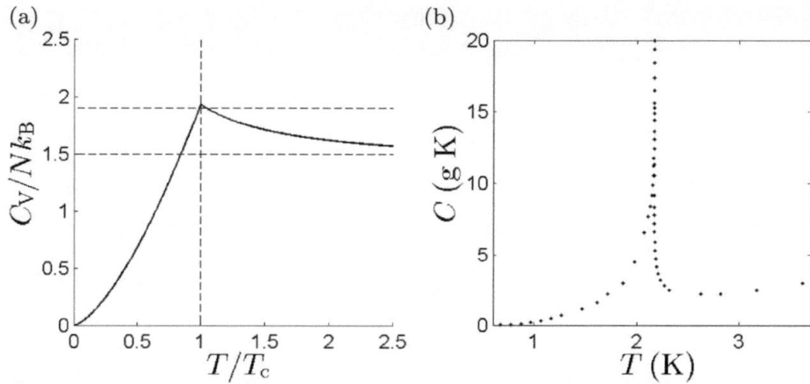

Fig. 2.11 **a** Heat capacity C_V of the ideal Bose gas as a function of temperature T. **b** Experimentally-measured heat capacity of liquid Helium about the λ-point of 2.2 K. Both curves show a similar cusped structure. Data reprinted with permission from [6]. Copyright 1961, North-Holland Publishing Co. All rights reserved

2.5.9   Heat Capacity

The heat capacity of a substance is the energy required to raise its temperature by unit amount. At constant volume it is defined as,

$$C_V = \left(\frac{\partial U}{\partial T}\right)_V.$$

(2.26)

From Eq. (2.25) we find,

$$C_V = \begin{cases} 1.93\, N k_{\mathrm{B}} T^{3/2} & \text{for } T < T_{\mathrm{c}}, \\ \dfrac{3}{2} N k_{\mathrm{B}} & \text{for } T \gg T_{\mathrm{c}}. \end{cases}$$

(2.27)

A more precise treatment, describing the dependence at intermediate temperatures, can be found in Ref. [7]. The form of $C_V(T)$ is depicted in Fig. 2.11, showing a cusp-like dependence around T_{c}. In general, discontinuities in the gradient of $C_V(T)$ are signatures of phase transitions between distinct states of matter. The similarity of this prediction to measured heat capacity curves for Helium about the λ-point was key evidence in linking helium II to Bose-Einstein condensation.

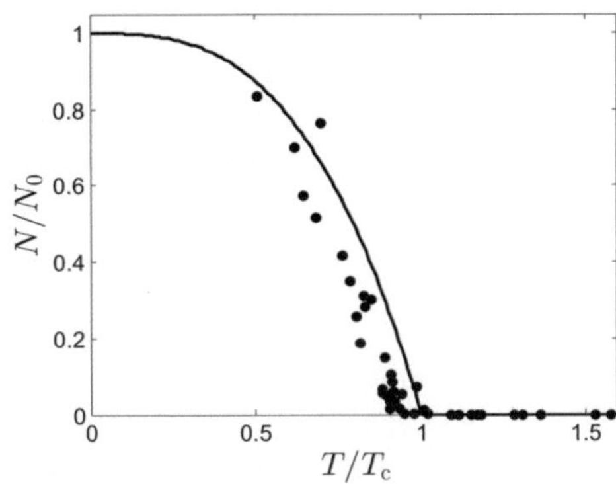

Fig. 2.12 Variation of condensate fraction N_0/N with temperature for a harmonically-trapped BEC, with the ideal-gas predictions (solid line) compared to experimental measurements (points), with $T_c = 280$ nK. Data reprinted with permission from [10]. Copyright 1996, American Physical Society

2.5.10 Ideal Bose Gas in a Harmonic Trap

2.5.10.1 Critical Temperature and Condensate Fraction

In almost all experiments, atomic Bose-Einstein condensates are confined by harmonic (quadratic) potentials, rather than boxes,[5] with the general form,

$$V(x, y, z) = \frac{1}{2}m\left(\omega_x^2 x^2 + \omega_y^2 y^2 + \omega_z^2 z^2\right),\tag{2.28}$$

where m is the atomic mass, and ω_x, ω_y and ω_z are trap frequencies which characterise the strength of the trap in each direction. Here the density of states is modified, being $g(E) = E^2/(2\hbar^3 \omega_x \omega_y \omega_z)$ in 3D. This leads, for example, to a critical temperature of the form,

$$T_c = \frac{\hbar}{k_B}(\omega_x \omega_y \omega_z)^{1/3}\left[\frac{N}{\zeta(3)}\right]^{1/3},\tag{2.29}$$

and for the condensate fraction to vary with temperature as,

$$\frac{N_0}{N} = 1 - \left(\frac{T}{T_c}\right)^3.\tag{2.30}$$

These predictions agree well with experimental measurements of harmonically-trapped atomic BECs, as seen in Fig. 2.12. This is despite the fact that atomic BECs are not *ideal* but feature significant interactions between atoms.

2.5.10.2 Density Profile

We can deduce the density profile of the (non-interacting) condensate in a harmonic trap as follows. The ground quantum state in a harmonic trap is the ground harmonic oscillator state. For simplicity, assume a spherically-symmetric trap with

---

[5] Box-like traps [8,9] are also possible, and allow the condensate to have uniform density, facilitating comparison with the theory of homogeneous condensates.

$\omega_x = \omega_y = \omega_z \equiv \omega_r$. A particle in the ground state (condensate) then has wavefunction $\psi(r) = \left(\frac{m\omega}{\pi\hbar}\right)^{3/4} e^{-m\omega r^2/2\hbar}$. The quantity $|\psi(r)|^2$ represents the probability of finding the particle at position r. For a condensate of N_0 such particles, with $N_0 \gg 1$, the particle density profile will follow as,

$$n(r) = N_0|\psi|^2 = N_0(\hbar\ell_r^2)^{-3/2}e^{-r^2/\ell_r^2}, \tag{2.31}$$

where we have introduced the harmonic oscillator length $\ell_r = \hbar/m\omega_r$ which characterises the width of the density distribution.

We can also deduce the density profile of the thermal gas. Taking the classical limit, the atoms will be distributed over energy according to the Boltzmann distribution $N(E) \propto e^{-E/k_B T}$. The trapping potential $V(r)$ allows us to map energy (potential) to position, leading to a spatial particle distribution,

$$n(r) = N_{\text{ex}}(\hbar\ell_{r,\text{th}}^2)^{-3/2}e^{-r^2/\ell_{r,\text{th}}^2}, \tag{2.32}$$

where $\ell_{r,\text{th}} = \sqrt{2k_B T/m\omega_r^2}$ characterises the width of the thermal gas and the profile has been normalised to N_{ex} atoms. For increased temperature, the atoms have higher average energy and climb further up the trap walls, leading to a wider profile. While the profiles of the ideal condensate and ideal thermal gas are both Gaussian in space, their widths have different functional forms. In particular, the width of the thermal gas depends on temperature, whereas the condensate width does not.

The typical experimental protocol to form a BEC proceeds by cooling a relatively warm gas towards absolute zero. Above T_c the gas has a broad thermal distribution, which shrinks during cooling. As T_c is under-passed, the condensate distribution forms. In typical atomic BEC experiments, $\ell_r \ll \ell_{r,\text{th}}$, such that this is distinctly narrower than the thermal gas, and the combined density profile is bi-modal. Under further cooling, the condensate profile grows (with fixed width) at the expense of the thermal profile, and for $T \ll T_c$ the thermal gas is negligible. In reality, atomic interactions modify the precise shapes of the density profiles but this picture *qualitatively* describes what is observed in experiments (see Figs. 1.3 and 1.5).

2.6   Ideal Fermi Gas

We outline the corresponding behaviour of the ideal Fermi gas. Since (identical) fermions are restricted to up to one per state, Bose-Einstein condensation is prohibited, and the Fermi gas behaves very differently as $T \to 0$. At $T = 0$ the Fermi-Dirac distribution (2.7) reduces to a step function,

$$f_{\text{FD}}(E) = \begin{cases} 1 & \text{for } E \leq E_F, \\ 0 & \text{for } E > E_F. \end{cases} \tag{2.33}$$

All states are occupied up to an energy threshold E_F, termed the *Fermi energy* (equal to the $T = 0$ chemical potential). With this simplified distribution it is straightforward to integrate the number of particles,

$$N = \int N(E)\, \mathrm{d}E = \int_0^{E_F} g(E)\, f_{\mathrm{FD}}(E)\, \mathrm{d}E = \frac{4\pi V}{3} \left(\frac{2\,m\,E_F}{h^2} \right)^{3/2}, \qquad (2.34)$$

where we have used the density of states (2.13). Note that the continuum approximation $N = \int g(E)\, N(E)\, \mathrm{d}E$ holds for $N \gg 1$ fermions since the unit occupation of the ground state is always negligible. Rearranging for the Fermi energy in terms of the particle density $n = N/V$ gives,

$$E_F = \frac{\hbar^2}{2\,m} \left(6\pi^2 n \right)^{2/3}. \qquad (2.35)$$

From this we define the Fermi momentum $p_F = \hbar k_F$ where $k_F = (6\pi^2 n)^{1/3}$ is the Fermi wavenumber. In momentum space, all states are occupied up to momentum p_F, termed the *Fermi sphere*.

Similarly, the total energy of the gas at $T = 0$ is,

$$U = \int N(E)E\, \mathrm{d}E = \frac{4\pi V}{5} \left(\frac{2\,m}{h^2} \right)^{3/2} E_F^{5/2} = \frac{3}{5} N E_F. \qquad (2.36)$$

From the pressure relation for an ideal gas, $P = 2U/3V$, the pressure of the ideal Fermi gas at $T = 0$ is,

$$P = \frac{2}{5} n E_F. \qquad (2.37)$$

This pressure is finite even at $T = 0$, unlike the Bose and classical gases, and does not arise from thermal agitation. Instead it is due to the stacking up of particles in energy levels, as constrained by the quantum rules for fermions. This *degeneracy pressure* prevents very dense stars, such as neutron stars, from collapsing under their own gravitational fields.

As temperature is increased from zero, the step-like Fermi-Dirac distribution becomes broadened about $E = E_F$, representing that some high energy particles become excited to energies exceeding E_F. It is useful to define the *Fermi temperature* $T_F = E_F/k_B$. At low temperatures $T \sim T_F$, only particles in states close to E_F can be excited out of the Fermi sphere, and the system is still dominated by the stacking of particles. For high temperatures $T \gg T_F$, there is significant excitation of most particles, thermal effects dominate, and the system approaches the classical Boltzmann result. The Fermi temperature is associated with the onset of degeneracy, i.e. when quantum effects dominate the system. These regimes are depicted in Fig. 2.13.

Now consider the Fermi gas to be confined in a harmonic trap. For $T \gg T_F$ the gas will have a broad, classical profile. As T is decreased, the profile will narrow but

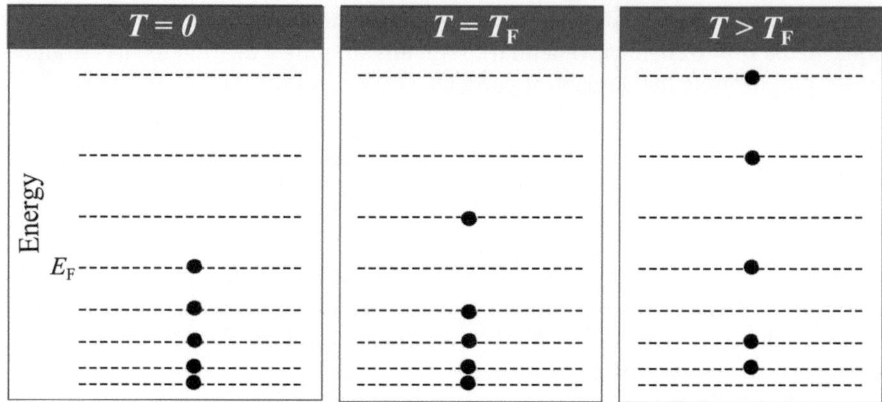

Fig. 2.13 Energy level occupations for an ideal Fermi gas. At $T = 0$, there is unit occupation of states up to the Fermi energy. At $T = T_F$, there is some excitation of states around $E = E_F$. For $T \gg T_F$, the system approaches the classical limit, with particles occupying many high-energy states

eventually saturates below T_F due to degeneracy pressure. The width of the Fermi gas at zero temperature is proportional to $N^{1/6} \ell_r$ [11], such that, for $N \gg 1$, this cloud is much wider than its classical and Bose counterparts. This picture is confirmed by the experimental images in Fig. 1.5.

Problems

2.1 *Boltzmann distribution I:*

Consider a system with 6 classical particles, total energy of 6ϵ, and 7 cells with energies 0, ϵ, 2ϵ, 3ϵ, 4ϵ, 5ϵ and 6ϵ. Complete the table below by entering the cell populations for each macrostate, the statistical weighting for each macrostate W, and the average population per cell $\bar{N}(E)$ (averaged over macrostates). What is the most probable macrostate? Plot $\bar{N}(E)$ versus E. It should be evident that the average distribution approximates the Boltzmann distribution, despite the small number of particles.

	Macrostates			
Cell energy E	1	...	11	$\bar{N}(E)$
6ϵ	?	...	?	?
5ϵ	?	...	?	?
$\vdots$	$\vdots$	...	$\vdots$	$\vdots$
ϵ	?	...	?	?
0	?	...	?	?
Statistical weighting W	?	...	?	

2.2 *Boltzmann distribution II:*

Consider a system with N classical particles distributed over 3 cells (labelled $1, 2,$ and 3) of energy 0, ϵ and 2ϵ. The total energy is $E = 0.5N\epsilon$.

(a) Obtain an expression for the number of microstates in terms of N and N_3, the population of cell 3.
(b) Plot the number of microstates as a function of N_2 (which parameterises the macrostate) for $N = 50$. Repeat for $N = 100$ and 500. Note how the distribution changes with N. What form do you expect the distribution to tend towards as N is increased to much larger values?

2.3 *Bose-Einstein condensation in 2D:*

Consider an ideal gas of bosons in two dimensions, confined within a two-dimensional box of volume $\mathcal{V}_{2D}$.

(a) Derive the density of states $g(E)$ for this two-dimensional system.
(b) Using this result show that the number of particles can be expressed as,

$$N_{\text{ex}} = \frac{2\pi m \mathcal{V}_{2D} k_B T}{h^2} \int_0^\infty \frac{ze^{-x}}{1 - ze^{-x}} dx,$$

where $z = e^{\mu/k_B T}$ and $x = E/k_B T$. Solve this integral using the substitution $y = ze^{-x}$.
(c) Obtain an expression for the chemical potential μ and thereby show that Bose-Einstein condensation is possible only at $T = 0$.

2.4 *Internal energy and heat capacity:*

Equation (2.25) summarizes how the internal energy of the boxed 3D ideal Bose gas scales with temperature. Derive the full expressions for the internal energy for the two regimes (a) $T < T_c$ (for which $z = 1$), and (b) $T \gg T_c$ (for which $z \ll 1$). Extend your results to derive the expressions for the heat capacity given in Eq. (2.27).

2.5 *Bose gas in a harmonic trap:*

Bose-Einstein condensates are typically confined in harmonic trapping potentials, as given by Eq. (2.28). Using the corresponding density of states provided in Sect. 2.5.10.1:

(a) Derive the expression for the critical number of particles.
(b) Derive the expression (2.29) for the critical temperature.
(c) Determine the expression (2.30) for the variation of condensate fraction N_0/N with T/T_c.

(d) In one of the first BEC experiments, a gas of 40, 000 Rubidium-87 atoms (atomic
mass 1.45×10^{-25} kg) underwent Bose-Einstein condensation at a temperature
of 280 nK. The harmonic trap was spherically-symmetric with with $\omega_r = 1130$
Hz. Calculate the critical temperature according to the ideal Bose gas prediction.
How does this compare to the result for the boxed gas (you may assume the
atomic density as 2.5×10^{18} m^{-3}).

2.6 *Compressibility of the Bose gas*:

The compressibility β of a gas, a measure of how much it shrinks in response to a
compressional force, is defined as,

$$\beta = -\frac{1}{\mathcal{V}}\frac{\partial \mathcal{V}}{\partial P}.$$

Determine the compressibility of the ideal gas for $T < T_c$.

Hint: Since T_c is a function of $\mathcal{V}$, you should ensure the full $\mathcal{V}$-dependence is
present before differentiating.

References

1. F. Mandl, *Statistical Physics*, 2nd edn. (Wiley, Chichester, 1988)
2. L.D. Landau, E.M. Lifshitz, *Statistical Physics*, 3rd edn. (Elsevier, Oxford, 1980)
3. S.N. Bose, Z. Phys. **26**, 178 (1924)
4. A. Einstein, Kgl. Preuss. Akad. Wiss. 261 (1924)
5. A. Einstein, Kgl. Preuss. Akad. Wiss. 3 (1925)
6. M.J. Buckingham, W.M. Fairbank, *The Nature of the Lambda-Transition in Liquid Helium*, in
   Progress in Low Temperature Physics, vol. 3, ed. by Gorter, C.J. (North Holland, Amsterdam,
   1961)
7. L.P. Pitaevskii, S. Stringari, *Bose-Einstein Condensation (International Series of Monographs
   on Physics)* (Oxford Science Publications, Oxford, 2003)
8. A.L. Gaunt, T.F. Schmidutz, I. Gotlibovych, R.P. Smith, Z. Hadzibabic, Phys. Rev. Lett. **110**,
   200406 (2013)
9. L. Chomaz et al., Nat. Comm. **6**, 6162 (2015)
10. J.R. Ensher, D.S. Jin, M.R. Matthews, C.E. Wieman, E.A. Cornell, Phys. Rev. Lett. **77**, 4984
    (1996)
11. D.A. Butts, D.S. Rokhsar, Phys. Rev. A **55**, 4346 (1997)

Gross-Pitaevskii Model of the Condensate

3

Abstract

The Gross-Pitaevskii equation (GPE) is a highly-successful and well-established model for describing an atomic Bose-Einstein condensate. Here we introduce this model, along with its derivation and assumptions. Throughout the rest of the chapter we explore its properties and key time-independent solutions.

3.1 The Gross-Pitaevskii Equation

We consider the system to be a gas of N identical and interacting bosonic atoms, each of mass m. The quantum state of the system at time t is specified by the many-body wavefunction, $\boldsymbol{9}(\mathbf{r}_1, \mathbf{r}_2, ...\mathbf{r}_N, t)$, where $\mathbf{r}_1, \mathbf{r}_1, ...\mathbf{r}_N$ are the instantaneous coordinates of the atoms. This wavefunction obeys the many-body Schrödinger equation,

$$i\hbar\frac{\partial \boldsymbol{9}}{\partial t} = \left[-\frac{\hbar^2}{2m}\left(\nabla_1^2 + \nabla_2^2 + ... + \nabla_N^2\right) + V_{\text{tot}}(\mathbf{r}_1, \mathbf{r}_1, ...\mathbf{r}_N, t)\right]\boldsymbol{9}, \qquad (3.1)$$

where ∇_i is the derivative with respect to $\mathbf{r}_i$ and V_{tot} is the combined potential acting on the system arising from particle interactions and any externally-applied potential. This equation exactly specifies the behaviour of the quantum many-body wavefunction. However, the complexity of this approach makes it intractable for modelling more than a few particles, let alone the thousands or millions typical of an atomic BEC. We will proceed to outline the physical assumptions and mathematical steps that lead from this exact description to a more tractable model, the Gross-Pitaevskii equation [1–4].

The above many-body description can be reduced to a more manageable form by working within the formalism of second quantization, whereby the many-body wavefunction is mapped onto a set of occupation numbers. We introduce the Bose field operators $\hat{\Psi}(\mathbf{r}, t)$ and $\hat{\Psi}^\dagger(\mathbf{r}, t)$, which account for the annihilation and creation, respectively, of an atom at position $\mathbf{r}$ and time t. Within this formalism the Hamiltonian operator for the system is,

C. F. Barenghi et al., *Quantum Fluids, Solitons, and Vortices*, Lecture Notes in Physics 1050, https://doi.org/10.1007/978-3-032-20171-3_3

$$\hat{H} = \int d^3\mathbf{r}\, \hat{\Psi}(\mathbf{r}, t) \left[-\frac{\hbar^2}{2m} \left(\nabla^2 + V(\mathbf{r}, t) \right) \right] \hat{\Psi}(\mathbf{r}, t)$$

$$+ \frac{1}{2} \iint d^3\mathbf{r}\, d^3\mathbf{r}'\, \hat{\Psi}^\dagger(\mathbf{r}, t)\, \hat{\Psi}^\dagger(\mathbf{r}', t)\, U(\mathbf{r} - \mathbf{r}')\, \hat{\Psi}(\mathbf{r}, t)\, \hat{\Psi}(\mathbf{r}', t). \tag{3.2}$$

where we have introduced the decomposition of the potential into the interaction potential between two atoms $U(\mathbf{r} - \mathbf{r}')$ and the external potential $V(\mathbf{r}, t)$.

We assume that the gas is sufficiently dilute that the interatomic interactions are dominated by low energy, two-body s-wave collisions. These interactions can be described by the contact (hard-sphere) interaction potential,

$$\mathcal{U}(\mathbf{r} - \mathbf{r}') = g\delta(\mathbf{r} - \mathbf{r}'), \tag{3.3}$$

where δ is Dirac's delta function and the coefficient g is given by,

$$g = \frac{4\pi \hbar^2 a_s}{m}, \tag{3.4}$$

where a_s is the *s-wave scattering length*. This approximation is valid providing the $a_s \ll d$, where d is the average interparticle distance (or, equivalently, $na_s^3 \ll 1$); we can interpret this regime as when the detailed shape of the interatomic potential becomes unimportant. The s-wave scattering length characterises both the strength and the nature of the contact interaction, being attractive for $a_s < 0$ and repulsive for $a_s > 0$. Further information regarding interatomic potentials and s-wave scattering can be found in Ref. [5].

Introducing the contact interaction into the Hamiltonian operator (3.2), one can integrate out $\mathbf{r}'$ to obtain a simpler form of the operator,

$$\hat{H} = \int d^3\mathbf{r}\, \hat{\Psi}^\dagger(\mathbf{r}, t) \left[-\frac{\hbar^2}{2m} \nabla^2 + V(\mathbf{r}, t) \right] \hat{\Psi}(\mathbf{r}, t)$$

$$+ \frac{g}{2} \int d^3\mathbf{r}\, \hat{\Psi}^\dagger(\mathbf{r}, t)\, \hat{\Psi}^\dagger(\mathbf{r}, t)\, \hat{\Psi}(\mathbf{r}, t)\, \hat{\Psi}(\mathbf{r}, t). \tag{3.5}$$

Under the second quantization formalism, the Hamiltonian operator evolves according to Heisenberg's equation,

$$i\hbar \frac{\partial \hat{\Psi}(\mathbf{r}, t)}{\partial t} = \left[\hat{\Psi}(\mathbf{r}, t), \hat{H} \right]. \tag{3.6}$$

Inserting the Hamiltonian operator (3.5) into Heisenberg's equation and using the known commutation relations of the bosonic field operators,

$$\left[\hat{\Psi}(\mathbf{r}, t), \hat{\Psi}^\dagger(\mathbf{r}', t) \right] = \delta(\mathbf{r} - \mathbf{r}'), \tag{3.7}$$

$$\left[\hat{\Psi}^\dagger(\mathbf{r}, t), \hat{\Psi}^\dagger(\mathbf{r}', t) \right] = \left[\hat{\Psi}(\mathbf{r}, t), \hat{\Psi}(\mathbf{r}', t) \right] = 0, \tag{3.8}$$

we obtain the time evolution equation for the Bose field operator,

$$i\hbar\frac{\partial \hat{\Psi}(\mathbf{r}, t)}{\partial t} = \left[\hat{H}_0 + g\hat{\Psi}^\dagger(\mathbf{r})\hat{\Psi}(\mathbf{r})\right]\hat{\Psi}^\dagger(\mathbf{r}). \tag{3.9}$$

We know that the many-body state of the system involves a macroscopic occupation of the condensate state. This motivates the decomposition of the Bose field operator into a mean-field term for the condensate atoms $\Psi(\mathbf{r}, t) = \langle\hat{\Psi}(\mathbf{r}, t)\rangle$ and a remaining operator describing the non-condensate atoms $\hat{\Psi}_{\mathrm{nc}}(\mathbf{r}, t)$,

$$\hat{\Psi}(\mathbf{r}, t) = \Psi(\mathbf{r}, t) + \hat{\Psi}_{\mathrm{nc}}(\mathbf{r}, t). \tag{3.10}$$

Inserting into the time evolution equation (3.9) and taking only the leading order terms in Ψ leads to,

$$i\hbar\frac{\partial \Psi(\mathbf{r}, t)}{\partial t} = -\frac{\hbar^2}{2m}\nabla^2\Psi(\mathbf{r}, t) + V(\mathbf{r}, t)\Psi\mathbf{r}, t) + g|\Psi\mathbf{r}, t)|^2\Psi(\mathbf{r}, t). \tag{3.11}$$

This is the *time-dependent Gross-Pitaevskii equation*(GPE), describing the evolution of the mean-field state of the condensate $\Psi(\mathbf{r}, t)$. The GPE was derived independently by Gross [6] and Pitaevskii [7] circa 1960 to study vortex lines in Bose gases.

By neglecting higher-order terms, we are neglecting the non-condensate atoms arising from thermal fluctuations and quantum fluctuations. The former assumption is valid providing the temperature of the system is much less than the critical temperature for condensation, $T \ll T_{\mathrm{c}}$. The latter is a valid approximation providing the atomic interactions (which drive quantum fluctuations) are sufficiently weak, characterised by the requirement $a_{\mathrm{s}} \ll \lambda_{\mathrm{dB}}$, where λ_{dB} is the typical de Broglie wavelength of the atoms. Indeed, providing these criteria are satisfied we find that the GPE gives excellent agreement with experimental observations of BECs across a wide range of static and dynamical scenarios.

The function Ψ is a complex field which is analogous to a single-particle wavefunction but represents all of the condensate atoms. For this reason Ψ is often termed the *macroscopic wavefunction* of the condensate. This aligns with the physical picture of the condensate as a giant matter wave (see Sect. 2.5.6).

We can write the macroscopic wavefunction as,

$$\Psi(\mathbf{r}, t) = \sqrt{n(\mathbf{r}, t)}\exp\left[iS(\mathbf{r}, t)\right], \tag{3.12}$$

where n and S are the density and phase distributions of the condensate, and is normalised to N atoms, i.e.,

$$\int |\Psi|^2\,\mathrm{d}^3\mathbf{r} = N. \tag{3.13}$$

Formally speaking ψ is not a wavefunction but rather the *order parameter* for the condensate, representing the degree of condensation. Note that the GPE conserves the norm of the macroscopic wavefunction. Energy is also conserved, providing the potential is time-independent.

The GPE resembles the single-particle Schrödinger equation, with the first two terms accounting for kinetic and potential energy, respectively. The key difference is the cubic term $g|\Psi|^2\Psi$, which arises directly from the atomic interactions and makes the equation *nonlinear*. The physical interpretation of the nonlinear term is that, at a given point in space, there is an energy contribution arising from the mean-field interactions of all the atoms in the immediate vicinity. Thus, although the interactions within a condensate are weak, they can have a significant affect on the solutions and dynamics, as we will explore in due course. Similar *Nonlinear Schrödinger Equations (NLSEs)* arise in optics, plasma physics and water waves. In one spatial dimension, the NLSE has special mathematical properties, such as soliton solutions and infinite conservation laws (see Chap. 5).

The GPE can also be extended to take thermal and quantum effects into account, but this topic is outside the scope of this book. Further information can be found in the tutorial of Proukakis and Jackson [8] and elsewhere [9, 10].

Although the GPE has analytic solutions in certain specific cases (as we will see), in general its solutions must be obtained by solving the equation numerically, for example, to determine its ground state and excited states, or to simulate dynamics. We outline a numerical approach to solving the GPE in one-dimension in the Appendix.

3.1.1 Atomic Parameters

Table 3.1 presents the mass and s-wave scattering length for some example atomic species commonly used to form Bose-Einstein condensates. The majority of BEC experiments feature atoms with positive s-wave scattering length; accordingly, this is the most well-studied regime. This is because BECs with negative a_s are more fragile, being prone to a collapse instability., as we will see later in this chapter.

Experimentalists can also control sign and magnitude of g using Feshbach resonances. Here magnetic fields are used to couple the two-body scattering to a bound state; when this coupling is close to some resonant magnetic field, huge changes in the two-body scattering properties are possible [5]. This can allow the s-wave scattering length to be tuned to particular values, ranging from strongly negative to strongly positive, and including the case where a_s is tuned to approximately zero to realise non-interacting condensates. This mechanism also allows for the scattering length to be tuned in time on timescales which are relatively fast compared to the typical condensate timescales, thereby enabling effectively instantaneous interaction quenches.

3.1.2 Mass, Energy and Momentum

The total mass of the condensate is $M = mN$, where N is provided by the normalization condition on Ψ, Eq. (3.13).

Table 3.1 Mass m and s–wave scattering length for some atomic species that are commonly used to form Bose-Einstein condensates. The scattering lengths are the natural values away from any Feshbach resonances

Atom	m (10^{-27}) kg	a_s (nm)
^{4}He	6.6	16.98
^{7}Li	11.7	−1.44
^{23}Na	38.2	3.28
^{41}K	68.1	4.13
^{87}Rb	144.4	5.29
^{133}Cs	220.8	14.82

The energy is,

$$E = \int \left[\frac{\hbar^2}{2m} |\nabla \Psi|^2 + V|\Psi|^2 + \frac{g}{2}|\Psi|^4 \right] d^3\mathbf{r} = E_{\text{kin}} + E_{\text{pot}} + E_{\text{int}}. \qquad (3.14)$$

The terms represent (from left to right) *kinetic energy* E_{kin}, *potential energy* E_{pot} and *interaction energy* E_{int}. Providing that the potential V is independent of time, then the energy $E = E_{\text{kin}} + E_{\text{pot}} + E_{\text{int}}$ is conserved during the time evolution of the condensate.

It can be useful, particularly when determining the energy numerically, to define $\Psi = \Psi_r + i\Psi_i$, where Ψ_r and Ψ_i are the real and imaginary parts of the wavefunction. Then, the $|\nabla\Psi|^2$ term in the energy can be expressed in a more convenient form, $|\nabla\Psi|^2 = (\nabla\Psi_r)^2 + (\nabla\Psi_i)^2$.

Meanwhile the momentum of the condensate is,

$$\mathcal{P} = \frac{i\hbar}{2} \int \left(\Psi\nabla\Psi^* - \Psi^*\nabla\Psi \right) d^3\mathbf{r}. \qquad (3.15)$$

3.2   Time-Independent GPE

Time-independent solutions of the GPE satisfy,

$$\Psi(\mathbf{r}, t) = \psi(\mathbf{r})e^{-i\mu t/\hbar}, \qquad (3.16)$$

where μ is a constant called the *chemical potential*. The exponential term represents the freedom for the phase to freely evolve with time, uniformly across the system, while the density $n(\mathbf{r}, t) = |\psi(\mathbf{r})|^2$ is unaffected. Inserting Eq. (3.16) into Eq. (3.11), we obtain the *time-independent GPE* for the time-independent wavefunction $\psi(\mathbf{r})$,

$$\mu\psi = -\frac{\hbar^2}{2m}\nabla^2\psi + V(\mathbf{r})\psi + g|\psi|^2\psi. \qquad (3.17)$$

Note that the potential V must be independent of time here. Solutions of the time-independent GPE are stationary solutions of the system, and the lowest energy solution is the ground state of the BEC. $\psi(\mathbf{r})$ is real for the simple solutions that we discuss in this Chapter.

The chemical potential is the eigenvalue of time-independent GPE, and direct integration leads to the expression,

$$\mu = \frac{1}{N}(E_{\text{kin}} + E_{\text{pot}} + 2E_{\text{int}}). \tag{3.18}$$

In the absence of interactions, this reduces to the energy per particle, consistent with the eigenvalue of the time-independent Schrödinger equation. More generally, the chemical potential is defined as $\mu = \partial E/\partial N$.

3.3   Fluid Dynamics Interpretation

There is a deep link between the GPE and fluid dynamics. Indeed, we can picture the condensate as a fluid, characterised by its density and velocity distributions. From the earlier relation, $\Psi(\mathbf{r}, t) = \sqrt{n(\mathbf{r}, t)}e^{iS(\mathbf{r},t)}$ (known in this context as the *Madelung transform*) the number density follows as $n(\mathbf{r}, t) = |\Psi(\mathbf{r}, t)|^2$. From this relation we have also the *mass density* $\rho(\mathbf{r}, t)$ conventionally used in fluid dynamics, $\rho(\mathbf{r}, t) = m\, n(\mathbf{r}, t)$.

The fluid velocity field $\mathbf{v}(\mathbf{r}, t)$ is defined from the phase via,

$$\mathbf{v}(\mathbf{r}, t) = \frac{\hbar}{m}\nabla S(\mathbf{r}, t). \tag{3.19}$$

Using the Madelung transform $\Psi = \sqrt{n}e^{iS}$ and the above velocity relation, we find that the energy integral of Eq. (3.14) can be written as,

$$E = \int \left[\frac{\hbar^2}{2m}\left(\nabla\sqrt{n}\right)^2 + \frac{mnv^2}{2} + Vn + \frac{gn^2}{2} \right] d^3\mathbf{r}. \tag{3.20}$$

The first two terms comprise the kinetic energy. The first of these is the quantum kinetic energy. It arises due to the zero-point motion of confined particles, and vanishes for a uniform system. The second term is the conventional kinetic energy associated with the flow of the fluid.

Inserting the Madelung transform into the GPE, and separating real and imaginary terms we obtain two equations. The first is the classical *continuity equation*,

$$\frac{\partial n}{\partial t} + \nabla \cdot (n\mathbf{v}) = 0. \tag{3.21}$$

The continuity equation expresses conservation of the number of atoms (or, when written in terms of $\rho(\mathbf{r}, t)$, conservation of mass). By integrating the equation over a given volume, we see that, if the number of atoms changes in that volume, it is because fluid has moved in or out of it.

The second equation is,

$$m\frac{\partial \mathbf{v}}{\partial t} = -\nabla \cdot \left(\frac{1}{2}mv^2 + V + gn - \frac{\hbar^2}{2m}\frac{\nabla^2\sqrt{n}}{\sqrt{n}}\right). \tag{3.22}$$

The $\nabla^2\sqrt{n}/\sqrt{n}$ term is termed the *quantum pressure* term (see below). With some manipulation, we can write this in the equivalent form,

$$mn\left(\frac{\partial \mathbf{v}}{\partial t} + (\mathbf{v}\cdot\nabla)\mathbf{v}\right) = -\nabla(P + P') - n\nabla V, \tag{3.23}$$

where P and P' are respectively the *pressure* and the *quantum pressure*,

$$P = \frac{gn^2}{2}, \qquad P' = -\frac{\hbar^2}{4m}n\nabla^2(\ln n). \tag{3.24}$$

Equations (3.21) and (3.22) (or, equivalently, Eqs. (3.21) and (3.23)) are known as the *superfluid hydrodynamic equations*. They can also be written in index notation[1]

Notice that the pressure depends only on the density. This property makes the condensate a *barotropic fluid*; as a consequence, surfaces of constant pressure are also surfaces of constant density. The quantum pressure is a pure quantum effect, and vanishes if we set Planck's constant equal to zero. It has the same origin as the quantum kinetic energy, i.e. zero point motion, which creates a pressure that opposes any 'squashing' or 'bending' of the condensate. In a uniform condensate the quantum pressure is zero because n is constant.

Equation (3.23) is very similar to the classical Euler equation for an inviscid fluid. To understand the relation between condensates and classical fluids, we compare the relative importance of pressure and quantum pressure. Using Eqs. (3.24), we estimate that the order of magnitude of P and P' are respectively $P \sim gn^2$ and $P' \sim \hbar^2 n/m\xi^2$, where ξ is the length scale of the variations of n. Then $P'/P \sim \hbar^2/(mng\xi^2)$, and hence the quantum pressure becomes negligible ($P' \ll P$) in the limit of length scales larger than ξ. If in addition, the trapping potential is absent ($V = 0$) then Eq. (3.23) become negligible, and the equation reduces to the classical Euler equation, which describes the motion of a classical fluid without viscosity.

---

[1] In index notation, Eqs. (3.21) and (3.23) are $\dfrac{\partial n}{\partial t} + \dfrac{\partial(nv_j)}{\partial x_j} = 0$ and $mn\left(\dfrac{\partial v_k}{\partial t} + v_j\dfrac{\partial v_k}{\partial x_j}\right) = -\dfrac{\partial P}{\partial x_k} - \dfrac{\partial P'_{jk}}{\partial x_j} - n\dfrac{\partial V}{\partial x_k}$, where v_j is the jth Cartesian component ($j = 1, 2, 3$) of the velocity $\mathbf{v}$, we have assumed summation over repeated indices, and where the components P'_{jk} of the quantum stress tensor P' are $P'_{jk} = -\dfrac{\hbar^2}{4m}n\dfrac{\partial^2(\ln n)}{\partial x_j \partial x_k}$.

The lengthscale in question is provided by the *healing length*, defined as,

$$\xi = \frac{\hbar}{\sqrt{gmn}}.$$ (3.25)

The typical value of the healing length in atomic BECs is $\xi \sim 10^{-6}$ m; for superfluid helium (^{4}He) the healing length is much smaller, $\xi \sim 10^{-10}$ m.

3.4   Stationary Solutions in Infinite or Semi–infinite Homogeneous Systems

In experiments, atomic condensates are confined by bowl-like trapping potentials $V(\mathbf{r})$. Condensates are therefore small (typically of the order of 10^{-5} or 10^{-4} m) and inhomogeneous (the density depends on the position). However, many general properties of atomic condensates can be understood from the simpler scenario of a homogeneous condensate in an infinitely-sized or semi-infinitely-sized system. The homogeneous condensate is also a useful model of superfluid helium, as the sizes of the samples of ^{4}He typically used in experiments range from 10^{-2} to 10^{-1} m, many orders of magnitude larger than the healing length. A homogeneous condensate would not be stable for $g < 0$ (as we see later) and so we consider $g > 0$ for now.

3.4.1   Uniform Condensate

For $V = 0$ (uniform condensate of infinite extent), the stationary solution is uniform, and the time-independent GPE becomes,

$$\mu\psi = g|\psi|^2\psi.$$ (3.26)

The solution is then,

$$\psi = \psi_0 = \sqrt{\mu/g}.$$ (3.27)

The corresponding number and mass densities are, respectively,

$$n = n_0 = |\psi_0|^2 = \mu/g, \qquad \rho = \rho_0 = m\mu/g.$$ (3.28)

3.4.2   Condensate Near a Wall

Consider a *one-dimensional hard wall* defined by,

$$V(x) = \begin{cases} \infty & \text{for } x < 0, \\ 0 & \text{for } x \geq 0. \end{cases}$$

No atoms exist in the region $x < 0$ (since this would require infinite energy), and so the boundary condition at $x = 0$ is $\psi(0) = 0$. Away from the wall (in the positive x direction) the condensate must recover its bulk form, giving the second boundary condition that $\psi(x) \to \psi_0 = \sqrt{\mu/g}$ for $x \to \infty$. In the semi-infinite region $x \geq 0$ the one-dimensional (1D) time-independent GPE is,

$$\mu\psi = -\frac{\hbar^2}{2m}\frac{\partial^2 \psi}{\partial x^2} + g|\psi|^2\psi. \tag{3.29}$$

The solution of this equation which satisfies the boundary conditions is,

$$\psi(x) = \psi_0 \tanh\left(\frac{x}{\xi}\right). \tag{3.30}$$

The meaning of the healing length ξ is now apparent: it is the characteristic minimal distance over which ψ changes spatially. The 'healing' profile is supported at a wall by the balance between the kinetic energy term in the GPE and the interaction term. Denoting the spatial scale of the variation in the wavefunction as ξ, these terms are of the order of $\hbar^2/m\xi^2$ and gn_0, respectively. Equating these terms and rearranging leads to $\xi = \hbar/\sqrt{mn_0 g}$, the healing length as defined in Eq. (3.25). Note that the healing length is sometimes defined with a $\sqrt{2}$ in the denominator.

In an infinite square well of width L_0, which is much wider than the healing length ($L_0 \gg \xi$), we then expect the wavefunction to 'heal' at each boundary, according to Eq. (3.30), and reach the bulk value in the centre of the well. This is shown in Fig. 3.1.

It is interesting to compare this to the case of $g = 0$, for which the ground state is given by the well-known solution of the Schrödinger equation for a particle in an infinite well, $\psi \sim \sin(\pi x/L_0)$. Clearly the interactions between the atoms broaden and flatten the density profile by increasing the energetic cost of concentrating atoms in one place.

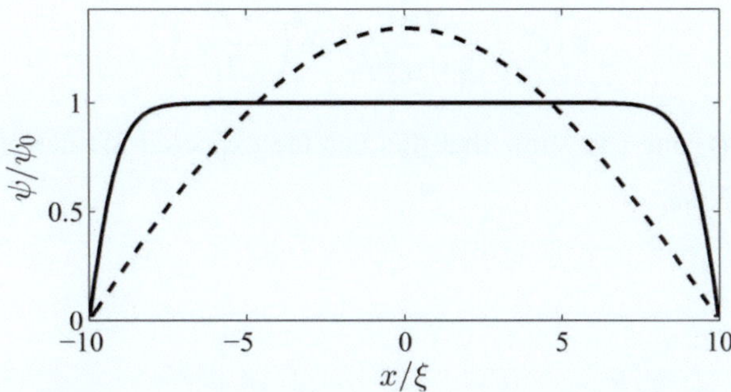

Fig. 3.1 Condensate wavefunction ψ (in units of ψ_0) as a function of position x (in units of ξ) within a 1D infinite square well of width (here with width 20ξ). Shown are the profiles for a non-interacting condensate ($g = 0$) and a repulsively-interacting ($g > 0$) condensate. Note how the wavefunction "heals" at each boundary according to Eq. (3.30), recovering its bulk density at a distance from the wall of the order of few times the healing length ξ

3.5   Stationary Solutions in Harmonic Potentials

Atomic condensates are typically confined by harmonic potentials which may, in general, be anisotropic in space. For simplicity here we start by considering a spherically-symmetric harmonic trap,

$$V(\mathbf{r}) = \frac{m}{2}\omega_r^2 r^2. \tag{3.31}$$

where $r^2 = x^2 + y^2 + z^2$. The characteristic length scale of this potential is the *harmonic oscillator length*,

$$\ell_r = \sqrt{\hbar/m\omega_r}. \tag{3.32}$$

There is no general analytic solution for the ground state (lowest energy) solution of the BEC in a harmonic trap; usually the ground state is found by numerically solving Eq. (3.17). However, there exist useful analytic results for certain regimes which we describe below. It is useful to work in terms of the *interaction parameter*,[2] Na_s/ℓ_r. Below we distinguish the following cases: no interactions, strong repulsive interactions ($Na_s/\ell_r \gg 1$) and weak interactions ($|Na_s/\ell_r| \ll 1$).

3.5.1   No Interactions

In the absence of atomic interactions ($g = 0$) the time-independent GPE reduces to the Schrödinger equation,

$$\mu\psi = -\frac{\hbar^2}{2m}\nabla^2\psi + \frac{m\omega_r^2 r^2}{2}\psi. \tag{3.33}$$

The ground state harmonic oscillator solution is well-known to be a three-dimensional Gaussian wave function,

$$\psi(\mathbf{r}) = \frac{N^{1/2}}{\pi^{3/4}\ell_r^{3/2}} \exp\left(-\frac{r^2}{2\ell^2}\right). \tag{3.34}$$

Using Eq. (3.14), one can show that this has the expected 3D harmonic oscillator energy $E = \frac{3}{2}N\hbar\omega_r$.

---

[2] More generally, for an anisotropic harmonic trap, the corresponding interaction parameter is $Na_s/\bar{\ell}$, where $\bar{\ell} = \sqrt{\hbar/m\bar{\omega}}$ and $\bar{\omega} = (\omega_x\omega_y\omega_z)^{1/3}$ is the geometric mean of the trap frequencies.

3.5.2 Strong Repulsive Interactions

Let the interactions be strongly repulsive, satisfying $Na_s/\ell_r \gg 1$. We expect a condensate profile which is significantly broadened and flattened due to the repulsive interactions. An analytic solution is found if we neglect the $\nabla^2\psi$-term in the GPE; this is known as the *Thomas-Fermi approximation*. The time-independent GPE simplifies to,

$$\mu\psi = g|\psi|^2\psi + V\psi. \tag{3.35}$$

Substituting $n = |\psi|^2$ and $V(r) = \frac{1}{2}m\omega^2r^2$, we obtain $n(r) = (2\mu - m\omega_r^2r^2)/2g$. Density cannot be negative, so we assume that $n(r) = 0$ if $2\mu \leq m\omega_r^2r^2$. The last equality defines the *Thomas-Fermi radius* R_r, which satisfies,

$$\mu = \frac{1}{2}m\omega_r^2R_r^2. \tag{3.36}$$

We conclude that the Thomas-Fermi density profile is,

$$n(r) = \begin{cases} \dfrac{\mu}{g}\left(1 - \dfrac{r^2}{R_r^2}\right) = \dfrac{m\omega_r^2(R_r^2 - r^2)}{2g} & \text{if } r \leq R_r, \\ 0 & \text{if } r > R_r, \end{cases} \tag{3.37}$$

and has the shape of an inverted parabola. Provided that $Na_s/\ell_r \gg 1$, the Thomas-Fermi solution is an excellent approximation of the solution of the GPE determined numerically, and compares well with experimental data, as shown in Fig. 3.2. Note, however, the slight deviation from the true numerical solution close to the condensate's edge; here the gradient terms, neglected within the Thomas-Fermi model, become significant.

The application of the normalization condition, Eq. (3.13), to the above solution and manipulation of the resulting expression leads to useful relations for the chemical potential and the energy of the condensate in terms of the number of atoms N,

$$\mu = \frac{\hbar\omega_r}{2}\left(\frac{15\,Na_s}{\ell_r}\right)^{2/5}, \quad E = \frac{5}{7}\mu N. \tag{3.38}$$

The latter is obtained from the relation $\mu = \partial E/\partial N$. Since $Na_s/\ell_r \gg 1$, it is evident that in the Thomas-Fermi regime the chemical potential and energy per particle are considerably greater than the typical trap energy $\hbar\omega_r$.

In the more general case where the harmonic potential is anisotropic in space, $V(x, y, z) = m(\omega_x^2x^2 + \omega_y^2y^2 + \omega_z^2z^2)/2$, the Thomas-Fermi boundary is an ellipsoidal surface satisfying the equation

$$\frac{x^2}{R_x^2} + \frac{y^2}{R_y^2} + \frac{z^2}{R_z^2} = 1, \tag{3.39}$$

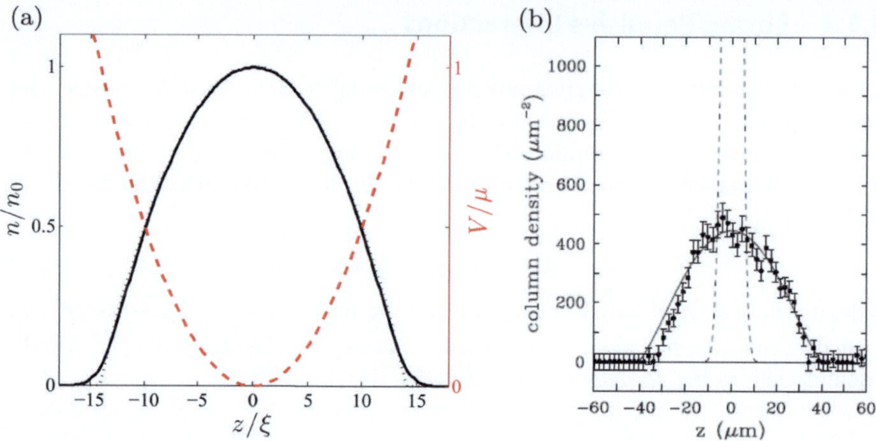

Fig. 3.2 a Density profile $n(z)$ plotted versus position z (in units of the healing length ξ). The agreement between the analytic Thomas-Fermi density profile (dotted black line) and the numerically-determined solution of the GPE (solid black line) is so good that the lines overlap everywhere but in the tails near $x \approx \pm 15\xi$. The harmonic trapping potential $V(z)$ is indicated by the dashed red line. **b** An experimental density profile, compared to the Thomas-Fermi prediction (solid line) and the non-interacting prediction (dashed line). Reprinted figure with permission from [11]. Copyright 1999 by the American Physical Society

where the three Thomas-Fermi radii R_x, R_y and R_z satisfy,

$$\mu = \frac{1}{2}m\omega_x^2 R_x^2 = \frac{1}{2}m\omega_y^2 R_y^2 = \frac{1}{2}m\omega_z^2 R_z^2. \tag{3.40}$$

In this anisotropic case, it is most convenient to write the density profile as,

$$n(x, y, z) = \begin{cases} \dfrac{\mu}{g}\left(1 - \dfrac{x^2}{R_x^2} - \dfrac{y^2}{R_y^2} - \dfrac{z^2}{R_z^2}\right), & \text{within the ellipsoid,} \\ 0 & \text{elsewhere.} \end{cases} \tag{3.41}$$

From this the anisotropic versions of the chemical potential and energy, Eq. (3.38), can be determined.

3.5.3 Weak Interactions

The following variational approach determines an approximate solution of the time-independent GPE in a harmonic potential when the interactions (either positive or negative) are weak, that is $|Na_s/\ell_r| < 1$.

In the limiting case $g = 0$ we know that the exact wavefunction is the Gaussian harmonic oscillator ground state, Eq. (3.34). For weak interactions we assume the following trial wavefunction, or Ansatz, which is Gaussian in shape but has variable width $\sigma\ell_r$,

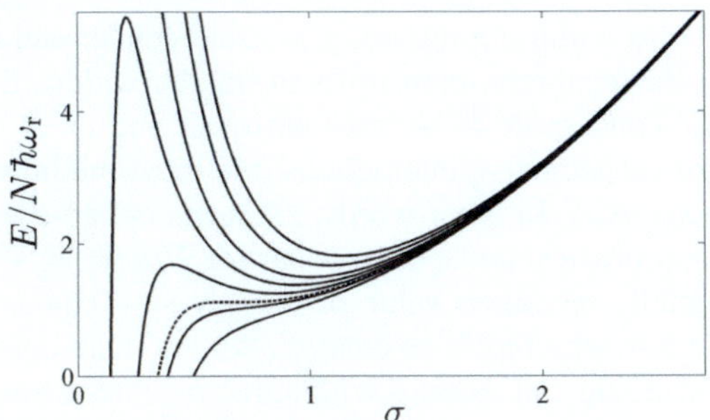

Fig. 3.3 Energy E (in units of $N\hbar\omega_r$) versus σ according to Eq. (3.43) for various values of the interaction parameter Na_s/ℓ_r corresponding, from top to bottom, to $[-1, -0.75, -0.67, -0.5, -0.25, 0, 0.5, 1]$. $Na_s/\ell_r = -0.67$ (dashed line) marks the critical point for the onset of collapse

$$\psi(\mathbf{r}) = \left(\frac{N}{\pi^{3/2}\sigma^3\ell_r^3}\right)^{1/2} \exp\left(-\frac{r^2}{2\sigma^2\ell_r^2}\right). \tag{3.42}$$

where σ is our *variational parameter*. If $g = 0$ then $\sigma = 1$, i.e. we recover the exact non-interacting result.

Using the energy integral (3.14), the energy of the Ansatz is,

$$E(\sigma) = \hbar\omega_r N \left[\frac{3}{4\sigma^2} + \frac{3\sigma^2}{4} + \frac{1}{\sqrt{2\pi}}\left(\frac{Na_s}{\ell_r}\right)\frac{1}{\sigma^3}\right]. \tag{3.43}$$

From left to right, the terms in the bracket represent kinetic energy, potential energy and interaction energy. For a given system (i.e. for specific values of N, ω, a_s and ℓ_r), Eq. (3.43) tells us how the energy varies with σ. The variational solution is defined as the variational state with the lowest energy, i.e. the minimum of $E(\sigma)$; the corresponding width is denoted $\sigma_{\min}$. Figure 3.3 plots $E(\sigma)$ for various values of the interaction parameter Na_s/ℓ. The behaviour is different depending on whether the interactions are repulsive or attractive:

- For **repulsive interactions** ($g > 0$), $E(\sigma)$ diverges to infinity for both $\sigma \to 0$ (due to the positive kinetic and interaction energies) and $\sigma \to \infty$ (due to the potential energy), with a global minimum in-between, corresponding to the variational ground state. If $g = 0$, $\sigma_{\min} = 1$, corresponding to the non-interacting Gaussian solution. For increasing g, $\sigma_{\min}$ increases, i.e., the condensate becomes wider.
- For **attractive interactions** ($g < 0$), $E(\sigma)$ now diverges to minus infinity as $\sigma \to 0$. This is due to the dominance of the negative interaction energy in this limit. The lowest energy solution is thus a wavepacket of zero width, i.e. an unstable collapsed state![3] However, for small $|Na_s/\ell_r|$, a *local minimum* exists

---

[3] In reality, the BEC does not quite collapse to zero width; at high densities, repulsive inter-atomic forces kick-in which cause the condensate to then explode outwards, an effect termed the *bosenova*.

in $E(\sigma)$ at non-zero width, representing a stable condensate of finite size. For larger $|Na_s/\ell_r|$, the local minimum shifts to smaller widths; the attractive interactions cause the condensate to become narrower and more peaked. However, beyond some critical attractive interactions, the local minimum disappears and no stable solutions exist. In other words, all states collapse to zero width. The variational method predicts collapse to occur for $Na_s/\ell_r \leq -0.67$; this is close to the experimentally measured value of $Na_s/\ell_r \leq -0.64$. This tendency to collapse is the reason why repulsive condensates are more common and why we have avoided discussing condensates with attractive interactions so far.

Note that the above-assumed Gaussian profile is just an approximation. In the presence of repulsive interactions, the true condensate profile (e.g. as obtained by numerical solution of the GPE) is broader than a Gaussian (becoming more Thomas-Fermi like for increasing repulsive interactions), while for attractive interactions the shape is narrower and more peaked.

3.5.4 Anisotropic Harmonic Potentials and Condensates of Reduced Dimensionality

The shape of the condensate is determined by the shape of the trapping potential. A spherical harmonic potential induces a spherical condensate. It is also common to encounter elongated, or *cigar-shaped*, condensates and flattened, or *pancake-shaped*, condensates. The former case is achieved if the condensate is more tightly trapped in two directions, e.g. $\omega_x, \omega_y > \omega_z$, and the latter case, if it is more tightly trapped in one direction, e.g. $\omega_z > \omega_x, \omega_y$. These shapes are illustrated in Fig. 3.4.

By making these trap anisotropies more extreme, it is possible to engineer condensates of reduced dimensionality. Consider first a highly elongated trap ($\omega_x, \omega_y \gg \omega_z$). If the transverse trapping potential (which is of energy $\hbar(\omega_x\omega_y)^{1/2}$) is much larger than the condensate energy scale (the chemical potential, μ), then excitations of the condensate in the x and y directions are highly suppressed, and the only significant dynamics occur in the z direction. The system has become effectively one-dimensional. An effectively two-dimensional condensate can be realized for $\omega_x, \omega_y \ll \omega_z$ and $\hbar\omega_z \gg \mu$.

In these limits the condensate can be described by suitable one-dimensional and two-dimensional GPEs. The reduction of the full three-dimensional GPE to these forms is straightforward, as we now outline for a one-dimensional system. Assuming the above criteria for an effectively one-dimensional condensate, we take the following Ansatz for the condensate wavefunction,

$$\psi(x, y, z, t) = \psi_z(z, t)G_x(x)G_y(y). \tag{3.44}$$

In other words, we have decomposed ψ into independent components along x, y and z. Under the criterion $\hbar(\omega_x\omega_y)^{1/2} \gg \mu$ then the x and y component will be "locked" into the respective ground harmonic oscillator states, which are represented by the Gaussian functions,

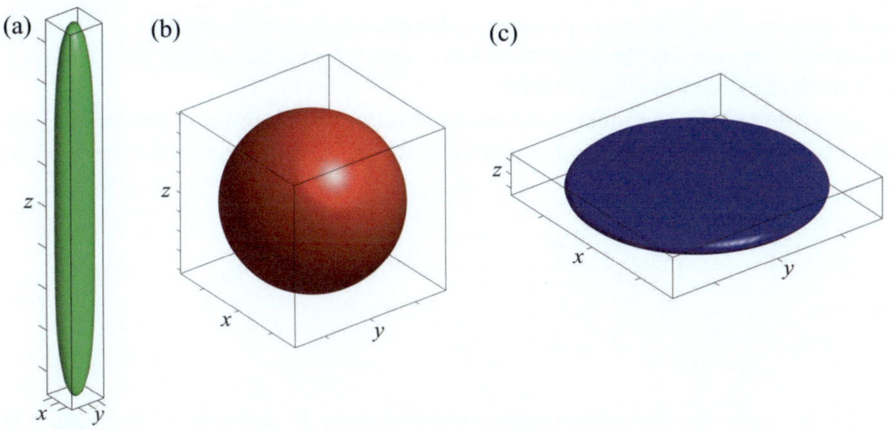

Fig. 3.4 The three most common condensate shapes that can be formed in an axisymmetric harmonic potential: **a** a cigar condensate ($\omega_x, \omega_y > \omega_z$), **b** a spherical condensate ($\omega_x = \omega_y = \omega_z$), and **c** a pancake condensate ($\omega_x, \omega_y < \omega_z$)

$$G_x(x) = \frac{1}{(\pi \ell_x^2)^{1/4}} e^{-x^2/2\ell_x^2}, \qquad G_y(y) = \frac{1}{(\pi \ell_y^2)^{1/4}} e^{-y^2/2\ell_y^2}, \tag{3.45}$$

where $\ell_x = \sqrt{\hbar/m\omega_x}$ and $\ell_y = \sqrt{\hbar/m\omega_y}$ denote the harmonic oscillator lengths along x and y. The time-dependence now only appears in the axial wavefunction, ψ_z. Note that ψ_z is normalised to the number of atoms, i.e. $\int |\psi_z|^2 \, dz = N$; as a result the transverse wavefunctions are both normalised to unity (leading to their pre-factors).

To obtain a 1D GPE, one proceeds by inserting the wavefunction Ansatz (3.44) into the 3D GPE and manipulating. Since,

$$\frac{d^2 G_1(x)}{dx^2} = \left(\frac{x^2}{\ell_x^4} - \frac{1}{\ell_x^2} \right) G_1(x), \tag{3.46}$$

and similarly for $G_y(y)$, each term in the GPE acquires a $G_x(x)G_y(y)$ factor. To eliminate these factors, one multiplies the equation through by $G_x^* G_y^*$ (where * denotes complex conjugate) and integrates over all x and y. It is helpful to note that $\int_{-\infty}^{\infty} e^{-x^2} dx = \sqrt{\pi}$. This leads to the following one-dimensional GPE for $\psi_z(z)$,

$$\mu_{1D}\psi_z = -\frac{\hbar^2}{2m} \frac{d^2\psi_z}{dz^2} + g_{1D}|\psi_z|^2\psi_z + \frac{1}{2} m\omega_z^2 z^2 \psi_z. \tag{3.47}$$

Here g_{1D} and μ_{1D} are the effective one-dimensional interaction strength and chemical potential, defined as,

$$g_{1D} = \frac{g}{2\pi \ell_x \ell_y}, \qquad \mu_{1D} = \mu - \frac{\hbar\omega_x}{2} - \frac{\hbar\omega_y}{2}. \tag{3.48}$$

Note that the trap geometries are often cylindrically symmetric, with $\omega_x = \omega_y$; this symmetry can simplify the integration steps. In the Appendix we introduce a numerical method to simulate this 1D GPE.

Following similar arguments for an effectively two-dimensional condensate, one obtains the effective two-dimensional GPE for the two-dimensional wavefunction $\psi_{xy}(x, y, t)$,

$$\mu_{2\mathrm{D}}\psi_\perp = -\frac{\hbar^2}{2m}\left(\frac{\mathrm{d}^2\psi_\perp}{\mathrm{d}x^2} + \frac{\mathrm{d}^2\psi_\perp}{\mathrm{d}y^2}\right) + g_{2\mathrm{D}}|\psi_\perp|^2\psi_\perp + \frac{1}{2}m(\omega_x^2 x^2 + \omega_y^2 y^2)\psi_\perp,$$

$$g_{2\mathrm{D}} = \frac{g}{\sqrt{2\pi}\,\ell_z}, \qquad \mu_{2\mathrm{D}} = \mu - \frac{\hbar\omega_z}{2}.$$

In this case the two-dimensional wavefunction is normalised according to $\int |\psi_\perp|^2 \,\mathrm{d}x\mathrm{d}y = N$.

In these one- and two-dimensional cases, the system energy is still described according to Eq. (3.14), with the gradient operator replaced by its one- and two-dimensional equivalents, and the integration taken over one and two dimensions, respectively. Moreover, the same analysis techniques presented for three-dimensional stationary solutions, e.g. the Thomas-Fermi approximation and the Gaussian variational approach, can be employed. In particular, the 1D GPE provides a simplified platform to study many generic properties of condensates, and, for example, its stationary solutions under hard-wall and periodic boundaries are well-established [12, 13].

Note that the system stability can be significantly affected by the dimensionality of the system, for example, collapse under attractive interactions does not occur within the 1D GPE, as will be discussed further in Chap. 5. Note that it is possible to derive modified forms of effective low-dimensional GPE which retain more of the three-dimensional physics of the system; one example in the non-polynomial Gross-Pitaevskii equation, which we will also meet in Chap. 5.

3.6   Imaging and Column-Integrated Density

The most common approach to image a condensate is via optical absorption imaging. The condensate is illuminated by an uniform light beam from one side. The atoms absorb a proportion of the light such that a two-dimensional shadow is cast behind the condensate; this is recorded by camera as shown in Fig. 3.5, forming an *absorption image* of the condensate. Examples are the images in Fig. 1.5. Importantly, the darkness of the shadow is proportional to the atomic density, integrated along the direction light is travelling in[4]; we call this the *column-integrated density*.

---

[4] Fortunately, the atomic density is so low that scattering of the light beam is negligible and so the light effectively takes a direct path through the condensate.

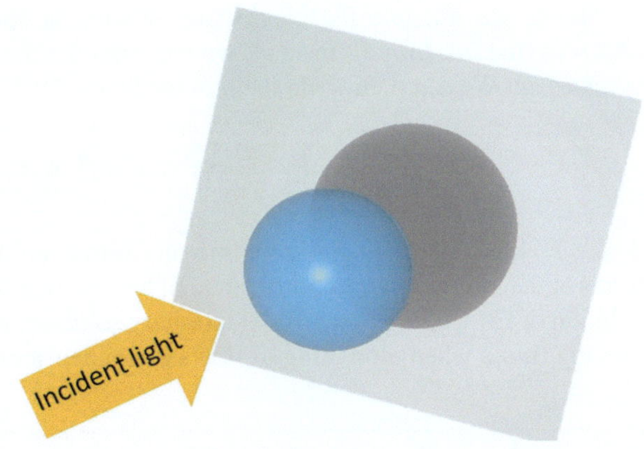

Fig. 3.5 Absorption imaging of a BEC. Laser light incident upon the BEC creates a shadow behind it, whose darkness is proportional to the column-integrated density of the BEC

To enable comparison between experimental absorption images and theoretical models, one must relate three-dimensional wavefunctions to the corresponding two-dimensional column-integrated density profiles. Assuming imaging in the z-direction, the column-integrated density n_{CI} is,

$$n_{\mathrm{CI}}(x, y) = \int_{-\infty}^{\infty} n(x, y, z)\, \mathrm{d}z. \tag{3.49}$$

3.7   Galilean Invariance and Moving Frames

A condensate in a homogeneous ($V = 0$) system satisfies the GPE,

$$i\hbar \frac{\partial \Psi}{\partial t} = \left(-\frac{\hbar^2}{2m}\nabla^2 + g|\Psi|^2 \right) \Psi. \tag{3.50}$$

The stationary solution is $\Psi_0 = \sqrt{n_0}\exp\left[-i\mu t/\hbar\right]$, corresponding to a static ($\mathbf{v} = 0$) condensate. Now let us imagine, instead, that this condensate is moving with uniform velocity v_0 in the positive x direction, say. We can construct this moving solution as,

$$\Psi = \Psi_0(x - v_0 t, y, z)\exp\left[i\frac{m v_0 x}{\hbar} - \frac{m v_0^2 t}{2\hbar} \right]. \tag{3.51}$$

Note that the density remains n_0 throughout. This is a demonstration of *Galilean invariance*, i.e. that the laws of physics are the same in all *inertial frames* (frames moving at fixed relative speed to each other). This is true only if the system is translationally invariant, i.e. the potential is the same everywhere.

Above, we imagine the condensate flowing at speed v_0 relative to the static observer (the *lab frame*). Instead, we can take the observer to be moving with the condensate. We can then write the *moving frame GPE*,

$$i\hbar \frac{\partial \Psi}{\partial t} = \left(-\frac{\hbar^2}{2m} \nabla^2 + g|\Psi|^2 + i\hbar v_0 \frac{\partial}{\partial \tilde{x}} \right) \Psi, \tag{3.52}$$

where $\tilde{x}$ is the x-coordinate in the moving frame and the Laplacian is evaluated in terms of the moving frame coordinates. In this moving frame, the flowing condensate solution of Eq. (3.51) is actually a stationary solution. It can be useful to work in the moving frame when modelling of flows of condensates.

3.8　Dimensionless Variables

The typical numbers which appear in the GPE equation are very small and cumbersome, for example the reduced Planck's constant is $\hbar = 1.055 \times 10^{-34}$ J s. When numerically solving the GPE to model a condensate, it would be better if the numbers which we compute were of order unity; this minimises the role of floating point errors which are inherent to modern digital computation. Another problem is that not all the parameters which appear in the GPE are independent: identifying the truly independent parameters reduces the number of numerical simulations which are needed to understand the nature of the solution. It is therefore useful to introduce *dimensionless variables* and write the GPE in simpler *dimensionless from.* To illustrate the procedure, we consider two examples: homogeneous and harmonically-trapped condensates.

Before we start, we notice for the sake of generality that we are free to introduce the chemical potential μ in the time-dependent GPE by letting, in analogy with Eq. (3.16),

$$\Psi(\mathbf{r}, t) = \psi(\mathbf{r}, t) e^{-i\mu t/\hbar}, \tag{3.53}$$

where now $\psi(\mathbf{r}, t)$ depends also on t; in other words, the exponential term takes care of part of (but not all of) the time dependence of the wavefunction. The resulting time-dependent GPE is,

$$i\hbar \frac{\partial \psi}{\partial t} = -\frac{\hbar^2}{2m} \nabla^2 \psi + g|\psi|^2 \psi + V\psi - \mu\psi. \tag{3.54}$$

3.8.1　Homogeneous Condensate

In the absence of trapping ($V = 0$), the governing equation is

$$i\hbar \frac{\partial \psi}{\partial t} = -\frac{\hbar^2}{2m} \nabla^2 \psi + g|\psi|^2 \psi - \mu\psi. \tag{3.55}$$

We have seen that the wave function of a uniform condensate at rest is $\psi_0 = \sqrt{\mu/g}$, corresponding to the number density $n_0 = \mu/g$. We have also seen that the characteristic minimum distance over which the wavefunction varies is the healing length $\xi = \hbar/\sqrt{m\mu}$. Therefore the quantities n_0 and ξ are convenient units of density and length. Similarly, it is apparent from Eqs. (3.16) or (3.53) that $\tau = \hbar/\mu$ is the natural unit of time. These remarks suggest the introduction of the following dimensionless variables (hereafter denoted by primes):

$$x' = \frac{x}{\xi}, \qquad y' = \frac{y}{\xi}, \qquad z' = \frac{z}{\xi}, \tag{3.56}$$

(in other words, $\mathbf{r}' = \mathbf{r}/\xi$), and

$$t' = \frac{t}{\tau}, \qquad \psi' = \frac{\psi}{\psi_0}. \tag{3.57}$$

To begin substituting these new variables into the GPE, we need to develop relations for their derivatives. Using the chain rule,

$$\frac{\mathrm{d}}{\mathrm{d}x} = \frac{1}{\xi}\frac{\mathrm{d}}{\mathrm{d}x'}, \quad \frac{\mathrm{d}}{\mathrm{d}y} = \frac{1}{\xi}\frac{\mathrm{d}}{\mathrm{d}y'}, \quad \frac{\mathrm{d}}{\mathrm{d}z} = \frac{1}{\xi}\frac{\mathrm{d}}{\mathrm{d}z'}, \quad \frac{\mathrm{d}}{\mathrm{d}t} = \frac{1}{\tau}\frac{\mathrm{d}}{\mathrm{d}t'}. \tag{3.58}$$

Hence the gradient and Laplacian operators acting on the primed variables are defined as,

$$\nabla = \frac{1}{\xi}\nabla', \quad \nabla^2 = \frac{1}{\xi^2}\nabla'^2. \tag{3.59}$$

Introducing these relations, Eq. (3.55) becomes the following *dimensionless GPE*,

$$i\frac{\partial\psi'}{\partial t'} = -\frac{1}{2}\nabla'^2\psi' + |\psi'|^2\psi' - \psi'. \tag{3.60}$$

This equation contains no parameters—it has been simplified to its mathematical essence.[5] These units are often termed *natural* or *healing length* units.

3.8.2 Harmonically-Trapped Condensate

Here we assume that the condensate is confined by a spherical harmonic trap $V = m\omega_r^2 r^2/2$. Then the governing equation is,

$$i\hbar\frac{\partial\psi}{\partial t} = -\frac{\hbar^2}{2m}\nabla^2\psi + g|\psi|^2\psi + V\psi - \mu\psi. \tag{3.61}$$

---

[5] In the literature, after transforming the GPE into dimensionless form, it is common to drop the primes.

In this case the natural units of length and time are based on the harmonic oscillator length $\ell_r = \sqrt{\hbar/m\omega_r}$ and the inverse of the trap frequency, ω_r^{-1}. We set $\mathbf{r}' = \mathbf{r}/\ell$ (that is to say $x' = x/\ell$, $y' = y/\ell$ and $z' = /\ell$) and $t' = t/\tau$, where $\tau = 1/\omega_r$ respectively.

It is conventional with these units to define the dimensionless wavefunction ψ' as being normalised to unity, i.e.,

$$\int |\psi'|^2 \mathrm{d}^3\mathbf{r}' = 1. \tag{3.62}$$

Comparing to Eq. (3.13) and noting that $\mathrm{d}^3\mathbf{r} = \ell_r^3 \mathrm{d}^3\mathbf{r}'$, it follows that $\psi = (N/\ell_r^3)^{1/2}\psi'$. Introducing these relations into Eq. (3.61) we arrive at the dimensionless form,

$$i\frac{\partial\psi'}{\partial t'} = -\frac{1}{2}\nabla'^2\psi' + C|\psi'|^2\psi' + \frac{r'^2}{2}\psi' - \mu'\psi'. \tag{3.63}$$

where $\mu' = \mu/\hbar\omega_r$ and,

$$C = \frac{4\pi a_s N}{\ell}. \tag{3.64}$$

is a dimensionless interaction parameter. These units are often term *harmonic oscillator units*. For anisotropic harmonic traps, the harmonic units can be defined instead in terms of one of the trap frequencies or their geometric mean, $\bar{\omega} = (\omega_x\omega_y\omega_z)^{1/3}$.

3.9   Extensions of the Gross-Pitaevskii Equation

We will explore key extensions of the GPE to two-component systems, condensates with dipolar interactions and the beyond-mean-field regime in later chapters. Below we summarise some additional extensions of the GPE.

3.9.1   Dissipative GPE

Thermal effects can be approximately modelled in a simple phenomenological way by introducing the small *dissipation parameter* $\gamma \ll 1$ at the left hand side of Eq. (3.11). We thus obtain the *damped GPE*,

$$(i - \gamma)\hbar\frac{\partial\Psi}{\partial t} = -\frac{\hbar^2}{2m}\nabla^2\Psi + V(\mathbf{r}, t)\Psi + g|\Psi|^2\Psi. \tag{3.65}$$

This equation leads to a decay of energy and excitations in the system, as well as a decay in the norm of the wavefunction (unless the wavefunction is suitably renormalised). Typically-used values of γ to model atomic BECs are of the order of 1%. Indeed, Choi et al. [14] fitted the model to experimental results for thermally-decaying oscillations of a condensate and inferred a value of $\gamma = 0.03$ for that particular experiment. The dissipative term can be derived from theories of finite-temperature Bose gases, with the value of γ being proportional to temperature [8].

One can extend this dissipative model by making the dissipation parameter dependent on position $\gamma = \gamma(\mathbf{r})$. This approach can be used to simulate a local sink of atoms from the condensate (for example, due to laser-induced outcoupling of atoms), inhomogeneous thermal damping (arising from an inhomogeneous thermal cloud), or to create damped, reflectionless boundary conditions [15].

3.9.2 Modelling Three-Body Loss

Atomic gases suffer from three-body loss in which an inelastic three-body collision leads to a highly-excited atom being ejected from the condensate. Three-body loss can be modelled within the time-dependent GPE by the inclusion of a phenomenological imaginary loss term,

$$
i\hbar\frac{\partial\Psi}{\partial t} = -\frac{\hbar^2}{2m}\nabla^2\Psi + V(\mathbf{r}, t)\Psi + g|\Psi|^2\Psi - i\frac{\hbar}{2}K_3|\Psi|^4\Psi, \tag{3.66}
$$

where K_3 is the three-body loss rate. Example values are $K_3 = 7.2 \times 10^{-30}\text{cm}^6/\text{s}$ for ^{87}Rb [16] and $K_3 = 1.1 \times 10^{-30}\text{cm}^6/\text{s}$ for ^{23}Na [17]. The term leads to the loss of atoms at a rate which is proportional to the atomic density cubed. Three-body losses are particularly important in systems with high atomic densities, such as encountered during the collapse of an attractively-interacting BEC or within a quantum droplet.

3.9.3 Optical Lattices and the Discrete GPE

Condensates may be subjected to static periodic potentials called *optical lattices*. These are formed by creating an optical standing wave with counter-propagating laser beams [18,19] and the potential experienced by the atoms is proportional to the intensity of the light field. A one-dimensional optical lattice, taken to be along x, has the form,

$$
V_{\text{OL}}(x) = V_0 \sin^2(k_x x), \tag{3.67}
$$

where k_x is the wavevector of the light along the x direction. For counter propagating beams $k_x = 2\pi/\lambda$, where λ is the laser wavelength. The effective wavevector of the lattice k_x can be increased beyond this value by introducing a relative angle between the laser beams. The depth of the optical lattice potential V_0 is controlled through the intensity of the laser beams. Two and three dimensional optical lattices can be formed by combining such potentials along two and three axes, respectively.

When the amplitude of the lattice potential is very large, the atoms become strongly localized in the lattice wells, creating a crystalline structure. In the mean-field description it is then appropriate to apply a tight-binding approximation, whereby the condensate wavefunction is written as a sum of the ground state wave functions from each well. Let us consider this for a quasi-one-dimensional condensate, as introduced in Sect. 3.5.4, which is aligned along x and experiences a strong transverse harmonic potential. The axial potential $V(x)$ includes the above optical lattice potential but may include additional contributions, e.g. a harmonic potential to provide overall confinement to the atoms. The condensate is parameterised by an axial wavefunction $\psi(x, t)$ which is normalised to the total number of condensate atoms $\int |\psi|^2 \, dx = N$ and obeys the 1D GPE,

$$i\hbar\frac{\partial \psi}{\partial t} = -\frac{\hbar^2}{2m}\frac{\partial^2 \psi}{\partial x^2} + V(x)\psi + g_{1D}|\psi|^2\psi, \tag{3.68}$$

where g_{1D} is the modified nonlinear coefficient given in Sect. 3.5.4.

Under the tight-binding model we assume that the mini-condensate localised within each well is in the ground state. We denote the ground state in the j'th well as $\psi_j(x - x_j)$, where x_j is the centre position of the j'th well. These local wavefunctions are normalised to unity and it is assumed that they do not depend on the instantaneous number of atoms in the well. These assumptions are valid providing the lattice depth V_0 is much larger than the chemical potential of the local condensates within each well. The full axial wavefunction is then constructed through the Ansatz [20],

$$\psi(x, t) = \sum_j u_j(t)\psi_j(x - x_j). \tag{3.69}$$

The time-dependent amplitude in the j'th well has the form $u_j(t) = \sqrt{N_j(t)}\exp(i\phi_j(t))$, where N_j and ϕ_j are the norm and phase of the site, respectively. The norm of the system is given by $N = \sum |u_j|^2$.

Inserting the above Ansatz into the 1D GPE leads to the *discrete GPE*,

$$\hbar i \frac{\partial u_j}{\partial t} = -K\left(u_{j-1} + u_{j+1}\right) + \epsilon_j u_j + g'|u_j|^2 u_j. \tag{3.70}$$

Here $K = -\int \left[\frac{\hbar^2}{2m}\nabla\psi_j^*\nabla\psi_{j+1} + \psi_j V(x)\psi_{j+1}\right]dx$ is the tunneling rate between adjacent sites, $\epsilon_j = \int \left[\frac{\hbar^2}{2m}(\nabla\psi_j)^2 + V(x)\psi_j^2\right]dx$ is the on-site energy, and $g' = g_{1D}N\int \psi_j^4\, dx$ is the effective nonlinear coefficient.

A rich set of solutions and dynamical states of the envelope function u_j exist depending on the system parameters [20]. Note that this discrete equation is a form of the discrete NLSE [21], which also has applications in optical waveguide arrays.

Problems

3.1 *Dimensions:*

(a) Using the normalization condition, determine the dimensions of the wavefunction Ψ in S.I. units (metres, kilograms, seconds).
(b) Verify that all terms of the GPE have the same dimension.
(c) Show that $g|\Psi|^2$ has dimension of energy.

3.2 *Spherical BEC:* Consider a BEC confined within a three-dimensional spherical harmonic trap.

(a) Normalise the wavefunction, and hence determine an expression for the Thomas-Fermi radius R_r in terms of N, a_s and ℓ_r.
(b) Determine an expression for the peak density in terms of N and R_r.
(c) Find an expression for the ratio R_r/ℓ_r, and comment on its behaviour for large N.
(d) What is the energy of the condensate?

3.3 *Variational solutions:* Derive the expression for the variational energy of a three-dimensional trapped condensate, Eq. (3.43). Repeat in two dimensions (for a potential $V(x, y) = m\omega_r^2(x^2 + y^2)/2$) and in one dimension (for a potential $V(x) = m\omega_r^2 x^2/2$). For each case plot $E/N\hbar\omega_r$ versus the variational width σ, for some different values of the interaction parameter Na_s/ℓ_r. What effect does dimensionality have on the shape of the curves? How do this change the qualitative behaviour described in Sect. 3.5.3?

3.4 *Numerical ground states:* Use the code provided in the Appendix (or your own code) to numerically solve the 1D GPE and thereby obtain the ground-state density profiles for a 1D condensate under harmonic confinement with i) no interactions, ii) repulsive interactions and iii) attractive interactions. Compare ii) with the corresponding Thomas-Fermi profile.

3.5 *Absorption image:* Consider a BEC in the non-interacting limit with wavefunction

$$\psi(x, y, z) = \sqrt{n_0}\, e^{-x^2/2\ell_x^2} e^{-y^2/2\ell_y^2} e^{-z^2/2\ell_z^2}, \tag{3.71}$$

where n_0 is the peak density and ℓ_x, ℓ_y and ℓ_z are the harmonic oscillator lengths in three Cartesian directions. The BEC is imaged along the z-direction. Determine the form of the column-integrated density $\psi_{\mathrm{CI}}(x, y)$. Hint: $\int_0^\infty e^{ax^2} = \frac{1}{2}\sqrt{\pi/a}$.

3.6 *Static and moving uniform BEC:* Consider a 1D uniform static condensate with $V(x) = 0$. Obtain an expression for the energy E in a length L of the condensate, in terms of n_0, g and L.

Now consider the condensate to be flowing with uniform speed v_0, by constructing a solution according to Eq. (3.51). Show that the solution satisfies the 1D GPE, and confirm that the velocity field of this solution is indeed $v(x) = v_0$. What is the corresponding energy for the flowing condensate, and how does it differ from the static result? Finally, what is its momentum?

3.7 *Dimensionless GPE:* Consider a homogeneous condensate. Identify dimensionless variables so that the dimensionless GPE is,

$$i\frac{\partial \psi'}{\partial t'} = -\nabla'^2 \psi' + |\psi'|^2 \psi' - \psi', \tag{3.72}$$

i.e., without the $1/2$ factor as in Eq. (3.60).

3.8 *Moving frame:* Consider a one-dimensional condensate. Let x, t be the space and time coordinates in the laboratory frame of reference. The condensate satisfies the following GPE:

$$i\hbar\frac{\partial \psi}{\partial t} = -\frac{\hbar^2}{2m}\frac{\partial^2 \psi}{\partial x^2} + g|\psi|^2\psi + V\psi.$$

Let x', t' be the space and time coordinates in a second frame of reference which moves with constant velocity v_x with respect to the first frame:

$$x' = x - v_x t,$$
$$t' = t.$$

Express the GPE in the moving frame of reference.

3.9 *Three-body loss:* Consider the GPE with the additional term to model three-body losses, Eq. (3.66). Show that the rate of decay of the number of condensate atoms N is proportional to the cube of the atomic density $n(\mathbf{r})$ as,

$$\frac{\partial N}{\partial t} = -\int K_3\, n^3(\mathbf{r})\, d\mathbf{r}. \tag{3.73}$$

References

1. J.F. Annett, *Superconductivity, Superfluids and Condensates* (Oxford University Press, Oxford, 2004)
2. C.J. Pethick, H. Smith, *Bose-Einstein Condensation in Dilute Gases* (Cambridge University Press, Cambridge, 2008)
3. L.P. Pitaevskii, S. Stringari, *Bose-Einstein Condensation.* International Series of Monographs on Physics (Oxford Science Publications, Oxford, 2003)

4. M. Tsubota, K. Kasamatsu, *Quantum Hydrodynamics and Turbulence* (Oxford University Press, 2025)
5. C. J. Foot, *Atomic Physics* (Oxford University Press, 2005)
6. E.P. Gross, Phys. Rev. **106**, 161 (1957)
7. L.P. Pitaevskii, Sov. Phys. JETP **13**, 451 (1961)
8. N.P. Proukakis, B. Jackson, J. Phys. B **41**, 203002 (2008)
9. P.B. Blakie, A.S. Bradley, M.J. Davis, R.J. Ballagh, C.W. Gardiner, Adv. Phys. **57**, 363 (2008)
10. N.P. Proukakis, S.A. Gardiner, M. Davis, M. Szymańska (eds.), *Quantum Gases: Finite Temperature and Non-Equilibrium Dynamics* (Imperial College Press, 2013)
11. F. Dalfovo, S. Giorgini, L.P. Pitaevskii, S. Stringari, Rev. Mod. Phys. **71**, 463 (1999)
12. L.D. Carr, C.W. Clark, W.P. Reinhardt, Phys. Rev. A **62**, 063610 (2000a)
13. L.D. Carr, C.W. Clark, W.P. Reinhardt, Phys. Rev. A **62**, 063611 (2000b)
14. S. Choi, S.A. Morgan, K. Burnett, Phys. Rev. A **57**, 4057 (1998)
15. M.T. Reeves, T.P. Billam, B.P. Anderson, A.S. Bradley, Phys. Rev. Lett. **114**, 155302 (2015)
16. D.M. Stamper-Kurn, M.R. Andrews, A.P. Chikkatur, S. Inouye, H.-J. Miesner, J. Stenger, W. Ketterle, Phys. Rev. Lett. **80**, 2027 (1998)
17. E.A. Burt, R.W. Ghrist, C.J. Myatt, M.J. Holland, E.A. Cornell, C.E. Wieman, Phys. Rev. Lett. **79**, 337 (1997)
18. I. Bloch, Nat. Phys. **1**, 23 (2005)
19. O. Morsch, M. Oberthaler, Rev. Mod. Phys. **78**, 179 (2006)
20. A. Trombettoni, A. Smerzi, Phys. Rev. Lett. **86**, 2353 (2001)
21. P.G. Kevrekidis, *The Discrete Nonlinear Schrödinger Equation*. Springer Tracts in Modern Physics (Springer, Berlin, 2009)

Waves and Excitations

4

Abstract

In the previous chapter we considered the shape of steady-state condensates, either homogeneous or confined by trapping potentials. We have seen that the condensate described by the GPE is a special kind of fluid, similar to the idealized Euler fluid without viscosity that appears in classical fluid dynamics textbooks. Not surprisingly for a fluid, the dynamics of the condensate exhibits a variety of interesting time-dependent phenomena. This includes sound waves to shape oscillations, which we explore in this chapter. Excitations in the form of solitons and vortices are explored in later chapters.

4.1 Linearity and Dispersion

When considering wave phenomena, we must keep in mind the distinction between *linear* and *nonlinear* waves, and between *dispersive* and *dispersionless* waves. Linear waves are small-amplitude perturbations of the background such that the governing differential equation is linear, hence the superposition principle applies (the sum of two wave solutions is another wave solution). For example, small ripples on a pond are linear waves, whereas a tsunami is a nonlinear wave. Dispersive waves are waves such that their speed depends on the wavelength of the wave. For example, under the action of gravity, waves at the surface of water travel faster if their wavelength is longer; on the contrary, sound waves in air travel at the same speed irrespective of their wavelength.

Another important concept is that of *dispersion relation*, which is the mathematical relation between the wave's angular velocity, $\omega = 2\pi/\tau$ (where τ is the period), and the wave's wavenumber $k = 2\pi/\lambda$ (where λ is the wavelength).

© The Author(s), under exclusive license to Springer Nature Switzerland AG 2026

C. F. Barenghi et al., *Quantum Fluids, Solitons, and Vortices*, Lecture Notes in Physics 1050, https://doi.org/10.1007/978-3-032-20171-3_4

4.1.1 Bogoliubov Dispersion Relation

Of particular importance in the study of quantum fluids is the behaviour of small perturbations to the ground state of the condensate (either for a homogeneous or trapped system). This includes *sound waves*, i.e., small-lengthscale density perturbations of the ground state which oscillate periodically, as illustrated in Fig. 4.1. We now derive the behaviour of these perturbations for a general BEC.

Starting with the time-dependent GPE, Eq. (3.54), we assume that our wavefunction takes the form $\psi(\mathbf{r}, t) = \psi_0(\mathbf{r}) + \delta\psi_j(\mathbf{r}, t)$, where ψ_0 is the ground-state (stationary) solution to the GPE, and $\delta\psi_j$ is the form of the jth perturbation. At this stage, we make no assumptions about the perturbation itself, other than that it is *small*, meaning that terms of order $\delta\psi_j^2$ or higher can be ignored. Substituting this into the left-hand side of the GPE gives

$$i\hbar\frac{\partial\psi(\mathbf{r}, t)}{\partial t} = i\hbar\frac{\partial\psi_0(\mathbf{r})}{\partial t} + i\hbar\frac{\partial\delta\psi_j(\mathbf{r}, t)}{\partial t}\,, \tag{4.1}$$

where we have made use of the stationary solution. On the right-hand side, we have the linear parts of the Hamiltonian, for which we introduce the notation $H_0 = -\frac{\hbar^2\nabla^2}{2m} + V$ to denote the single-particle effects, giving

$$[H_0 - \mu]\,\psi(\mathbf{r}, t) = [H_0 - \mu]\,(\psi_0(\mathbf{r}) + \delta\psi_j(\mathbf{r}, t))\,, \tag{4.2}$$

and the contact interactions (dropping the function variables),

$$\begin{aligned}
g|\psi|^2\psi &= g|\psi_0 + \delta\psi_j|^2(\psi_0 + \delta\psi_j) \\
&= g(\psi_0 + \delta\psi_j^*)(\psi_0 + \delta\psi_j)(\psi_0 + \delta\psi_j) \\
&= g(\psi_0^2 + \psi_0(\delta\psi_j + \delta\psi_j^*) + |\delta\psi_j|^2)(\psi_0 + \delta\psi_j) \\
&= g(\psi_0^3 + 2\psi_0^2\delta\psi_j + \psi_0^2\delta\psi_j^*) + \psi_0(\delta\psi_j + \delta\psi_j^*)\delta\psi_j\,, \tag{4.3}
\end{aligned}$$

where we have linearised the small perturbations by removing higher-order terms in $\delta\psi_j$ and made use of the purely real ground state.

We can now introduce the Bogoliubov–de Gennes Ansatz, which assumes perturbations of the form

$$\delta\psi_j(\mathbf{r}, t) = u_j(\mathbf{r})e^{-i\omega_j t} + v_j^*(\mathbf{r})e^{i\omega_j t}\,, \tag{4.4}$$

where u_j and v_j are the particle and hole amplitudes, respectively, for a quasiparticle with energy $\hbar\omega_j$, normalised such that

$$\int (|u(\mathbf{r})|^2 - |v(\mathbf{r})|^2)\mathrm{d}^3\mathbf{r} = 1\,. \tag{4.5}$$

So far, we have derived excitations around the linearised GPE, remembering that $(4.1) = (4.2) + (4.3)$. First, substituting this Ansatz into Eq. (4.1) gives

$$i\hbar\frac{\partial\delta\psi_j(\mathbf{r}, t)}{\partial t} = \hbar\omega_j u_j(\mathbf{r})e^{-i\omega_j t} - \hbar\omega_j v_j^*(\mathbf{r})e^{i\omega_j t}\,. \tag{4.6}$$

The linear part, Eq. (4.2), is straightforward, but the nonlinear part (4.3) requires some calculation:

$$g(\psi_0^3 + 2\psi_0^2\delta\psi_j + \psi_0^2\delta\psi_j^*) = g\psi_0^3 + 2g\psi_0^2\left[u_j(\mathbf{r})e^{-i\omega_j t} + v_j^*(\mathbf{r})e^{i\omega_j t}\right]$$
$$+ g\psi_0^2\left[u_j^*(\mathbf{r})e^{i\omega_j t} + v_j(\mathbf{r})e^{-i\omega_j t}\right]\,. \tag{4.7}$$

Summing all of the contributions, this becomes

$$\hbar\omega_j\left[u_j(\mathbf{r})e^{-i\omega_j t} - v_j^*(\mathbf{r})e^{i\omega_j t}\right] = \left[H_0 + g\psi_0^2 - \mu\right]\psi_0(\mathbf{r})$$
$$+ [H_0 - \mu]\left(u_j(\mathbf{r})e^{-i\omega_j t} + v_j^*(\mathbf{r})e^{i\omega_j t}\right)$$
$$+ 2g\psi_0^2\left[u_j(\mathbf{r})e^{-i\omega_j t} + v_j^*(\mathbf{r})e^{i\omega_j t}\right]$$
$$+ g\psi_0^2\left[u_j^*(\mathbf{r})e^{i\omega_j t} + v_j(\mathbf{r})e^{-i\omega_j t}\right]\,, \tag{4.8}$$

where the first term is simply the stationary GPE. Collecting terms with $e^{i\omega_j t}$ and $e^{-i\omega_j t}$ gives two equations:

$$e^{-i\omega_j t} : \qquad \hbar\omega_j u_j(\mathbf{r}) = [H_0 - \mu]u_j(\mathbf{r}) + 2g\psi_0^2 u_j(\mathbf{r}) + g\psi_0^2 v_j(\mathbf{r}) \tag{4.9}$$

$$e^{i\omega_j t} : \qquad -\hbar\omega_j v_j(\mathbf{r}) = [H_0 - \mu]v_j(\mathbf{r}) + 2g\psi_0^2 v_j(\mathbf{r}) + g\psi_0^2 u_j(\mathbf{r})\,, \tag{4.10}$$

where we have taken the complex conjugate of the second equation. These can equivalently be written in matrix form:

$$\epsilon_j\begin{pmatrix} u_j \\ v_j \end{pmatrix} = \begin{pmatrix} [H_0 - \mu] + 2g\psi_0^2 & g\psi_0^2 \\ -g\psi_0^2 & -[H_0 - \mu] - 2g\psi_0^2 \end{pmatrix}\begin{pmatrix} u_j \\ v_j \end{pmatrix} \tag{4.11}$$

where we have labelled the excitation energy $\epsilon_j = \hbar\omega_j$.

To find the general excitation spectrum of the GPE, one must first numerically compute the ground state ψ_0 with its corresponding chemical potential μ. One then constructs the matrix in Eq. (4.11), and numerically diagonalises it to find the eigenvalues ϵ_j, corresponding to the jth excitation energy, and the eigenvectors u_j and v_j, which give the mode structure in real space according to Eq. (4.4).

For the homogeneous condensate more analytic progress can be made. We note that in the continuous limit the excitation energy is also continuous, and that momentum is a good quantum number. In this case, $V = 0$, $\mu = g\psi_0^2$, and we can assume that the excitations are plane waves such that $u_j(\mathbf{r}) \to u_k e^{i\mathbf{k}\cdot\mathbf{r}}$ and $v_j(\mathbf{r}) \to v_k e^{i\mathbf{k}\cdot\mathbf{r}}$, and $\omega_j \to \omega(\mathbf{k})$. Note that

$$H_0 u_j(\mathbf{r}) = -\frac{\hbar^2 \nabla^2}{2m} u_j(\mathbf{r}) \to \frac{\hbar^2 k^2}{2m} u_k e^{i\mathbf{k}\cdot\mathbf{r}}, \tag{4.12}$$

and the same for v_j. Equations (4.9) and (4.10) then reduce to

$$\hbar\omega(\mathbf{k}) u_k = \frac{\hbar^2 k^2}{2m} u_k + g\psi_0^2 (u_k + v_k), \tag{4.13}$$

$$-\hbar\omega(\mathbf{k}) v_k = \frac{\hbar^2 k^2}{2m} v_k + g\psi_0^2 (u_k + v_k). \tag{4.14}$$

Adding and subtracting these gives:

$$\hbar\omega(\mathbf{k})(u_k + v_k) = \frac{\hbar^2 k^2}{2m}(u_k - v_k), \tag{4.15}$$

$$\hbar\omega(\mathbf{k})(u_k - v_k) = \left[\frac{\hbar^2 k^2}{2m} + 2g\psi_0^2\right](u_k + v_k) \tag{4.16}$$

Finally, multiplying (4.16) by $\hbar^2 k^2/2m$ and substituting (4.15) into it gives the homogeneous energy spectrum

$$\hbar\omega(k) = \sqrt{\frac{\hbar^2 k^2}{2m}\left(\frac{\hbar^2 k^2}{2m} + 2g\psi_0^2\right)}, \tag{4.17}$$

noting that for a homogeneous BEC this only depends on $k = |\mathbf{k}|$. This is often called the *Bogoliubov dispersion relation* after Nikolay Bogoliubov who first derived it. It relates the wave's angular frequency ω to its wavenumber k, or equivalently, its period $2\pi/\omega$ to its wavelength $2\pi/k$.

Notice that the phase velocity of the wave[1] $v_{\text{ph}} = \omega/k$ depends, in general, on k. Suppose that the initial condition at $t = 0$ is a generic wave packet, i.e., the superposition of different plane waves with different amplitudes and phases; since these waves move at different phase velocities, the wave-packet spreads out as it propagates, or *disperses*.

Consider the behaviour of the dispersion relation $\omega(k)$ for different regimes of interactions, as plotted in Fig. 4.1b.

- In the **absence of interactions** ($g = 0$) the dispersion relation reduces to $\hbar\omega = \hbar^2 k^2/2m$, in other words the wave behaves like a free particle of momentum $p = \hbar k$ and energy $\hbar\omega$. Note that ω is real; then the exponential terms in the solution of Eq. (4.4) have imaginary exponents and so describe a temporally-oscillating solution.

---

[1] The phase velocity of a wave is the rate at which its phase propagates in space.

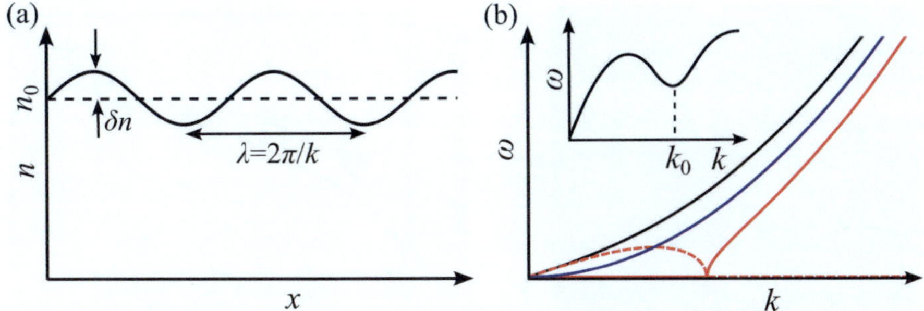

Fig. 4.1 **a** One dimensional sound waves, that is, sinusoidal perturbations of the background density n_0, of wavelength λ and amplitude $\delta n_0(x, t) \ll n_0$. **b** The dispersion relation $\omega(k)$ of the homogeneous (weakly-interacting) condensate, according to Eq. (4.17), for $g > 0$ (black line), $g = 0$ (blue line) and $g < 0$ (red line). Solid lines plot the real part of ω and dashed lines plot the imaginary part. For $g < 0$, ω becomes imaginary for small k; everywhere else ω is real. Inset: In helium II, the dispersion relation has a different and distinct shape, featuring a maxon (local maximum) and roton (local minimum). The roton wavenumber is indicated as k_0

- For **repulsive interactions** ($g > 0$), this free-particle behaviour ($\omega \sim k^2$) is recovered in the limit of large k/short waves. However, for low k/long waves, the dispersion relation is linear in k This linear behaviour is characteristic of sound waves—see below. As for $g = 0$, the angular frequencies are real.
- For **attractive interactions**, the situation is fundamentally different. For $g < 0$ and in the regime of sufficiently small k, ω^2 is negative and, correspondingly, ω becomes complex. Then these exponential terms develop *real* and *positive* exponents, such that they exponentially increase in amplitude over time. This signifies the dynamical instability of the homogeneous attractively-interacting condensate—small perturbations are not stable and grow out of control. In fact, this instability is due to the collapse instability we've already described for an attractive condensate; here the condensate prefers to collapse rather than stay as a uniform density profile.

In helium II the shape of the dispersion curve is somewhat different (see Fig. 4.1(inset)), due to the strong inter-atomic interactions in the system. The dispersion relation is linear for small k, but then features a local maximum, termed the *maxon*, and a local minimum, termed the *roton*. At even higher k the dispersion relation flattens off.

4.1.2 Sound Waves

For repulsive interactions ($g > 0$) and in the limit of small k/long waves, the above dispersion relation predicts waves whose angular frequency increases linearly with wavenumber. This is characteristic of sound waves. The phase velocity of these waves is $v_{\mathrm{ph}} = \omega/k \approx \sqrt{n_0 g/m}$, which is approximately constant for all wavelengths. This

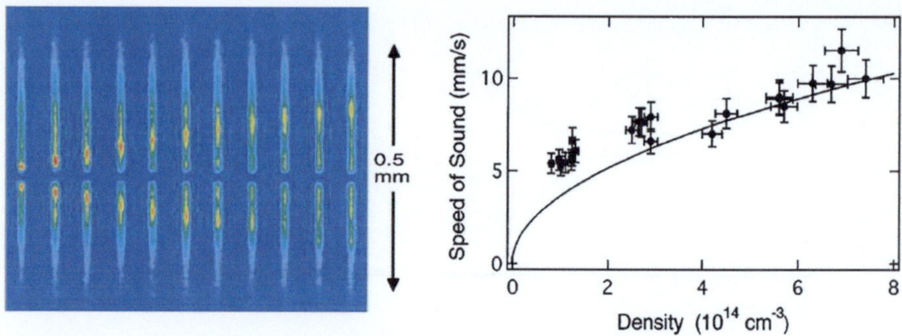

Fig. 4.2 Left: Experimental creation of sound waves after suddenly generating a 'hole' in the centre of the condensate using a laser beam. Shown is the column-integrated density profile of the condensate taken at regular intervals of time. Yellow, red and blue are respectively large, medium and small values of the density. Right: The measured speed of sound c (points) as a function of the background number density, in agreement with Eq. (4.18) (solid line). Reprinted figures with permission from [1]. Copyright 1997 by the American Physical Society

defines the speed of sound,

$$c = \sqrt{\frac{n_0 g}{m}}. \tag{4.18}$$

The physical interpretation of these waves is readily obtained using the Madelung transform. By perturbing the state of uniform density $n = n_0 = \mu/g$, we obtain the one-dimensional wave equation,

$$\frac{\partial^2}{\partial t^2} \delta n = c^2 \frac{\partial^2}{\partial x^2} \delta n, \tag{4.19}$$

where $\delta n(x, t) \ll n_0$ are density perturbations about the background condensate (here taken to be the homogeneous condensate), as shown in Fig. 4.1. The wave solution which we have found is one-dimensional—the wave propagates along x— but can be easily generalized to two and three dimensions.

In a trapped condensate ($V \neq 0$) the speed of sound will vary with the position due to the spatial dependence of the density. The speed of sound is less near the edge of the condensate BEC where the density tends to zero.

The prediction of sound waves was tested experimentally in Ref. [1] by using a laser beam to initially and suddenly "punch" a hole in the density at the centre of a condensate, much like a stone being thrown into a pond. This generated low amplitude ripples, i.e., sound waves, which travelled outwards along the condensate, as shown in Fig. 4.2a. The speed of the waves was found to follow the square-root of the density, as seen in Fig. 4.2b and in agreement with the prediction of Eq. (4.18).

In a trapped condensate of finite size, sound waves should have a wavelength considerably smaller than the condensate size (or, equivalently, the angular frequency of the wave should be considerably larger than the trap frequency); this, for example, is clearly satisfied in the experimental images in Fig. 4.2(left). However, if the wavelength of the density perturbations becomes of the order of the condensate size,

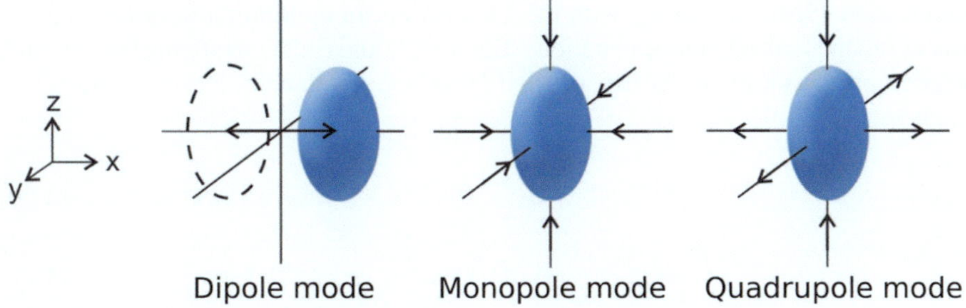

Fig. 4.3 The three common collective modes of a harmonically-trapped condensate

then these excitations involve a motion of the whole system. These are the *collective modes*.

There is a wide family of collective modes which are supported under harmonic trapping. Here we consider the simplest and most common types, illustrated in Fig. 4.3:

- The **dipole mode** corresponds to an oscillation of the condensate's centre-of-mass about the trap centre. In a harmonic trap, this oscillation occurs at the trap frequency in the respective direction. This mode is not affected by g since, in a harmonic trap, the centre-of-mass motion is decoupled from the internal dynamics. For this reason, this mode is often excited experimentally to measure the trap frequency.
- The **monopole mode** involves contraction-expansion oscillations of the condensate, which are in-phase across the directions.
- The **quadrupole mode** also involves contraction-expansion oscillations, but where the oscillation in one direction is in anti-phase to that in the other directions. Both the quadrupole and monopole modes are sensitive to interactions.

4.1.3 Scaling Solutions

Experimentally, the collective modes are typically induced by forming a stationary condensate, and then suddenly changing the harmonic trap. To induce the dipole mode, the trap can be suddenly translated in space; the condensate finds itself up the trap wall and begins to undergo centre-of-mass oscillations about the trap centre. To induce the monopole, quadrupole or similar modes, the trap frequencies can be suddenly changed in time. This scenario, in the absence of centre-of-mass motion, is the one we consider here. A similar methodology can be used to account for the centre-of-mass dynamics.

We consider a condensate which is at equilibrium at $t = 0$, with the trap frequencies suddenly changed for $t > 0$. We follow the approach introduced in Ref. [2]. We can model the ensuing oscillations of the condensate through the hydrodynamical description (Sect. 3.3), along with the Thomas-Fermi approximation (Sect. 3.5.2). Recalling the hydrodynamic equations, Eqs. (3.21) and (3.23), dropping terms which depend on the gradients of density, and introducing a general harmonic potential, leads to,

$$\frac{\partial n}{\partial t} + \nabla \cdot (n\mathbf{v}) = 0, \qquad (4.20)$$

$$m\frac{\partial \mathbf{v}}{\partial t} + \nabla \cdot \left(\frac{1}{2}mv^2 + \frac{1}{2}m(\omega_x^2 x^2 + \omega_y^2 y^2 + \omega_z^2 z^2) + gn \right) = 0. \qquad (4.21)$$

The equilibrium solution of the condensate at $t = 0$ is found by setting $\mathbf{v}$ and the time-derivatives to zero; then the second equation reduces to,

$$\nabla \cdot \left(\frac{1}{2}m(\omega_x^2 x^2 + \omega_y^2 y^2 + \omega_z^2 z^2) + gn \right) = 0. \qquad (4.22)$$

Integrating over space and rearranging gives,

$$n = \frac{2\mu - m(\omega_x^2 x^2 + \omega_y^2 y^2 + \omega_z^2 z^2)}{2g}, \text{ for } n \geq 0, \qquad (4.23)$$

where μ arises as the integration constant. This is the equilibrium density profile for the condensate in the Thomas-Fermi limit, as obtained in Sect. 3.5.2. We may also write this in the form,

$$n = n_0 \left(1 - \frac{x^2}{R_{x,0}^2} - \frac{y^2}{R_{y,0}^2} - \frac{z^2}{R_{z,0}^2} \right), \text{ for } n(x, y, z) \geq 0, \qquad (4.24)$$

where $R_{j,0} = \sqrt{2\mu/m\omega_j}$, with $j = x, y, z$, are the Thomas-Fermi radii and n_0 is the central density of the condensate at $t = 0$. Applying the usual normalization condition that $\int n(x, y, z)\, \mathrm{d}^3\mathbf{r} = N$ gives the expression for the central density,

$$n_0 = \frac{15\,N}{8\pi R_{x,0} R_{y,0} R_{z,0}}. \qquad (4.25)$$

Following a sudden change in the trap frequencies, $\omega_j \rightarrow \omega_j(t)$, the condensate profile becomes time-dependent. We consider the time-dependent density to maintain the same general shape throughout but where its dimensions become scaled over time. This is accounted for by making the radii time-dependent, $R_j \rightarrow R_j(t)$. If we further introduce *scaling parameters* $b_j(t) = R_j(t)/R_{j,0}$ then we can write the time-dependent profile as,

$$n(x, y, z, t) = \frac{n_0}{b_x b_y b_z} \left(1 - \frac{x^2}{b_x^2 R_{x,0}^2} - \frac{y^2}{b_y^2 R_{y,0}^2} - \frac{z^2}{b_z^2 R_{z,0}^2} \right), \qquad (4.26)$$

This is known as the *scaling solution*. The modified pre-factor accounts for the time-dependence of the central density. The initial conditions of the dynamics are,

$$b_j(t=0) = 1, \quad \dot{b}_j(t=0) = 0. \tag{4.27}$$

To satisfy the continuity equation, the velocity field which matches this density must be of the form,

$$\mathbf{v}(\mathbf{r}, t) = \frac{1}{2}\nabla \cdot \left[\alpha_x(t)x^2 + \alpha_y(t)y^2 + \alpha_z(t)z^2\right], \quad \alpha_j = \frac{\dot{b}_j}{b_j}. \tag{4.28}$$

One proceeds (although the derivation is beyond our scope) to introduce the time-dependent density and velocity distributions into the Thomas-Fermi hydrodynamic Eqs. (4.20) and (4.21). This leads to three coupled equations of motion for the scaling variables $b_j(t)$,

$$\ddot{b}_j + \omega_j(t)^2 b_j - \frac{\omega_{j,0}^2}{b_j b_x b_y b_z} = 0. \tag{4.29}$$

Remarkably, these equations involve the scaling variables b_i and the trap frequencies, only. What is also remarkable is that the same scaling equations of motion arise for a Gaussian Ansatz, a justifiable approximation for weak interactions. As such these scaling equations have a much wider coverage than the strongly-interacting Thomas-Fermi limit.

For a cylindrically symmetric trap $V(r, z) = m(\omega_r^2 r^2 + \omega_z^2 z^2)/2$, where $r^2 = x^2 + y^2$, this description reduces to two equations of motion,

$$\ddot{b}_r + \omega_r(t)^2 b_r - \frac{\omega_{r,0}^2}{b_r^3 b_z} = 0, \quad \ddot{b}_z + \omega_z(t)^2 b_z - \frac{\omega_{z,0}^2}{b_r^2 b_z^2} = 0. \tag{4.30}$$

To demonstrate the collective mode dynamics, we solve these two ordinary differential equations numerically for Thomas-Fermi condensate initially confined to a spherically-symmetric trap with $\omega_r = \omega_z = 2\pi \times 50$Hz. To induce a monopole mode, we reduce the trap frequencies by 10% for $t > 0$. As seen in Fig. 4.4a, the widths increase initially, and continue to oscillate around a new, larger equilibrium width. Characteristic of a monopole mode, the oscillations are in phase along r and z. Meanwhile, a quadrupole mode is generated by simultaneously increasing ω_z and decreasing ω_r (both by 10%). As seen in Fig. 4.4b, the condensate initially expands radially and shrinks axially, and continues to oscillate in anti-phase.

The above scaling equations of motion are valid for arbitrarily large mode amplitudes (providing the Thomas-Fermi approximation is maintained). In the limit of perturbatively small-amplitude modes (e.g. by linearizing about the equilibrium condensate, similar to Sect. 4.1 for a homogeneous system), one can determine the frequency of the collective modes analytically [3]. Under cylindrical symmetry, the mode frequencies obey,

$$\omega_{\mathrm{M,Q}}^2 = \omega_r^2\left(2 + \frac{3}{2}\lambda^2 \pm \frac{1}{2}\sqrt{16 - 16\lambda^2 + 9\lambda^4}\right), \tag{4.31}$$

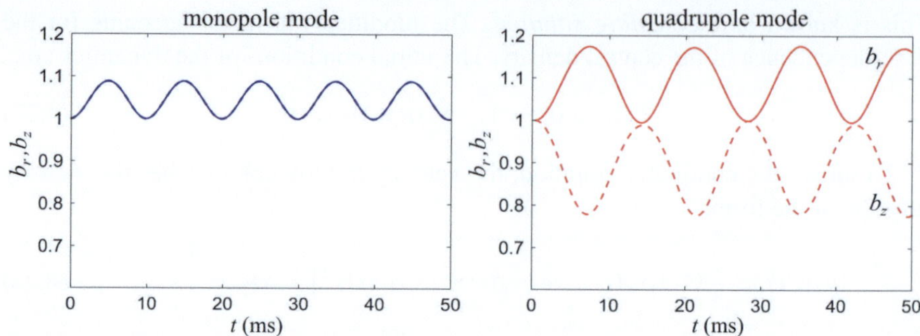

Fig. 4.4 A monopole mode and a quadrupole mode of a condensate in a cylindrically-symmetric harmonic trap. Shown are the radial and axial scaling parameters as a function of time, i.e. $b_r(t)$ and $b_z(t)$, as solved according to the Thomas-Fermi scaling equations of motion, Eq. (4.30). Note that for the monopole model, the two curves lie on top of each other

where the "+" refers to the monopole mode frequency, ω_M, and the "−" to the quadrupole mode frequency, ω_Q, and $\lambda = \omega_z/\omega_r$ is the *trap ratio*. For an approximately spherical trap ($\lambda \approx 1$) this gives,

$$\omega_M \approx \sqrt{5}\,\omega_r, \quad \omega_Q \approx \sqrt{2}\,\omega_r. \tag{4.32}$$

These are in close agreement with the frequencies of the oscillations in Fig. 4.4.

These scaling predictions give excellent agreement with the mode dynamics observed in experiments. These modes play an important role in this field. They are straightforward to generate experimentally and can be measured to high accuracy, and provide a versatile means to test theoretical models and assumptions. According to these predictions, the modes persist forever since the condensate has no viscosity. In reality, thermal dissipation causes the modes to decay over time, although usually on a much longer timescale than the oscillations themselves.

4.1.4   Expansion of the Condensate

A particular case of these scaling dynamics is when the trap is suddenly switched off and the condensate is allowed to expand freely. This is routinely performed in BEC experiments since some expansion of the gas is often necessary to enable imaging of small features such as dark solitons and vortices.

For $\omega_r(t) = \omega_z(t) = 0$ the cylindrically-symmetric scaling equations (4.30) reduce to,

$$\ddot{b}_r = \frac{\omega_{r,0}^2}{b_r^3 b_z}, \quad \ddot{b}_z = \frac{\omega_{z,0}^2}{b_r^2 b_z^2}. \tag{4.33}$$

Replacing the time variable with $\tau = \omega_z t$ and introducing the initial trap ratio, $\lambda = \omega_{z,0}/\omega_{r,0}$, gives,

$$\frac{d^2 b_r}{d\tau^2} = \frac{1}{b_r^3 b_z}, \quad \frac{d^2 b_z}{d\tau^2} = \frac{\lambda^2}{b_r^2 b_z^2}. \tag{4.34}$$

It is possible to obtain analytic expressions for $b_r(t)$ and $b_z(t)$ for the case of a cigar-shaped condensate, $\lambda \ll 1$ [2]. We proceed by expanding the solutions in powers of λ^2, i.e.,

$$b_r(\tau) = 1 + \alpha_r(\tau)\lambda^2 + \beta_r(t)\lambda^4 + ..., \quad b_z(\tau) = 1 + \alpha_z(\tau)\lambda^2 + \beta_z(\tau)\lambda^4 + ...$$

To lowest order in λ, the axial dynamics satisfy,

$$\frac{d^2 b_z}{d\tau^2} = 0 \quad \rightarrow \quad b_z(\tau) = 1. \tag{4.35}$$

In obtaining this solution for b_z we have applied the initial conditions in Eq. (4.27). Employing this result, we find the radial dynamics satisfy,

$$\frac{d^2 b_r}{d\tau^2} = \frac{1}{b_r^3} \quad \rightarrow \quad b_r(t) = \sqrt{1 + \tau^2}. \tag{4.36}$$

Continuing to second order, we find that the axial expansion satisfies,

$$b_z(\tau) = 1 + \lambda^2[\tau \arctan\tau - \ln\sqrt{1 + \tau^2}] + \mathcal{O}(\lambda^4). \tag{4.37}$$

The expansion develops very differently in the two directions. The radial size increases rapidly at first, whereas the axial spreading is weak, suppressed by the λ^2 factor. We define the aspect ratio of the condensate as $R_r/R_z = R_{r,0}b_r/R_{z,0}b_z$. Initially, $R_r/R_z = \omega_z/\omega_r = \lambda$. Over time the aspect ratio is modified by the scaling dynamics as $\lambda b_r/b_z$. Using the above analytic expressions, it can be shown that in the limit of large τ, the aspect ratio approaches the value,

$$\left(\frac{R_r}{R_z}\right)_{\tau \to \infty} = \frac{2}{\pi\lambda}. \tag{4.38}$$

In other words the condensate reverses its aspect ratio. These predictions agree accurately with experimental observations of condensate expansion, for example, see Fig. 18 of Ref. [3] and Fig. 4.5 of Ref. [4].

Problems

4.1 *Condensate oscillations:*

Consider a condensate of N atoms of mass m and interaction parameter g trapped by the harmonic potential $V(\boldsymbol{r}) = (1/2)m\omega_0^2 r^2$ where ω_0 is the trap frequency and $\boldsymbol{r} = (r, \theta, \phi)$ is the position in spherical coordinates (r, θ, ϕ). At length scales larger than the healing length, the condensate is described by the following hydrodynamic equations without the quantum pressure term:

$$\frac{\partial n}{\partial t} + \nabla \cdot (n\boldsymbol{v}) = 0,$$

$$m\frac{\partial \boldsymbol{v}}{\partial t} + \nabla\left(\frac{1}{2}mv^2 + gn + V\right) = 0,$$

where $n(\boldsymbol{r}, t)$ is the number density, $\boldsymbol{v}(\boldsymbol{r}, t)$ is the velocity and $v = |\boldsymbol{v}|$.

1. Determine the density profile $n(\boldsymbol{r} = n_{eq}(r)$ in the equilibrium state (no time dependence, $\boldsymbol{v} = \boldsymbol{0}$) under the normalization condition $N = \int n(\boldsymbol{r})\, d^3\boldsymbol{r}$. What are the the peak density $\hat{n}_0$ at the centre of the condensate and the radius R of the condensate in terms of the parameters N, m, g and ω_0?
2. Perturb the equilibrium state by assuming small perturbations (denoted by primes)

$$n(\boldsymbol{r}, t) = n_{eq}(r) + n'(r, \theta, \phi, t) + \cdots,$$
$$\boldsymbol{v}(\boldsymbol{r}, t) = \boldsymbol{v}'(r, \theta, \phi, t) + \cdots,$$

   hence derive linearized equations for the perturbations n' and $\boldsymbol{v}'$.
3. Assume that all perturbations depend on time as $e^{i\omega t}$, where ω is the angular frequency, hence show that the density perturbations obey

$$\omega^2 n' = -\frac{\omega_0^2}{2}(R^2 - r^2)\nabla^2 n' + \omega_0^2 r\frac{\partial n'}{\partial r}.$$

4. Assume that the density perturbations depend on r, ϕ, θ as $r^\ell e^{im\phi} P_\ell^m(\cos\theta)$ where ℓ and m are integers and P_ℓ^m are associated Legendre polynomials, hence show that the angular frequency of the perturbations is

$$\omega = \sqrt{\ell}\,\omega_0.$$

Useful formulae:

In spherical coordinates, the Laplacian of a scalar field $f(r, \theta, \phi)$ is

$$\nabla^2 f = \frac{\partial^2 f}{\partial r^2} + \frac{2}{r}\frac{\partial f}{\partial r} + \frac{1}{r^2 \sin \theta}\frac{\partial}{\partial \theta}\left(\sin \theta \frac{\partial f}{\partial \theta}\right) + \frac{1}{r^2 \sin^2 \theta}\frac{\partial^2 f}{\partial \phi^2}.$$

The associated Legendre polynomials satisfy the following associated Legendre equation:

$$(1 - x^2)\frac{d^2}{dx^2}P_\ell^m(x) - 2x\frac{d}{dx}P_\ell^m(x) + \left[\ell(\ell + 1) - \frac{m^2}{1 - x^2}\right]P_\ell^m(x) = 0,$$

4.2 *Simulating collective modes:*

Use the 1D GPE code provided in the Appendix, or your own code, to simulate the centre-of-mass modes and monopole modes of a harmonically-trapped, repulsively-interacting 1D condensate. Excite a centre-of-mass oscillation by instantaneously shifting the trap by some small distance at $t = 0$. Similarly, excite a monopole mode by slightly modifying the trap frequency at $t = 0$. Compare the frequencies to the analytic predictions.

4.3 *Generation and propagation of sound waves:*

Using the 1D GPE code provided in the Appendix, simulate the generation and propagation of sound waves in a trapped condensate. Follow the protocol of Andrews et al. [1] by suddenly introducing a repulsive Gaussian potential $V(x) = V_0 \exp(-x^2/\sigma^2)$ (from time $t = 0$ in the simulation) with width $\sigma = 2$ microns and amplitude $V_0 = 1 \times 10^{-31}$J. Verify that the observed speed of the outgoing wave agrees with the predicted speed of sound of the system.

4.4 *Quantum carpets:*

Use the 1D GPE code in the Appendix to simulate the so-called "quantum carpet" states, a highly-excited state of the condensate. Starting from the Gaussian harmonic oscillator ground state, release the non-interacting condensate into an infinite square well (achieve by setting the potential to a high value towards the edge of the box, and zero elsewhere). Repeat for repulsive and contrast with the non-interacting case.

Now simulate the longer-term behaviour. The wavefunction undergoes revivals, known as the Talbot effect, and forms a "quantum carpet" [5].

References

1. M.R. Andrews et al., Phys. Rev. Lett. **79**, 553 (1997)
2. Y. Castin, R. Dum, Phys. Rev. Lett. **77**, 5315 (1996)
3. F. Dalfovo, S. Giorgini, L.P. Pitaaevskii, S. Stringari, Rev. Mod. Phys. **71**, 463 (1999)
4. C.F. Barenghi, N.G. Parker, *A Primer on Quantum Fluids* (Springer, Berlin, 2016)
5. I. Marzoli et al., Acta Phys. Slov. **48**, 323 (1998)

Solitons

5

Abstract

In the previous chapter we described the linear waves that are supported in quantum fluids. Here we introduce nonlinear waves in the form of solitons, from their idealized mathematical form to how solitons manifest in experimental systems, where dimensionality, inhomogeneity, thermal and quantum effects all modify the behaviour.

5.1 Soliton Solutions

In one-dimension and for a homogeneous system (corresponding to the absence of an external potential), the GPE is,

$$i\hbar\frac{\partial\psi}{\partial t} = -\frac{\hbar^2}{2m}\frac{\mathrm{d}^2\psi}{\mathrm{d}x^2} + g|\psi|^2\psi, \tag{5.1}$$

where the variables and parameters take their 1D definitions. This is a form of the 1D *nonlinear Schrödinger equation*. This equation is well-studied in the context of nonlinear optics [1]. For $g < 0$ the NLSE is called "focussing" in the optics literature, as the attractive cubic nonlinearity leads to focusing of light pulses in optical fibres, while for $g > 0$ it is termed "defocussing" due to the fact that the repulsive nonlinearity promotes the broadening of the light pulses.

Equation (5.1) has the special property of being *integrable* such that its solutions possess an infinite set of conserved quantities (integrals of motion). The simplest of these quantities (and those with a clear physical interpretation for our system) are the norm N, the momentum $\mathcal{P}$ and the energy E,

$$N = \int\limits_{-\infty}^{+\infty} |\psi|^2 \mathrm{d}x, \tag{5.2}$$

$$P = \frac{i\hbar}{2} \int\limits_{-\infty}^{+\infty} \left(\psi \frac{\partial \psi^*}{\partial x} - \psi^* \frac{\partial \psi}{\partial x} \right) \mathrm{d}x, \tag{5.3}$$

$$E = \int\limits_{-\infty}^{+\infty} \left(\frac{\hbar^2}{2m} \left| \frac{\partial \psi}{\partial x} \right|^2 + \frac{g}{2} |\psi|^4 \right) \mathrm{d}x. \tag{5.4}$$

It is due to these special properties that Eq. (5.1) supports solutions known as *solitons*.

Solitons are nonlinear waves which arise in many areas of physics, from fluids to optics to plasmas [2]. Solitons have three characteristic properties [3]:

- They have a permanent, unchanging form.
- They are localized in space.
- They emerge unscathed from collisions with other solitons.

Their permanent form is due to wave dispersion being perfectly balanced by the nonlinearity; as a consequence, solitons propagate without spreading in space. This makes them analogous to particles, and motivated their particle-like name "solitons".

The soliton solutions of the nonlinear Schrödinger equation were obtained in the pioneering works of Zakharov and Shabat, using a technique called the *inverse scattering transform* (see Ref. [4] for more information). Depending on the sign of the nonlinearity/interactions g, *dark solitons* and *bright solitons* can be supported, as we see in this Chapter. Dark and bright solitons were first studied in the context of nonlinear optics; there they correspond to a dip and a peak in an optical intensity field, respectively, giving rise to their names.

In reality, condensates are three-dimensional and feature trapping potentials, and so "solitons" therein are not strictly solitons in the true mathematical definition. However, these solutions possess the key solitonic properties and so we continue to term them as "solitons". In some of the literature, these solutions are instead referred to as "solitary waves" to distinguish them from the strict mathematical soliton solutions.

5.2 Dark Solitons

5.2.1 Dark Soliton Solutions

Dark solitons are supported for repulsive interactions ($g > 0$). These nonlinear waves consist of a localized density dip with a phase jump across it, and propagate at speed u.

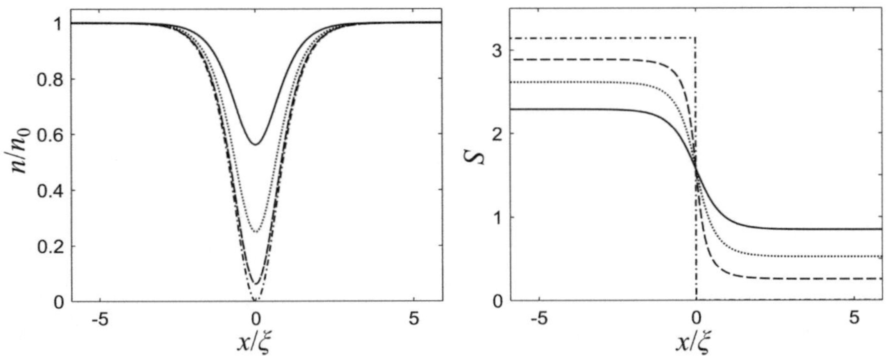

Fig. 5.1 Density $n(x)$ and phase $S(x)$ profiles of dark solitons with various speeds: $u/c = 0$ (dot-dashed line), $u/c = 0.25$ (dashed line), $u/c = 0.5$ (dotted line) and $u/c = 0.75$ (solid line). Density is scaled in terms of the background density of the homogeneous system, n_0, and position in terms of the healing length, ξ

The speed can exist in the range $0 < u \le c$, where c is the speed of sound. A broad review of dark solitons in condensates is given in Ref. [5].

The general dark soliton solution to Eq. (5.1) with $g > 0$ is,

$$\psi(x, t) = \sqrt{n_0} \left\{ B \tanh \left[\frac{B(x - ut)}{\xi} \right] + i \frac{u}{c} \right\} \exp \left(-\frac{i \mu t}{\hbar} \right), \qquad (5.5)$$

where $B = \sqrt{1 - u^2/c^2}$. The density and phase profiles of dark soliton solutions with various speeds u/c are shown in Fig. 5.1. Note that the soliton width is always of the order of the healing length ξ.

A dark soliton state is an excited state; the ground state is the soliton-free homogeneous density. In the lab frame (the frame at rest with the background condensate), the soliton is a moving solution. However, dark solitons become stationary solutions in the moving frame, e.g. the $u/c = 0.5$ soliton is a stationary (excited) state in the frame moving at the same speed.

The density depression and the phase profile of the dark soliton vary with its speed speed. If $u = 0$ then one obtains the stationary *black soliton*, whose density profile is,

$$n(x) = n_0 \tanh^2(x/\xi). \qquad (5.6)$$

The density of the black soliton goes to zero at its centre, and the phase jump is a sharp step of π. At the opposite speed extreme, the $u = c$ dark soliton has zero density depth and no phase slip, i.e. it is indistinguishable from the background.

For more general speeds, the *soliton depth*, that is, the maximum depth of the soliton density depression, follows from Eq. (5.5) as,

$$n_{\rm d} = n_0(1 - u^2/c^2). \qquad (5.7)$$

The phase slip across the soliton also varies smoothly with speed. We define the total phase slip ΔS as the difference between the phases at $\pm\infty$, i.e. $\Delta S = S(x = -\infty) -$

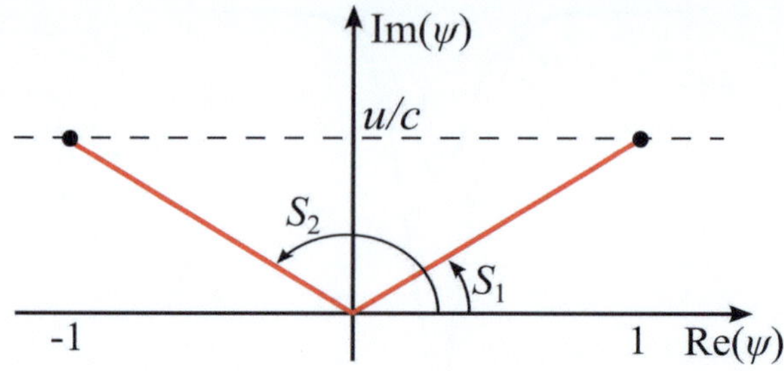

Fig. 5.2 Phase jump from $x \to \infty$ (at $(1, u/c)$) to $x \to -\infty$ (at $(-1, u/c)$) corresponding to a dark soliton

$S(x = \infty)$. The dark soliton solution becomes $\psi \to \sqrt{n_0}(1 + iu/c)$ for $x \to \infty$, and $\psi \to \sqrt{n_0}(-1 + iu/c)$ for $x \to -\infty$. Therefore, as we move from $x = \infty$, through the origin, to $x = -\infty$, the phase of ψ (that is, the angle S between $\mathrm{Re}(\psi)$ and $\mathrm{Im}(\psi)$) changes from $S = S_1$ at the point $(1, u/c)$ on the complex plane ψ, to $S = \pi/2$ at $(0, u/c)$, to $S = \pi - S_1$ at $(-1, u/c)$. Hence the change of the phase from $x = \infty$ to $x = -\infty$ is $\Delta S = 2\arccos(u/c)$. Taking the limit $u/c \to 0$, we conclude that the phase jump is $\Delta S = \pi$, as above A dark soliton is therefore a 1D *phase defect*: a discontinuity of the quantum mechanical phase (Fig. 5.2).

The energy of the dark soliton can be defined as,

$$E_\mathrm{s} = \frac{4}{3} n_0 \hbar c B^3, \tag{5.8}$$

and its momentum as,

$$\mathcal{P}_\mathrm{s} = -\frac{2\hbar n_0 u B}{c} + 2\hbar n_0 \arctan\left(\frac{Bc}{u}\right). \tag{5.9}$$

It is left as an exercise to derive these results (Problem 5.1).

5.2.2   Particle-Like Behaviour

Here we show that the dark soliton is analogous to a classical particle. Differentiating the soliton energy E_s and momentum $\mathcal{P}_\mathrm{s}$ with respect to speed u gives,

$$\frac{\mathrm{d}E_\mathrm{s}}{\mathrm{d}u} = -\frac{4n_0 \hbar u B}{c}, \quad \frac{\mathrm{d}P_\mathrm{s}}{\mathrm{d}u} = -\frac{4n_0 \hbar B}{c}. \tag{5.10}$$

Note that the energy and momentum both *decrease* with as the soliton gets faster. Using these results and the chain rule we can then form,

$$\frac{\mathrm{d}E_\mathrm{s}}{\mathrm{d}\mathcal{P}_\mathrm{s}} = \frac{\mathrm{d}E_\mathrm{s}}{\mathrm{d}u}\frac{\mathrm{d}u}{\mathrm{d}\mathcal{P}_\mathrm{s}} = u. \tag{5.11}$$

This result informs us that the dark soliton behaves like a classical particle with effective mass $m_s = \dfrac{dP_s}{du}$, which we can see from Eq. (5.10) is,

$$m_s = -\frac{4n_0\hbar B}{c}.$$ (5.12)

The dark soliton thus behaves as a classical particle with *negative mass*. This is not surprising given that the dark soliton is an absence of atoms. We can also estimate the ratio of the soliton mass to the atomic mass,

$$\frac{|m_s|}{m} \sim \frac{4\hbar n_0}{mc} \sim 4\xi n_0.$$ (5.13)

ξn_0 is the number of atoms within a ξ-sized length of the system; since typically $\xi n_0 \gg 1$ we conclude that the soliton is considerably more massive than an atom.

In the limit of slow solitons ($B \approx 1$) the energetics of the soliton reduces to a particularly simple form. Taking the Taylor expansion of the soliton energy, Eq. (5.4), about $v = 0$ and up to terms in v^2 gives,

$$E(v) = E(0) + \frac{dE}{du}\bigg|_0 u + \frac{1}{2}\frac{d^2E}{du^2}\bigg|_0 u^2 + \mathcal{O}(v^3)$$ (5.14)

$$\approx \frac{4}{3}n_0\hbar c - \frac{2n_0\hbar u^2}{c}.$$ (5.15)

Introducing the soliton mass, $m_s = -4n_0\hbar/c$, we obtain,

$$E(v) = E_0 + \frac{1}{2}m_s u^2,$$ (5.16)

which is the form for a classical particle moving in free space with *rest mass* E_0 and kinetic energy $m_s u^2/2$. This relation shows that, due to the negative effective mass, slower solitons have greater energy. Conversely, if the soliton loses energy (due to dissipative processes such as thermal dissipation) it will speed up.

5.2.3 Particle-Like Oscillations in a Harmonic Trap

We here consider how the particle-like behaviour of the dark soliton manifests within a harmonically-trapped condensate. Assuming that the background density is slowly-varying in space, we can define the soliton energy as per the homogeneus system, Eq. (5.8), but where the uniform density is replaced by its local value, $n(x)$. We then obtain,

$$E_s(u, x) = \frac{4}{3}\frac{\hbar}{\sqrt{mg}}\left(n(x)g - mu^2\right)^3.$$ (5.17)

For a harmonic potential $V(x) = m\omega^2 x^2/2$, we take the condensate profile to follow the Thomas-Fermi form, $n(x)g = \mu - V(x) = \mu - m\omega^2 x^2/2$. Inserting into the above equation gives,

$$E_s(u, x) = \frac{4}{3} \frac{\hbar}{\sqrt{mg}} \left(\mu - \frac{1}{2} m\omega_x^2 x^2 - mu^2 \right)^3. \tag{5.18}$$

We proceed to expand this expression for slow solitons ($u/c \ll 1$) and close to the origin, via a two-dimensional Taylor series,

$$\begin{aligned} E_s(u, x) = \ & E_s(0, 0) + x \frac{\partial E_s}{\partial x}\bigg|_{(0,0)} + u \frac{\partial E_s}{\partial u}\bigg|_{(0,0)} \\ & + \frac{1}{2} \left[u^2 \frac{\partial^2 E_s}{\partial u^2}\bigg|_{(0,0)} + 2ux \frac{\partial^2 E_s}{\partial x \partial u}\bigg|_{(0,0)} + x^2 \frac{\partial^2 E_s}{\partial x^2}\bigg|_{(0,0)} \right] + \mathcal{O}(x^3, u^3). \end{aligned}$$

This leads to a result of the form,

$$E_s(u, x) = E_0 + \frac{1}{2} m_s u^2 + \frac{1}{4} m_s \omega_x^2 x^2, \tag{5.19}$$

where $m_s = -4\hbar n_0/c$ and $E_0 = 4\hbar \mu^{3/2}/4\sqrt{mg}$ are the effective mass and rest mass in the low-speed limit. For a mass m obeying the classic harmonic oscillator $\ddot{x} + \omega_x^2 x = 0$, the corresponding expression is $E = E_0 + mu^2/2 + m\omega_x^2 x^2/2$. By comparison we see that the soliton behaves like an oscillator with effective frequency,

$$\omega_s = \frac{\omega_x}{\sqrt{2}}. \tag{5.20}$$

This result for dark soliton oscillations in a harmonic trap was first predicted in Ref. [6] and typically gives excellent agreement with experiments and GPE simulations providing the system is suitably quasi-one-dimensional, close to the Thomas-Fermi regime and that thermal dissipation is weak.

5.2.4 Longitudinal Instability of Dark Solitons (Infinite-Sized Systems)

As established in Sect. 5.1, the dark soliton is an exact one-dimensional solution in a homogeneous system, i.e. when propagating on a uniform background density. However the soliton becomes unstable when propagating through an inhomogeneous background density, leading to the decay of the soliton via the emission of sound waves and, in some cases, splitting into multiple smaller dark solitons [6–8]. From a mathematical point of view, the addition of a spatially-varying potential $V(x)$ to the homogeneous 1D GPE breaks the integrability of the system and the robustness of the true soliton solution.

This instability is a general occurence whenever a dark soliton moves longitudinally through an inhomogeneous density, as established for a soliton interacting with a range of inhomogeneities including optical lattices, linear ramps, localised potentials and micro-traps [6–9]. Since experimental condensates are inherently inhomogeneous, this instability is widespread. However, as we will see shortly, the impact on the soliton tends to be small in magnitude and can be masked by competing processes of sound absorption. Note that an exceptional case is for a soliton interacting with a hard wall–in this special case the soliton behaves as if colliding with its mirror image and reflects elastically.

The longitudinal instability is most clearly illustrated in an infinite-sized system for which the emitted sound waves propapage to infinity without the complication of re-interacting with the soliton, as we will explore in the following section.

Longitudinal Instability for a Step Potential

A simple scenario to illustrate the instability is that of a dark soliton incident upon a potential step of amplitude V_0, depicted in Fig. 5.3. In the step region the condensate density has a reduced value $n_0(1 - V_0/\mu)$ relative to the unperturbed density n_0. At the step edge the density smoothly heals between these values. Since the local speed of sound is related to the local condensate density, this system is the analog of the interface between optical media of different refractive index.

Figure 5.3 (bottom panel) illustrates the interaction of the soliton with the step according to simulations of the 1D GPE. There are three qualitative regimes of interaction:

- *Dissipative transmission*: For relatively low V_0, the soliton transmits over the step. To a first approximation, the soliton maintains its depth (as defined in Eq. (5.7)) and emerges with reduced speed (due to the reduced local speed of sound). As the soliton traverses the interface, counter-propagating sound pulses are emitted (with amplitudes typically in the range $1-10\%$). The loss of energy from the soliton contributes to an increase in its speed (due to its negative effective mass); however, the reduced ambient speed of sound has a bigger impact on the soliton speed and so overall the soliton speed is reduced.
- *Dissipative reflection*: For larger step heights (but not so large as to create a hard interface), the soliton reflects, with emission of sound waves.
- *Elastic reflection*: For high steps $V_0 \gg \mu$, the fluid is fully expelled from the step region such that the interface is essentially a hard wall; in this regime the soliton reflects elastically with no decay.

Qualitatively similar behaviour arises when dark solitons interact with other forms of sharply-varying potentials, for example, localised barriers [11].

For completeness we also comment on the case where the step potential is negative $V_0 < 0$; here the soliton either transmits with sound emission and an increase in speed (due to the increased ambient speed of sound in the step region) or breaks into multiple solitons [7].

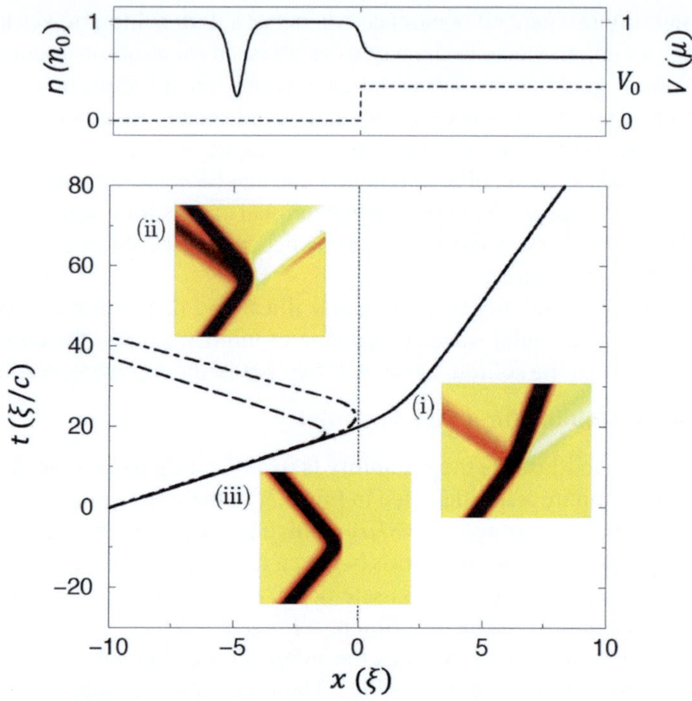

Fig. 5.3 Top panel: Schematic of a dark soliton (initial speed $u = 0.5c$) incident upon a potential step (of amplitude V_0) in an infinite-sized system. Bottom panel: Representative results for the three regimes of soliton-step interaction showing the path of the soliton: (i) *dissipative transmission* ($V_0 = 0.2\mu$, solid line), (ii) *dissipative reflection* ($V_0 = 0.55\mu$, dot-dashed line), (iii) *elastic reflection* ($V_0 \gg \mu$, dashed line). Insets are space-time plots of the time-dependent fluid density (having subtracted the ground state background density to highlight excitations). These are simulated results of the 1D GPE. Reprinted with permission from [10]. Copright 2004, The Author(s). All rights reserved

Longitudinal Instability for Slowly-Varying Potentials

For a slowly-varying background density, numerical experiments show that the power radiated from the soliton as sound waves P_s follows the relation,

$$P_s \equiv \frac{dE_s}{dt} = \kappa a_s^2, \tag{5.21}$$

where E_s is the soliton energy, a_s is the soliton acceleration (relative to the fluid[1]) and κ is a coefficient.

This instability and emission law has an analog in nonlinear optical systems, in which it is caused by spatial inhomogeneities of the nonlinearity within the optical

---

[1] Acceleration of the whole fluid induces no additional sound emission from the soliton, the only contribution comes from soliton acceleration relative to the ambient background.

medium, for example, due to saturation effects. By employing multiscale asymptotic techniques, it has been derived that the power radiated from the soliton follows an acceleration-squared relation but with a coefficient which is a function of the soliton speed and local field intensity (which is analogous to density in the context of the GPE) [12]. This result translates well to dark solitons in inhomogeneous condensates. A convenient scenario to demonstrate this is a condensate experiencing a potential which is harmonic up to some distance from the origin and flat beyond; this truncated potential allows emitted sound to propagate to infinity without re-interacting with the soliton. As shown in Fig. 5.4a, the dark soliton (to a first approximation) undergoes harmonic oscillations with frequency $\omega_s \approx \omega_x/\sqrt{2}$ in accordance with the particle-analogy prediction in Sect. 5.2.3. Counter-propagating sound waves (light/dark bands) are visibly emitted, with the power emission (Fig. 5.4b) oscillating in time. Note how the power radiated is zero when the soliton is at the trap centre,; here the background density is locally homogeneous and the soliton is locally stable. Maximal power emission occurs at the turning points of the soliton position, where the local acceleration and density inhomogeneity are maximal. The power emission is well captured by the acceleration-squared result, both with a fitted coefficient and with the coefficient provided by the nonlinear optics approach [12].

The loss of energy of the soliton, coupled with its negative effective mass, causes the oscillations to grow over time; this has been coined "anti-damping". In Fig. 5.4a the anti-damping causes a small but noticeable departure from conservative oscillations (white dashed line). Indeed, in this example, the growth in oscillation amplitude eventually allows the soliton to escape in inner harmonic region to the outer region. Since the outer region has uniform background density, the soliton propagates with constant speed and with no further sound emission.

Longitudinal Instability in Finite-Sized Systems

The outcome of the longitudinal instability becomes more complicated in finite-sized, bounded condensates due to the re-interaction and co-existence of the emitted sound with the soliton, and depends sensitively on the shape of the imposed potential.

For purely harmonic potentials, soliton decay is prohibited by the emitted sound being focussed back into the soliton to achieve a quasi-equilibrium. The steady state is that of the soliton oscillating within a background condensate which is excited with sound waves and low-lying collective modes. These excitations slightly perturb the soliton motion but, to a close approximation, the soliton oscillates with constant amplitude and at close to the predicted frequency $\omega_s \approx \omega_x/\sqrt{2}$, as illustrated in Fig. 5.5.

A further special case where dark soliton decay is prohibited is that of an infinite square well; the soliton propagates through the condensate with no sound emission and reflects elastically from the walls.

Deviations from these two special cases–the purely harmonic trap and the infinite square well–typically lead to a net decay of the soliton. This has been demonstrated for solitons in harmonic traps which are modified with a localised dimple [6,8], obstacle [11] or optical lattice [13,14], in square wells with "soft" walls [15], and potentials which are Gaussian or quartic [16]. An example in depicted in Fig. 5.6 for

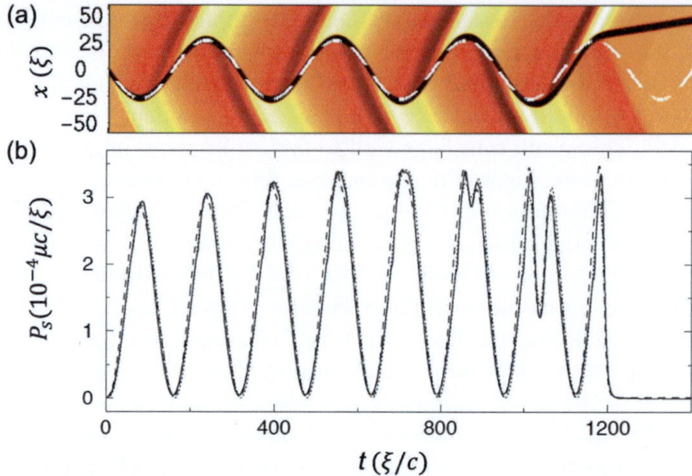

Fig. 5.4 Sound emission from a dark soliton oscillating in a finite harmonic trap embedded within an infinite-sized system. The soliton is initially positioned at the centre of the harmonic trap with a velocity in the negative x direction. Top panel: Space-time plot of the time-dependent condensate density (minus the ground state density). The dashed white line shows the expected soliton trajectory in the absence of sound emission. Sound emission is evident (bright/dark bands), with an amplitude of $\sim 2\% n_0$. **b** Power radiated from the oscillating soliton P_s. The acceleration-squared law with constant, fitted coefficient (dashed line) and with the derived coefficient in the context of nonlinear optics (dotted line) agree well with the power measured numerically from the simulations (solid line). Reprinted figure with permission from [8]. Copyright 2003 by the American Physical Society

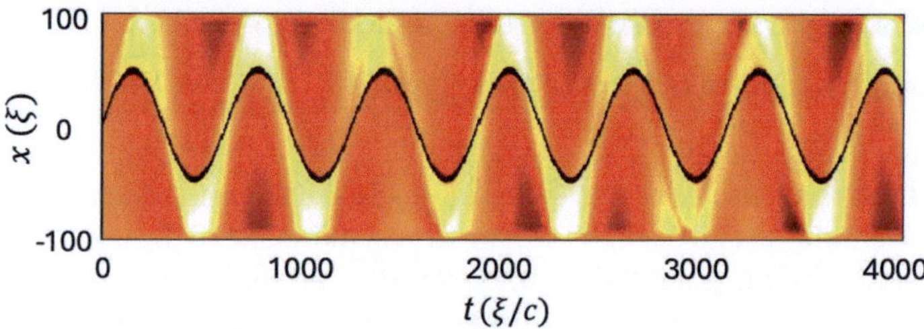

Fig. 5.5 A dark soliton oscillating in a harmonically-trapped condensate, simulated by the 1D GPE. Shown is a space-time plot of the time-dependent fluid density (having subtracted the ground state background density to highlight excitations). The soliton undergoes harmonic-like oscillations and excites the background condensate. Reprinted with permission from [10]. Copyright 2004, The Author(s). All rights reserved

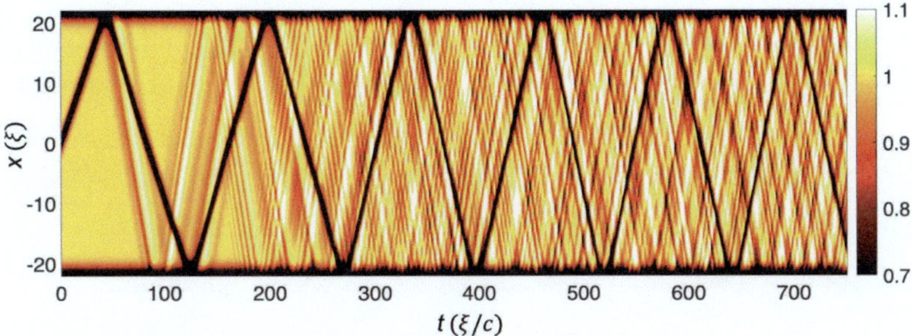

Fig. 5.6 A dark soliton oscillating in a box-like potential with "soft" walls, as simulated by the 1D GPE. The soliton emits sound waves each time it reflects from a wall, leading to gradual decay of the soliton and growth of the ambient sound field. Reprinted under CC-BY-4.0 license from [15]. Copyright 2017, The Author(s)

a soliton oscillating in a box potential with soft walls. Each time the soliton reflects from a wall, sound is emitted, leading to the gradual decay of the soliton and the growth of a sound field in the condensate. This is akin to the thermalization of the system. At late times the original soliton becomes indistinguishable from the sound field.

The soliton-sound interaction can be engineered to transfer energy from the sound field to the soliton, rather than vice-versa. This was demonstrated through simulations in which a time-dependent potential was used to drive energy into the sound field of the condensate which, in turn, led to an increase in the energy of a co-existent dark soliton [17].

5.2.5 Collisions

The 1D homogeneous GPE (5.1) possesses exact solutions for arbitrary numbers of solitons with arbitrary speeds, including their collisions [4]. The two-soliton solution can be found in Ref. [5]. The collisions of two dark solitons of equal speed are shown in Fig. 5.7 for different speeds. Notice how the solitons emerge from the collision with unchanged form and speed, one of the fundamental properties of solitons. The only effect on the outgoing solitons is a shift in their position; this is known as the *phase shift* during the collision.

There are two regimes of collision [5]. For high incoming speed ($u/c \geq 0.5$) the solitons pass through each other. However for low incoming speed ($u/c < 0.5$), the solitons reflect: as they begin to overlap, the solitons slow down and become deeper until they become "black" and stationary, before reversing direction and propagating away from each other to reach their original speeds. This interaction is analogous to a short-range repulsive interaction between two classical particles; indeed one can derive the approximate form for this interaction potential and show it decays exponentially with the soliton separation [5].

Due to this repulsive interaction between dark solitons, an infinite chain of suitably-spaced dark solitons is a stationary solution. Indeed, by using the particle-

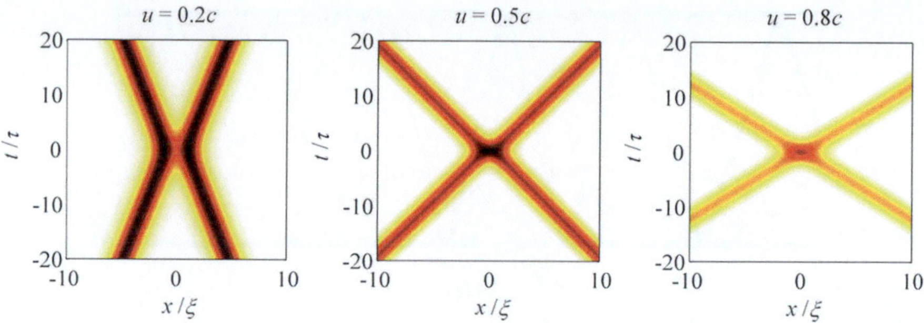

Fig. 5.7 Collisions of two dark solitons at different incoming speeds. White represents the background density n_0 and darker shades represents lower densities. Note that the time axis is centred on the soliton collision

analogy such a chain of dark solitons can be described by the Toda lattice, a simplified model from solid state physics for describing a 1D crystal. The Toda lattice system is itself integrable such that the envelope of the lattice excitations supports soliton solutions. Thus one can support solitons of chains of solitons, so-called *hypersolitons* [18].

5.2.6   Instability in Higher Dimensions: The Snake Instability

Dark soliton structures can be supported in real three-dimensional condensates providing there is sufficient transverse confinement to render the system effectively one-dimensional, for example, within a tight waveguide potential. The dark soliton is embedded as a plane of reduced density and with transversely-uniform phase; this is depicted in Fig. 5.8 (left panel).

The high-dimensional dark solitons are prone to a transverse instability, under which the plane is unstable to long-wavelength transverse perturbations. This is known as the *snake instability* due to the snake-like corrugation of the unstable soliton stripe. As the perturbation grows and the soliton bends, transverse velocity fields are triggered which tear the plane apart, with the product being one or more vortex rings (plus some transient sound waves). In two-dimensional systems, dark solitons are embedded as stripes and the products emerging from the snake instability are vortex-antivortex pairs.

This instability was first studied within the context of nonlinear optics and the homogeneous two-dimensional NLSE. According to linear stability analysis [21], the dark soliton stripe is unstable to transverse perturbations with wavenumbers below a critical value. p_c which satisfies,

$$p_c^2 = \frac{2\pi}{\xi}\left(B^2 - 2 + 2\sqrt{B^4 + B^2 + 1}\right), \tag{5.22}$$

where $B = \sqrt{1 - u^2/c^2}$ as introduced in Sect. 5.2.1. Figure 5.8(right) depicts how the critical wavenumber varies with soliton speed; for faster solitons the critical

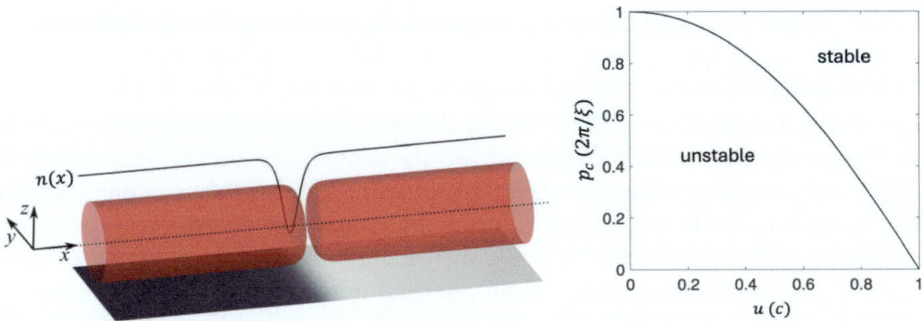

Fig. 5.8 Left panel: Schematic of a dark soliton plane embedded within a quasi-1D condensate. Shown is a low-density isosurface of the condensate (red surfaces), a 2D projection of the phase profile across the dark soliton (greyscale heat map) and the integrated 1D density profile (line). Right panel: The variation of the critical wavenumber, according to linear stability analysis, for the snake instability p_c of a dark soliton stripe in a homogeneous 2D system as a function of the soliton speed u

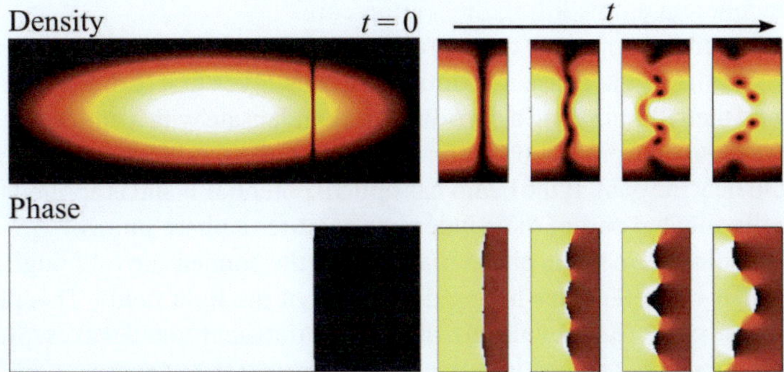

Fig. 5.9 Two-dimensional density and phase images through a 3D harmonically-confined condensate, which has a 3D dark soliton positioned initially to one side (left). The initial phase appears as a step profile. If the condensate is too wide, the soliton undergoes the snake instability (right), leading to its decay into vortex rings. Reprinted with permission from [10]. Copright 2004, The Author(s). All rights reserved

wavenumber decreases. In the unstable regime and close to the critical value, non-linear effects and radiation from the perturbed soliton stripe can halt the runaway growth of the transverse perturbations [22].

The snake instability of 3D dark solitons has been observed experimentally in trapped condensates with relatively weak transverse confinement [23,24]. The instability and ensuing decay is illustrated in Fig. 5.9, according to a GPE simulation. At the crossover between 1D and 3D geometries, it is possible to form *solitonic vortices*, a type of vortex which is squeezed by the system geometry and has some dark soliton-like properties. These states become energetically favoured over a dark soliton stripe when the transverse size of the system is greater than approximately 6 healing lengths [26] and have been observed experimentally [25]. We will revisit these excitations in Sect. 9.13.

5.2.7 Experimental Observation and Formation of Dark Solitons

The first experimentally-created dark solitons in condensates were short-lived due to high levels of thermal dissipation [19,20]. Thermal dissipation arises from atoms in the thermal fraction of the gas scattering off the soliton, causing it to gradually lose energy and eventually disappear. The rate of thermal decay of the soliton increases with temperature and cross-sectional area of the soliton. Later experiments, working in a highly-elongated, effectively one-dimensional geometry and at very cold temperature, generated dark solitons which persisted for several seconds, equivalent to tens of oscillations in the trap [27]. The dynamics were in excellent agreement with the 1D GPE.

Dark solitons extend only a few healing length in the longitudinal direction, which is typically less than the resolution limit of optical imaging, and so some time-of-flight expansion of the cloud is typically undertaken to enhance the size of the soliton core before imaging. Experimentally-imaged dark solitons have reduced density contrast compared to GPE predictions; this is due to a combination of thermal atoms and quantum-incoherent atoms filling the soliton core.

The most common approach to controllably generate a dark soliton, as used in the above experiments, is via phase imprinting. This involves first forming a condensate in the trap and then illuminating a portion of the condensate with a uniform laser beam (which is far-detuned from any optical transitions to avoid exciting atomic transitions which would heat the gas). If the beam has optical potential $\mathcal{V}$ and is applied for a time τ, then the illuminated part of the condensate acquires a phase jump of $\Delta S = \mathcal{V}\tau/\hbar$. Thus a longitudinal step in the phase is applied to the condensate (although note that the step itself is blurred somewhat by diffraction of the light field). This phase step evolves into one or more dark solitons, along with transient sound waves. Such phase imprinting can be also applied in numerical experiments to generate dark solitons; indeed, it is left as an exercise to simulate the production of dark solitons via this approach.

Dark solitons have also be created via collisions of two BECs, whereby the standing wave pattern evolves into one or more dark solitons [28,29], in the wake of an obstacle potential being moved relative to the fluid [30], and from phase defects formed spontaneously as the gas is rapidly cooled through Bose-Einstein condensation transition [25].

5.3 Bright Solitons

5.3.1 Bright Soliton Solutions

For attractive interactions ($g < 0$) the 1D GPE (5.1) supports *bright solitons*. In contrast to dark solitons, the bright solitons are self-trapped wavepackets in which attractive interactions overcome dispersion. We saw in Sect. 3 that the homogeneous condensate in 1D with attractive interactions is unstable. Actually the stable ground state in 1D is a bright soliton. However, dimensionality significantly modifies the solutions such that three-dimensional bright solitons are metastable and prone to a collapse instability, which we will explore further below. A detailed review of bright solitons in condensates can be found in Ref. [31].

The general solution for a single bright soliton, containing N atoms and moving at speed u, is,

$$\psi(x, t) = \sqrt{\frac{N}{2\xi_s}} \operatorname{sech}\left(\frac{x - ut}{\xi_s}\right) \exp\left[if(x, t) + i\phi\right], \tag{5.23}$$

where $\xi_s = 2\hbar^2/m|g|N$ characterises the soliton width, ϕ is a global phase offset, and

$$f(x, t) = \frac{mux}{\hbar} - \frac{t}{\hbar}\left(\frac{mu^2}{2} - \frac{\hbar^2}{2m\xi_s^2}\right), \tag{5.24}$$

is a time- and space-dependent phase factor. The soliton maintains a sech-squared density profile, shown in Fig. 5.10a, as it propagates. For stronger attractive interactions and/or more atoms, the soliton is narrower, indicative of a stronger binding effect. For a dark soliton, the density profile of the soliton is related to its speed and its speed is limited by the speed of sound; for bright solitons, the density profile and speed are decoupled, and a bright soliton can in principle take on any speed u.

The stationary ($u = 0$) bright soliton is the ground state solution of the $g < 0$ 1D GPE (5.1). To understand the manner is which this solution is supported we take a variational approach. As an Ansatz for the soliton solution, we adopt a 1D Gaussian wavepacket of width ℓ, normalised to N atoms,

$$\psi(x) = \frac{N^{1/2}}{\pi^{1/4}\ell^{1/2}} \exp\left(-\frac{x^2}{2\ell^2}\right). \tag{5.25}$$

Using the energy functional Eq. (5.4), the energy-per-particle E/N of this wavepacket is,

$$\frac{E(\ell)}{N} = \frac{\hbar^2}{4m\ell^2} + \frac{gN}{2\sqrt{2\pi}\ell}. \tag{5.26}$$

As illustrated in Fig. 5.10b, for $g < 0$ there exists a local minimum in $E(\ell)/N$, representing the ground state bright soliton solution. This minimum is formed by a balance between the dispersive term, which scales like $1/\ell^2$ and dominates for small ℓ, and the attractive nonlinear term, which scales like $-1/\ell$ and dominates elsewhere. Differentiating E/N with respect to ℓ, setting to zero and rearranging gives the width of the variational solution, $\ell_v = \frac{\sqrt{2\pi}\hbar^2}{m|g|N}$; this variational solution agrees well with the true solution–see Fig. 5.10a.

Meanwhile, for $g \geq 0$, E/N decreases monotonically with ℓ, informing us that the states are prone to expanding (due to disperson and, for $g > 0$, repulsion between the atoms) and there is no peaked stationary state, as expected for untrapped repulsive condensates (Sect. 3.4).

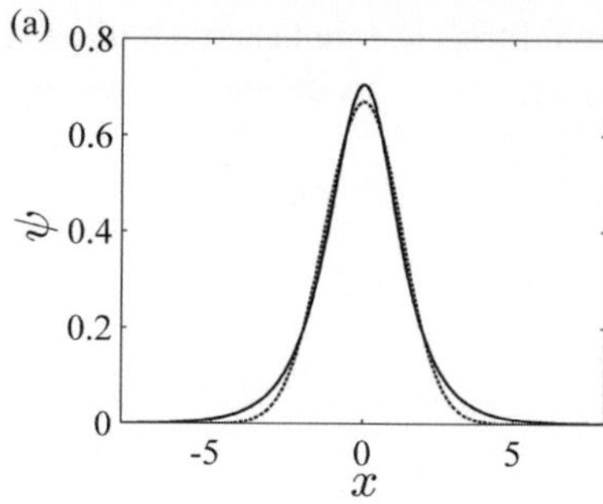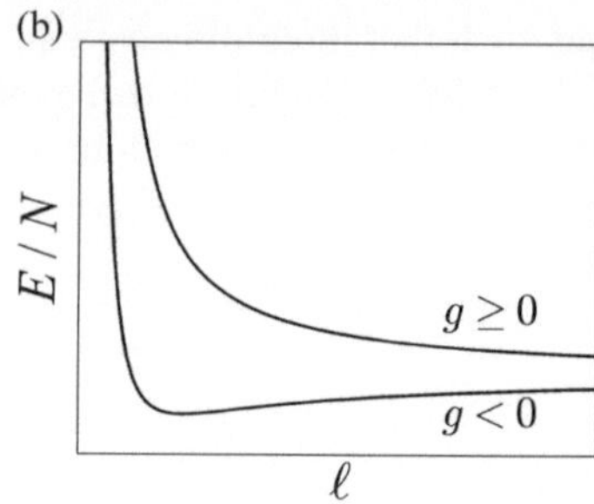

Fig. 5.10 a Density profile $n(x) = |\psi|^2$ of the bright soliton solution of Eq. (5.23) (solid line). The variational solution of the Gaussian Ansatz (5.25) is in good agreement (dashed line). **b** The energy-per-particle E/N of the Gaussian Ansatz, given by Eq. (5.26), versus the width of the Ansatz, ℓ, for two regimes of interactions

Under periodic or box-like boundary conditions, the stationary solitons of the 1D GPE consist of a line of stationary bright solitons [32], sometimes referred to as a *bright soliton train*. We will revisit soliton trains later when discussing methods to form bright solitons.

5.3.2 Collisions

Being a self-contained condensate, a bright soliton has a global phase ϕ and this significantly affects the manner in which bright solitons interact under collision. Dark solitons, in contrast, have no such phase freedom.

To gain insight into their collisions and the phase dependency, consider two bright solitons, each with N atoms. Soliton 1 begins at position $-x_0$ (with $x_0 > 0$) and propagates to the right with speed u, while soliton 2 begins at position x_0 and propagates to the left with the same speed. Their individual solutions are,

$$\psi_1(x, t) = \sqrt{\frac{N}{2\xi_s}} \, \text{sech} \left(\frac{x + x_0 - ut}{\xi_s} \right) e^{if(x,t)} e^{i\phi_1}, \qquad (5.27)$$

$$\psi_2(x, t) = \sqrt{\frac{N}{2\xi_s}} \, \text{sech} \left(\frac{x - x_0 + ut}{\xi_s} \right) e^{if(x,t)} e^{i\phi_2}. \qquad (5.28)$$

Note that due to the symmetric configuration, both solitons have the same time- and space-dependent phase factor $e^{if(x,t)}$, but we allow for different global phase offsets, ϕ_1 and ϕ_2.

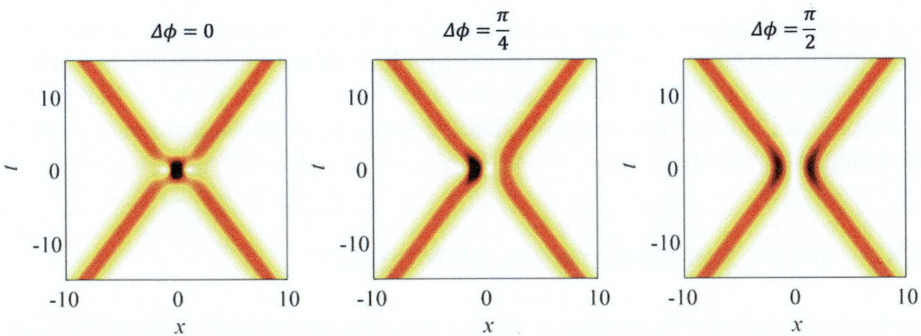

Fig. 5.11 **a** Density profile $n(x, t) = |\psi(x, t)|^2$ during the collision of two bright solitons, with speed $u = 0.2$ and for different relative phases

Assuming that the solitons are well-separated we can construct their superposition as $\psi' = \psi_1 + \psi_2$.[2] We proceed to calculate the density profile of this superposed state, $|\psi'| = |\psi_1 + \psi_2|^2$,

$$|\psi'(x, t)|^2 = \left| |\psi_1| e^{if(x,t)} e^{i\phi_1} + |\psi_2| e^{if(x,t)} e^{i\phi_2} \right|^2 \tag{5.29}$$

$$= |\psi_1|^2 + |\psi_2| + |\psi_1||\psi_2| \left(e^{i\Delta\phi} + e^{-i\Delta\phi} \right), \tag{5.30}$$

where we have introduced the *relative phase* $\Delta\phi = \phi_2 - \phi_1$. Using the identity $\cos\theta = (e^{i\theta} + e^{-i\theta})/2$ we obtain,

$$|\psi'(x, t)|^2 = |\psi_1|^2 + |\psi_2| + 2|\psi_1||\psi_2| \cos\Delta\phi. \tag{5.31}$$

Let us see how this affects the overlap of the two solitons by calculating the density at their midpoint (the origin). Introducing the form of ψ_1 and ψ_2 from Eq. (5.28) and setting $x = 0$ gives,

$$|\psi'(0, t)|^2 = \frac{N}{2\xi_s} \left[2\,\mathrm{sech}^2 \left(\frac{x_0 - ut}{\xi_s} \right) + 2\cos\Delta\phi\,\mathrm{sech}^2 \left(\frac{x_0 - ut}{\xi_s} \right) \right]. \tag{5.32}$$

If $\Delta\phi = 0$ then the density at the midpoint reinforces (constructive interference), in other words, the solitons overlap with each other. However, for $\Delta\phi = \pi$ the midpoint density is forced to zero (destructive interference), and the overlap of the solitons is prohibited. For intermediate values of the relative phase, the overlap varies smoothly between these extremes.

Figure 5.11 shows the collisions for different relative phases. True to their solitonic character, the solitons emerge unscathed from the collision, barring a shift. The

---

[2] The superposition theorem does not apply to the GPE, a nonlinear equation; constructing a superposition is only a valid approximation if the density is low. This condition is satisfied here since we are concerned with the weak overlap between well-separated solitons.

role of relative phase becomes clear: for $\Delta\phi = 0$ the solitons merge at the point of collision, while for $\Delta\phi = \pi$ overlap is prohibited and they appear to bounce. In between the collision becomes asymmetric. Despite these different behaviours during the collision, it is remarkable that the outgoing densities and position shifts are independent of $\Delta\phi$.

5.3.3   Bright Solitons Under Axial Potentials

As discussed earlier in the context of dark solitons, the presence of an inhomogeneous potential breaks the integrability of the 1D GPE such that the solutions are no longer bright solitons in the strictly integrable sense. For slowly-varying potentials, however, the states retain many of the key properties of solitons, for example, preservation of their shape and propagation without emission of radiation. Meanwhile, for sharply-varying potentials this behaviour breaks down. We will illustrate these regimes below through the examples of a bright soliton interacting with a step potential and under the influence of a harmonic trap.

Bright Soliton Incidential Upon a Step Potential

Consider a bright soliton incident upon a tanh-shaped potential step $V(x) = V_0[1 + \tanh(x/\sigma)]/2$, where the step width is relatively sharp (that is, the step width σ narrow compared to the soliton width ξ_s) and $V_0 > 0$ is the potential of the step. The outcome is illustrated in Fig. 5.12b in terms of the reflection coefficient R, which is the proportion of atoms reflected from the step.

A classical particle would transmit over the barrier when its kinetic energy is equal to or exceeds the step potential. We see a similar behaviour for the soliton - for low incident speeds the soliton fully reflects and for high speeds it fully transmits. The transition between these regimes is smooth and centred on the classical threshold. Here the incident soliton splits into a transmitted and a reflected soliton plus some additional low amplitude transmitted and reflected radiation. For wider steps the transition region narrows, evidencing that the soliton is behaving more like a classical particle.

Soliton splitting is similarly seen when a bright soliton is incident upon a narrow potential barrier. Moreover, the split solitons can be recombined if they synchronously re-interact with each other at the barrier. This splitting and recombination realizes the basic components of a matter-wave interferometer and has been demonstrated experimentally [33].

A bright soliton can also reflect from a negative potential $V_0 < 0$ [34]. This is an example of *quantum reflection*, a process in which a particle reflects from a spatially-varying potential without reaching a classical turning point, and is a consequence of the wave nature of the particle. Significant reflection occurs when the potential energy experienced by the particle changes abruptly over a lengthscale characterised by the particle's quantum mechanical wavelength, a regime which is achieved for sufficiently low incident speed and for a sufficiently sharp potential. This behaviour is confirmed in Fig. 5.12c, which presents the reflection coefficient for a bright soliton incident upon a negative tanh potential. As for a positive step, the soliton splits into a reflected and transmitted soliton, with enhanced reflection for low speeds and sharp steps. In contrast a classical particle would fully transmit. Quantum reflection of a BEC was first demonstrated experimentally for a trapped repulsively-interacting BEC incident upon a solid surface [35] (due to the sharp and attractive Casimir-Polder potential of the surface experienced by the atoms). Quantum reflection of bright solitons was later engineered at a narrow attractive potential generated by a tightly-focussed laser beam [36].

Dark solitons, in contrast, do not exhibit quantum reflection at a negative step–this is because the healing of the background condensate across the step ensures that the dark soliton experiences a sufficiently smooth effective potential.

Bright Soliton Within a Harmonic Potential

An axial harmonic trap makes the ground state more narrow and peaked than the $V(x) = 0$ soliton but nonetheless the state retains a sech-like profile providing the interaction energy dominates over the axial trapping energy. This limit is parametrized by $Na_s/l_x \gg 1$, where $l_x = \sqrt{\hbar/m\omega_x}$ is the axial harmonic oscillator length. In the opposite limit $Na_s/l_x \ll 1$ the state approaches the non-interacting Gaussian ground state with width l_x. One can demonstrate this by extending the variational approach from Sect. 5.3.1 to include a harmonic trap and considering how this modifies the variational solution (which is posed as an exercise at the end of this Chapter).

Despite the modified density profile, the solutions continue to show soliton-like behaviour. For example, if the ground state is set into motion in the trap (such as by being given a momentum kick or by being instantaneously positioned off-centre in the potential) it will propagate with a fixed profile and undergo centre-of-mass oscillations at the trap frequency, as per a classical particle. Note that this behaviour is a more general result–Morgan et al. [37] considered in generality the eigenstates solutions of various GPEs with arbitrary number of dimensions and with a general potential, and showed that the ground states can propagate in a rectilinear path with unchanging form. This is true providing the potential is time-independent and that the nonlinearity is decoupled from the absolute position, a requirement satisfied by the GPE.

In addition to propagating with constant form, multiple solitons in a harmonic trap pass through each other unscathed.

5.3.4 Particle Model

Due to their particle-like properties, it is natural to apply a particle model to capture the dynamics of one or more solitons. An isolated soliton will behave like a classical particle of mass mN, providing any imposed potential is slowly varying (to avoid regimes of soliton splitting discussed earlier): in a homogeneous system the soliton will be stationary or propagate with constant speed while under an inhomogeneous potential it will undergo classical centre-of-mass dynamics.

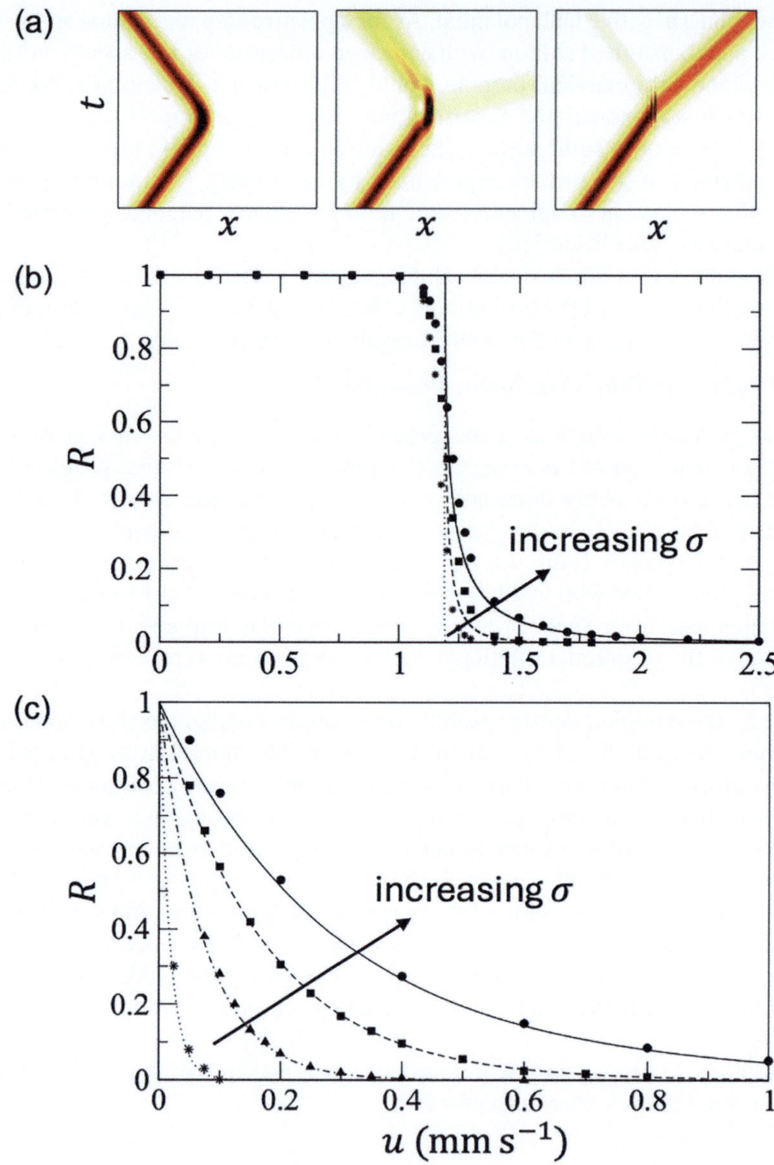

Fig. 5.12 Interaction of a bright soliton with a tanh-shaped step potential. Reprinted from [34] with permission from Elsevier. **a** Space-time density plots showing regimes of high, intermediate and low reflection coefficient. **b–c** Reflection coefficient R as a function of the soliton speed u when the step is (**b**) positive and (**c**) negative, according to 1D GPE simulations (points) and a non-interacting plane-wave approach (lines). Various step widths σ are considered, with arrows showing the direction of increasing step width σ

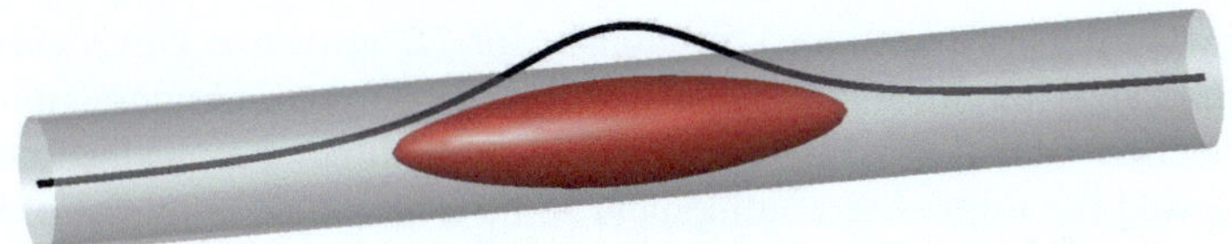

Fig. 5.13 Illustration of a 3D bright soliton in a quasi-1D waveguide. Shown is an isosurface (red) of the 3D density profile and the integrated 1D density profile along the axis (black line)

For multiple interacting solitons, one can define an appropriate interaction potential between pairs of solitons [31,38]. Let us consider first the homogeneous 1D system. Consider two interacting soliton particles of norm N_1 and N_2 and positions x_1 and x_2, then the interaction potential as,

$$
\mathcal{V}(x_1 - x_2) = -\frac{N_1 N_2 (N_1 + N_2)}{32 N^3} \operatorname{sech}^2 \left(\frac{N_1 N_2 (x_1 - x_1)}{2\xi_N N (N_1 + N_2)} \right), \tag{5.33}
$$

where $N = N_1 + N_2$ is the total norm. Recall that the interaction between two bright solitons leads to asymptotic position shifts only, with the soliton speed and mass unchanged. This potential exactly reproduces the required asymptotic position shift for equal norm solitons. Note how the interaction potential is attractive–this is due to the underlying attractive atomic interactions. Note also how the potential varies in space as sech^2–this is due to the overlap between the sech^2 soliton wavepackets.

For solitons of differing norms the asymptotic shift predicted by the particle model approaches the correct value for $|N_1 - N_2| \ll |u_1 - u_2|\hbar/2|g|N$ [39], i.e. for weak soliton interactions (small norms or interaction strengths) or short interaction times (collisions with high relative speed).

The introduction of an axial potential causes the asymptotic position shifts to deviate from those predicted by the above interaction potential. However the interaction potential remains a satisfactory approximation providing the external potential is approximately uniform over the region of the collision and the trapped states are bright-soliton-shaped. Then one can extend the particle model to include the influence of the axial potential on the centre-of-mass of the solitons. Note the particle is not suited for trains of solitons, where the solitons are never well separated or where the phase difference (which does not enter the model) becomes important. The particle model has been used to study the dynamics of multiple bright solitons oscillating and colliding in a harmonic trap [39]; here the dynamics for two solitons are regular, however for three or more solitons there are regimes of chaotic dynamics.

5.3.5 3D Bright Solitons in Waveguide Potentials

In the real 3D world, bright solitons can be formed within a waveguide potential featuring strong transverse trapping, as illustrated in Fig. 5.13. These states are self-trapped along the longitudinal direction, as per the 1D bright soliton. However, the additional dimensions introduces new physics beyond the 1D case, notably the collapse instability we introduced previously for 3D attractive BECs (Sect. 3.5.3) which significantly modifies the soliton stability, properties and interactions.

In this section we will consider the waveguide to be directed along the longitudinal direction x and the transverse confinement to be a harmonic potential of the form $V(x, y, z) = m\omega_\perp^2 (y^2 + z^2)/2$.

Time-Independent Solutions

In 3D integrability is lost and no exact analytic results exist for the 3D bright solitons. Identification of the solutions requires solving the 3D GPE numerically or employing approximate analytic methods. A convenient Ansatz for a stationary 3D bright soliton in a waveguide with transverse harmonic confinement is [40],

$$\psi(\mathbf{r}) = \sqrt{\frac{N}{\pi \sigma_x \sigma_\perp^2 \ell_\perp^3}} \operatorname{sech}\left(\frac{x}{\sigma_x \ell_\perp}\right) \exp\left(-\frac{y^2 + z^2}{2\sigma_\perp^2 \ell_\perp^2}\right), \qquad (5.34)$$

where $\sigma_\perp$ and σ_x are dimensionless variational parameters for the transverse and axial width of the wavepacket. The energy-per-particle of this state according to the energy integral (3.14) is,

$$\frac{E}{N} = \hbar\omega_\perp \left(\frac{1}{2\sigma_\perp^2} + \frac{\sigma_r^2}{2} + \frac{1}{6\sigma_x^2} + -\frac{k}{3\sigma_x \sigma_\perp^2}\right), \qquad (5.35)$$

where $k = N|a_s|/\ell_\perp$ is a dimensionless interaction coefficient.

For sufficiently weak interactions $0 \leq k \leq k_c$, the variational energy landscape $E(\sigma_x, \sigma_\perp)$ has the qualitative form depicted in Fig. 5.14a. The global energy minimum is a collapsed state of zero width, with the energy diverging to $-\infty$ as $\sigma_x, \sigma_\perp \to 0$; in this very high density regime, a higher-order description is required to capture the full physics (e.g. three-body loss and quantum fluctuations).

There exists a local minimum for finite widths $(\sigma_{x0}, \sigma_{\perp 0})$, representing the variational solution of the 3D bright soliton. Along $\sigma_\perp$, the energy rises rapidly on either side of the local minimum due to the transverse trapping potential. Along σ_x the behaviour is more complicated. On the low σ_x side, a saddle point separates the local minimum from the global minimum. This is formed due to a delicate playoff between all of the energy terms.

For $k > k_c$, however, the saddle point and local minimum merge, and the 3D bright soliton state is unstable to collapse. Importantly, the 1D approach does not account for the collapse instability; the 1D limit here corresponds to the $\sigma_\perp = 1$ line in the $(\sigma_x - \sigma_\perp)$ plane, such that collapse regions cannot be accessed. There exist analytic expressions for the variational bright soliton solutions; these can be obtained by identifying the stationary points of the energy landscape (5.35) and then isolating those corresponding to minima [40]. The solution widths decrease as the interaction parameter k increases [Fig. 5.14a] up to the critical value $k_{c,v} = 3^{-1/4} \approx 0.76$, beyond which no variational solutions exist. It is interesting to note that at the critical point the solution is isotropic. A purely Gaussian variational Ansatz gives a remarkably close result, $k_{c,v} \approx 0.78$.

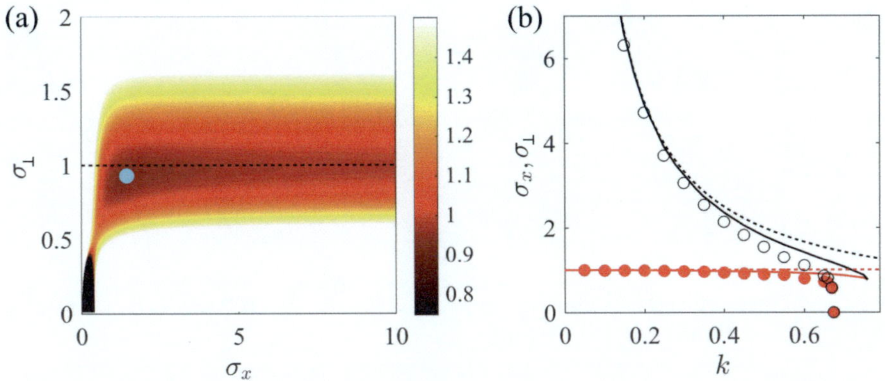

Fig. 5.14 (Color online) **a** Energy landscape of the variational Ansatz (5.34) in the absence of axial trapping ($\lambda = 0$) and for $k = N|a_s|/l_\perp = 0.6$, as a function of the axial and radial width parameters, σ_x and $\sigma_\perp$. The local minimum representing the metastable 3D bright soliton is shown by the light blue dot. **b** Dimensionless widths of the 3D bright soliton as a function of the interaction parameter k, based on numerical solutions of the 3D NLSE (points), the variational approach (solid lines) and the quasi-1D prediction (dashed lines)

The transition to 1D behaviour is confirmed in Fig. 5.14b: as k becomes increasingly small the variational soliton widths (solid lines) agree with their 1D counterparts (dashed lines), $\sigma_\perp = 1$ and $\sigma_x = \xi_s/2\ell_\perp = 1/k$. Thus the 1D model is only valid for low k.

Numerical solutions of the 3D NLSE (circles in Fig. 5.14c) confirm the qualitative predictions of the variational model (solid lines). However, the NLSE-derived widths decrease with k faster than predicted by the variational method; thus the solitons are more peaked in density for a given k and become unstable at a lower interaction parameter $k_c = 0.675 \pm 0.005$ [41].

Collisions

The collisions of 3D bright solitons can also deviate significantly from the 1D case [42]. Two 3D solitons, which are stable in isolation, i.e. each satisfying $k < k_c$, can undergo a catastrophic collapse instability during a collision if the raised densities are sufficiently high to trigger collapse. The relative phase between the colliding solitons, which governs the form of the collisional density as we saw in Sect. 5.3.2, becomes critically important, as illustrated in Fig. 5.15. For $\Delta\phi = \pi$, a central density node is suppressed during the collapse, promoting stability of the collision. For $\Delta\phi = 0$ however, a high density central node is formed during the collision, rendering the collision unstable to collapse.

The collisional speed also plays a non-trivial role; at low speeds the relative phase has its biggest effect, while for faster collisions the relative phase becomes less important as the collisions becoming more stable and approach the 1D result. This is a matter of timescales–for fast collisions the collapse instability has insufficient time to progress. Figure 5.15b presents the phase diagram demarkating stable versus unstable collisions as a function of incoming speed and interaction parameter k of the initial solitons, clearly showing the enhanced stability for out-of-phase collisions.

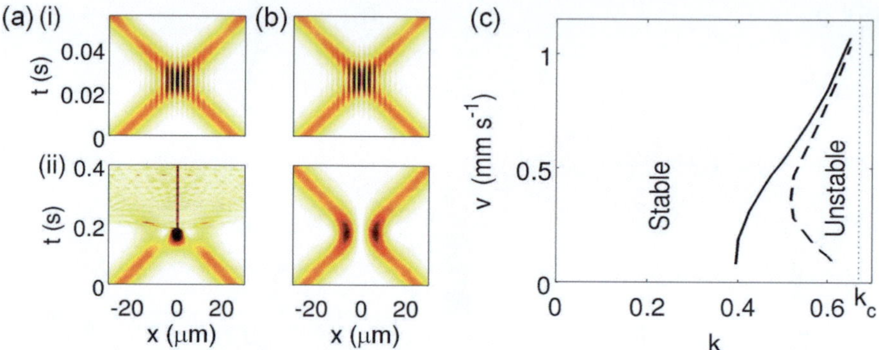

Fig. 5.15 Stability of 3D soliton collisions in an axially homogeneous waveguide potential. **a** Space-time density plots showing the collision between two $k = 0.4$ bright solitons with $\Delta\phi = 0$ for (i) high and (ii) low incident speed. **b** The same but for $\Delta\phi = \pi$. **c** Collisional stability diagram as a function of incident soliton speed v_i and interaction parameter k, where the solid (dashed) line marks the boundary between stable and unstable collisions for $\Delta\phi = 0(\pi)$. The vertical line denoted k_c indicates the critical interaction strength for collapse of a single, isolated 3D soliton (**b**). Adapted with permission from [42]. Copyright 2008, IOP Publishing Ltd. All rights reserved

For $\Delta\phi \neq 0, \pi$ (modulo 2π), the asymmetric collisional density can lead to irreversible population transfer between two 3D solitons. The amount of population transfer depends strongly on relative phase and is enhanced for collisional speeds [42] due to the increased interaction time. Population transfer can also trigger collapse in the receiving soliton. For $k \ll 1$ the quasi-1D regime is reached and the effects of population transfer and unstable collisions disappear.

5.3.6   Bright Solitons and the Non-polynomial GPE

It is possible to modify the 1D GPE such that it captures more of the 3D character of a 3D bright soliton, including the collapse instability, and giving it a wider range of validity than the bare 1D GPE. Salasnich et al. [43,44] factorised the 3D GPE wavefunction into a slowly-varying axial function, multiplied by a rapidly-varying transverse function. The transverse function depends on the axial function, unlike the quasi-1D reduction presented in Sect. 3.5.4 A variational calculation then yields the non-polynomial Schrödinger equation [43],

$$
i\hbar\frac{\partial\psi}{\partial t} = \left[-\frac{\hbar^2}{2m}\frac{\partial^2}{\partial x^2} + V(x) + g\frac{|\psi|^2}{A(x,t)} \right.
$$
$$
\left. + \frac{\hbar\omega_\perp}{2}\left(\frac{1}{A(x,t)} + A(x,t)\right) \right]\psi, \tag{5.36}
$$

where $A(x,t) = \sqrt{1 + 2|a_s||\psi|^2}$. When $|a_s||\psi|^2 \ll 1$ this reduces to the bare 1D GPE. The non-polynomial model not only captures the collapse of the 3D bright soliton, but it predicts a critical point for collapse of $k_c = 2/3$, which in close agreement with the full 3D result. Furthermore, the equation predicts soliton dynamics in excellent agreement with 3D simulations [45].

5.3.7 Experimental Formation and Observation of Bright Solitons

Bright solitons were first formed with condensates in 2002 [46,47]. They are typically generated using an atomic species whose interactions can be tuned via a magnetic Feshbach resonance. A stable repulsive condensate is first formed in a highly-elongated harmonic trap. The interaction strength is then tuned to being attractive by means of the Feshbach resonance. In these early experiments, the critical number of atoms for the system was exceeded, driving a collapse of the condensate. Out of the collapse, one or more bright solitons formed. More recent experiments have formed bright solitons by choosing final interaction parameters which avoid the collapse instability. The weak axial trap is either kept on, in which case the soliton/s oscillate axially, or is switched off, such that the solitons propagate freely.

An experimental proof of bright solitons is shown in Fig. 5.16. For repulsive interactions it was seen that the condensate expanded over time, while for attractive interactions it was seen to maintain its shape, characteristic of a soliton. More recent experiments have studied the collisions of bright solitons with each other [48] and with potential barriers [49].

A train of multiple bright solitons was experimentally created by beginning with an elongated repulsively-interacting condensate and then rapidly changing the scattering length to be attractive [50]. This triggers the modulational instability of the background condensate, whereby modulations grow exponentially in time [51]. The growth rate is largest for the lengthscale $\xi_{\mathrm{MI}} = 2\pi l_\perp/\sqrt{4|a_f|n_{\mathrm{1D}}}$, where a_f is the final scattering length and n_{1D} is the one-dimensional atomic density. This instability causes the condensate to segregate into localised wavepackets of size ξ_{MI}. If the final scattering length is not too attractive, then these wavepackets take the form of solitons. In this regime it was observed that the number of solitons formed is well captured by simply the original length of the condensate divided by the instability lengthscale ξ_{MI}. The experiment observed that the formed solitons interact repulsively with each other; this served to stabilise the train and avoid multiple solitons merging to undergo collapse. This repulsion is thought to be due to π-phase differences being present between the solitons. However for more attractive final scattering lengths, each soliton formed exceeded the critical threshold for collapse and undergoes a collapse, complicating the outcome.

5.3.8 Bright Solitons in Optical Lattices

We now consider how an imposed optical lattice modifies the properties of bright solitons. There are several distinct regimes which we outline below.

Solitons which are narrow relative to the lattice wavelength are only influenced by the local properties of the potential: they have a single-peaked profile which closely matches the unperturbed soliton and they propagate under the influence of the potential in analogy to a classical particle with mass Nm.

However when the optical lattice is relatively weak (that is, the amplitude of the lattice V_0 is much smaller than the chemical potential of the condensate μ) and the

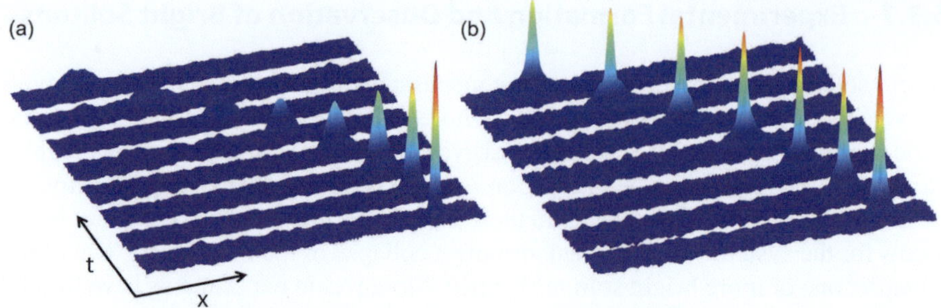

Fig. 5.16 a As a repulsive BEC travels along a waveguide (with tight transverse harmonic trapping
and very weak axial confinement) it spreads out. **b** For attractive interactions, the condensate is seen
to maintain its shape over time, characteristic of a bright soliton. Image courtesy of S. L. Cornish
(University of Durham)

soliton length is large compared to the lattice space (such that the soliton extends over
multiple lattice sites), the dispersion of the wave can become significantly modified
such that it can adopt an effective mass which is quite distinct from Nm. Indeed, it is
possible for the effective mass to become negative; in this regime bright solitons can
be supported with repulsive nonlinearity. These *gap solitons* have been generated
experimentally [52].

A final regime of solitons in optical lattices exists when the optical lattice is very
strong ($V_0 \gg \mu$) such that the atoms are tightly bound within each lattice site and
the system can become modelled by the discrete NLSE introduced in Sect. 3.9.3.
Within this system, solitons and breather solutions of the envelope function can be
supported [53]. Indeed, a recent experiment generated and observed such multi-site
bright solitons within a highly tunable optical lattice system [59], in good agreement
with 3D GPE results.

5.3.9 Breathers

Breathers are localised, periodic solutions of nonlinear equations They exist in many
nonlinear systems, for example in fiber optics [54], and are paradigms of rogue waves
in the oceans [55]. The homogeneous attractive 1D GPE supports a family of breather
solutions called the *Satsuma-Yajima breathers* or sometimes *higher-order solitons*.
These solutions may be considered as a bound state of multiple solitons which co-
exist locally and interfere to form a quasi-periodic localised solution [40,56]. The
solution appears to "breathe" in time. Just like their constituent solitons, the breathers
can be propagated in space and pass through each other unscathed during collisions.

A Satsuma-Yajima breather solution composed of N_s solitons can be formed from
a bright soliton when the nonlinearity is instantaneously changed by a factor of N_s^2.
In a BEC this scenario can be implemented by a quasi-instantaneous quench of the
interaction strength by Feshbach engineering of the atomic s-wave scattering length
of the atoms a_s. Indeed, breathers of attractive BECs have been created experimentally by this method [57,58]. If the interaction/nonlinearity is changed by a factor

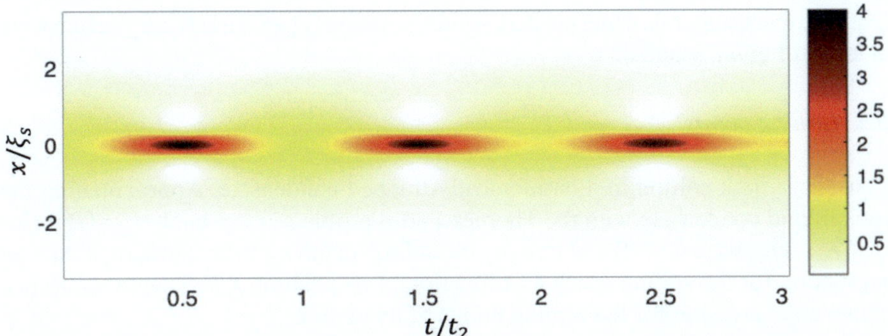

Fig. 5.17 The $N_s = 2$ breather solution of the homogeneous 1D $g < 0$ GPE. The colour bar represents the density of the solution, relative to the peak density of the initial state

which does not equal N_s^2, the system evolves to the nearest breather solution with additional radiation being released.

Consider the $N_s = 2$ breather in the homogeneous 1D GPE, triggered by a sudden four-fold increase in $|g|$. The exact analytical solution for the evolution of the density in space and time is [56],

$$n(x, t) = 4n_0 \left| \frac{\cosh\left[3x/(2\xi_s)\right] + 3\,\mathrm{e}^{-2\pi \mathrm{i}t/t_2}\cosh\left[x/(2\xi_s)\right]}{3\cos\left(2\pi t/t_2\right) + 4\cosh\left(x/\xi_s\right) + \cosh\left(2\,x/\xi_s\right)} \right|^2, \qquad (5.37)$$

where $t_2 = \dfrac{8\pi\hbar^3}{mN^2|g|^2}$ is the period of the $N_s = 2$ breather. The solution is presented in Fig. 5.17.

Problems

5.1 *Integrals of motion*

For a dark soliton, the integrals of motion in Eqs. (5.2), (5.3), (5.4) are renormalised so as to remove the contribution from the background and lead to finite values,

$$N_\mathrm{s} = \int_{-\infty}^{+\infty} (n_0 - |\psi|^2)\,\mathrm{d}x$$

$$\mathcal{P}_\mathrm{s} = \frac{i\hbar}{2} \int_{-\infty}^{+\infty} \left(\psi\frac{\partial\psi^*}{\partial x} - \psi^*\frac{\partial\psi}{\partial x}\right)\left(1 - \frac{n_0}{|\psi|^2}\right)\mathrm{d}x$$

$$E_\mathrm{s} = \int_{-\infty}^{+\infty} \left(\frac{\hbar^2}{2m}\left|\frac{\partial\psi}{\partial x}\right|^2 + \frac{g}{2}(|\psi|^2 - n_0)^2\right)\mathrm{d}x$$

Evaluate these integrals using the dark soliton solution Eq. (5.5), leaving your answers in terms of ξ, n_0, u and c.

5.2 *Turning points*

Consider a dark soliton in a harmonically-trapped condensate. Approximating the background condensate with the Thomas-Fermi profile $n(x) = n_0(1 - x^2/R_x^2)$ (for $x \leq R_x$, otherwise $n = 0$) and treating the soliton depth n_d to be constant, obtain an expression for the soliton speed as a function of its position x and depth n_d. Hence obtain an expression for the turning points of its motion.

5.3 *Simulating a dark soliton*

Using the 1D GPE solver providing in the Appendix (or your own solver), simulate the motion of a dark soliton within a harmonically-confined condensate where the initial condition is a $u = 0.5c$ dark soliton located at the trap centre. Assume a Rubidium-87 condensate with 50,000 atoms within a cylindrically-symmetric quasi-one-dimensional harmonic trap with axial frequency $\omega_x = 2\pi \times 40$Hz and transverse frequency $\omega_r = 2\pi \times 100$ Hz. Verifying that the period of the soliton oscillation agrees with the expected result.

To generate the initial condition you will need first need to obtain the soliton-free ground state and then multiply this by the appropriate dark soliton solution (accounting for the local healing length and speed of sound).

5.4 *Soliton and barrier*

Extending on the previous exercise, introduce a small repulsive Gaussian potential barrier of the form $V_0 \exp(-(x - x_0)^2/\sigma^2)$ to the system with amplitude $V_0 = 2 \times 10^{-31}$J, width $\sigma = 2 \times 10^{-6}$m and offset $x_0 = 5 \times 10^{-6}$m.

1. How does the presence of the obstacle affect the soliton path? Is any visible sound emitted when the soliton interacts with the barrier?
2. Double the amplitude of the potential. How does this modify the soliton path?
3. By considering the Thomas-Fermi prediction for the condensate density at the centre of the barrier, estimate the critical obstacle amplitude at which the soliton no longer transmits past the obstacle. Is this consistent with the two above simulations?

5.5 *Imprinting a soliton in imaginary time*

One can numerically create a black soliton solution at arbitrary position in the condensate by imposing a π-phase step during imaginary time propagation of the 1D GPE. The imaginary time method then evolves towards the lowest energy state featuring this phase step, i.e. a black soliton at the specific location.

1. Adapt the imaginary time routine to introduce a π-phase step and thereby generate a black soliton solution at the centre of the condensate. Propagate the state in real time to verify that it is evolves as you would expect.
2. Repeat but with the black soliton being imprinted off-centre in the trap at $x_0 = 10 \times 10^{-6}$m; how does the soliton evolves?
3. Using the same approach, construct an initial solution with two dark solitons positioned at $x_0 = \pm 10 \times 10^{-6}$m and evolve in real time to simulate dark soliton collisions. If you reduce the initial offset to $x_0 = \pm 1 \times 10^{-6}$ how does this qualitatively change the soliton dynamics and how can this be interpreted?

5.6 *Chemical potential of soliton*

Show that the static ($u = 0$) bright soliton solution, obtained from Eq. (5.23), is a solution to the 1D attractive time-independent GPE with $V(x) = 0$, i.e,

$$\mu\psi = -\frac{\hbar^2}{2m}\frac{\mathrm{d}^2\psi}{\mathrm{d}x^2} - |g||\psi|^2\psi, \tag{5.38}$$

and hence determine an expression for the chemical potential of the soliton.

5.7 *Soliton as a classical particle*

Using the general bright soliton solution, Eq. (5.23), evaluate the soliton integrals of motion according to Eqs. (5.2), (5.3) and (5.4). The soliton solution is already normalised to the number of atoms, N. Show that the bright soliton behaves as a classical particle with positive mass.

5.8 *Simulating a bright soliton*

In the experimental of Nguyen et al. [48] bright solitons were formed from around 28,000 Lithium-7 atoms within a waveguide potential with transverse harmonic trapping of frequency $\omega_\perp = 2\pi \times 254$ Hz and for a scattering length of $a_s = -3.2 \times 10^{-11}$ m.

1. Using the 1D GPE solver providing in the Appendix (or your own solver) and with zero axial trapping, introduce a stationary bright soliton solution corresponding to the above parameters as an initial condition and propagate in time to verify that it maintains a stationary self-trapped form.
2. Modify your initial condition to give the soliton a velocity kick by multiplying by the factor $\exp(imvx/\hbar)$, taking the imposed velocity as $v = 0.001$ m/s. Does the soliton behave as you would expect?
3. Repeat but with a weak axial harmonic potential $\omega_x = 2\pi \times 10$Hz introduced. Confirm that the soliton behaves like a classical particle under the imposed potential Does the soliton profile change in time and why is this the case?

5.9 *Bright solitons under axial trapping*

In Sect. 5.3.5 we presented the variational solutions for 3D bright solitons confined within a waveguide potential with harmonic transverse confinement only.

1. Extend this analysis to consider the additional presence of harmonic trapping along x defined in terms of the trap ratio $\lambda = \omega_x/\omega_\perp$. Derive the corresponding contribution to the energy-per-particle.
2. Produce a two-dimensional heatmap of the energy-per-particle, as a function of the axial and transverse variational widths σ_x and $\sigma_\perp$ for a moderate interaction strength $k = 0.4$. How does the introduction of axial trapping modify the energy landscape and the variational solution?
3. In the bright soliton experiment of Khaykovich et al. [47] the axial potential was harmonic but inverted, i.e. $\lambda^2 < 0$. How does the presence of an inverted axial harmonic potential modify the variational solutions and their range of stability?

References

1. R.W. Boyd, *Nonlinear Optics*, 4th edn. (Academic Press, London, 2020)
2. T. Dauxois, M. Peyrard, *Physics of Solitons* (Cambridge University Press, Cambridge, 2006)
3. P.G. Drazin, R.S. Johnson, *Solitons: an introduction*, 2nd edn. (Cambridge University Press, Cambridge, 1989)
4. M.J. Ablowitz, H. Segur, *Solitons and the Inverse Scattering Transform* (SIAM, Philadelphia, 1981)
5. D.J. Frantzeskakis, J. Phys. A **43**, 213001 (2010)
6. T. Busch, J.R. Anglin, Phys. Rev. Lett. **84**, 2298 (2000)
7. G. Huang, J. Szeftel, S. Zhu, Phys. Rev. A **65**, 053605 (2002)
8. N.G. Parker, N.P. Proukakis, M. Leadbeater, C.S. Adams, Phys. Rev. Lett. **90**, 220401 (2003)
9. N.G. Parker, N.P. Proukakis, M. Leadbeater, C.S. Adams, J. Phys. B: At. Mol. Opt. Phys. **36**, 2891 (2003)
10. N.G. Parker, *Numerical Studies of Vortices and Dark Solitons in Atomic Bose-Einstein Condensates* (Durham University, 2004)
11. N.P. Proukakis, N.G. Parker, D.J. Franzeskakis, C.S. Adams, J. Opt. B **6**, S380 (2004)
12. D.E. Pelinovsky, Y.S. Kivshar, V.V. Afanasjev, Phys. Rev. E **54**, 2015 (1996)
13. P.G. Kevrekidis, R. Carretero-Gonzalez, G. Theocharis, D.J. Frantzeskakis, B.A. Malomed, Phys. Rev. A **68**, 035602 (2003)
14. N.G. Parker, N.P. Proukakis, C.F. Barenghi, C.S. Adams, J. Phys. B: At. Mol. Opt. Phys. **37**, S175 (2004)
15. M. Sciacca, C.F. Barenghi, N.G. Parker, Phys. Rev. A **95**, 013628 (2017)
16. N.G. Parker, N.P. Proukakis, C.S. Adams, Phys. Rev. A **81**, 033606 (2010)
17. N.P. Proukakis, N.G. Parker, C.F. Barenghi, C.S. Adams, Phys. Rev. Lett. **93**, 130408 (2004)
18. M. Ma, R. Navarro, R. Carretero-Gonzalez, Phys. Rev. E **93**, 022202 (2016)
19. S. Burger et al., Phys. Rev. Lett. **83**, 5198 (1999)
20. J. Denschlag et al., Science **287**, 97 (2000)
21. A. Kuznetsov, S.K. Turitsyn, Zh. Eksp. Teor. Fiz. **94**, 119 (1988). (Sov. Phys. JETP, **67**, 1583 (1988))
22. D.E. Pelinovsky, Y.A. Stepanyants, Y.S. Kivshar, Phys. Rev. E **51**, 5016 (1995)

23. B.P. Anderson, P.C. Haljan, C.A. Regal, D.L. Feder, L.A. Collins, C.W. Clark, E.A. Cornell, Phys. Rev. Lett. **86**, 2926 (2001)
24. Z. Dutton, M. Budde, C. Slowe, L.V. Hau, Science **293**, 663 (2001)
25. G. Lamporesi, S. Donadello, S. Serafini, F. Dalfovo, G. Ferrari, Nat. Phys. **9**, 656 (2013)
26. J. Brand, W.P. Reinhardt, Phys. Rev. A **65**, 043612 (2002)
27. C. Becker et al., Nat. Phys. **4**, 496 (2008)
28. A. Weller et al., Phys. Rev. Lett. **101**, 130401 (2008)
29. I. Shomroni, E. Lahoud, S. Levy, J. Steinhauer, Nat. Phys. **5**, 193 (2009)
30. P. Engels, C. Atherton, Phys. Rev. Lett. **99**, 160405 (2007)
31. T.P. Billam, A.L. Marchant, S.L. Cornish, S.A. Gardiner, N.G. Parker, Bright solitary matter waves: formation, stability and interactions, in *Spontaneous Symmetry Breaking, Self-Trapping and Josephson Oscillations*, ed. by B.A. Malomed (Springer, Berlin, 2013)
32. L.D. Carr, C.W. Clark, W.P. Reinhardt, Phys. Rev. A **62**, 063611 (2000)
33. O.J. Wales, A. Rakonjac, T.P. Billam, J.L. Helm, S.A. Gardiner, S.L. Cornish, Commun. Phys. **3**, 51 (2020)
34. S.L. Cornish, N.G. Parker, A.M. Martin, T.E. Jugg, R.G. Scott, T.M. Fromhold, C.S. Adams, Phys. D **238**, 1299 (2009)
35. T.A. Pasquini, Y. Shin, C. Sanner, M. Saba, A. Schirotzek, D.E. Pritchard, W. Ketterle, Phys. Rev. Lett. **93**, 223201 (2004)
36. A.L. Marchant, T.P. Billam, M.M.H. Yu, A. Rakonjac, J.L. Helm, J. Polo, C. Weiss, S.A. Gardiner, S.L. Cornish, Phys. Rev. A **93**, 021604(R) (2016)
37. S.A. Morgan, R.J. Ballagh, K. Burnett, Phys. Rev. A **55**, 4338 (1997)
38. R. Scharf, Chaos. Solitons Fract. **5**, 2527 (1995)
39. A.D. Martin, C.S. Adams, S.A. Gardiner, Phys. Rev. Lett. **98**, 020402 (2007)
40. L.D. Carr, Y. Castin, Phys. Rev. A **66**, 063602 (2002)
41. N.G. Parker, S.L. Cornish, C.S. Adams, A.M. Martin, J. Phys. B **40**, 3127 (2007)
42. N.G. Parker, A.M. Martin, S.L. Cornish, C.S. Adams, J. Phys. B **41**, 045303 (2008)
43. L. Salasnich, A. Parola, L. Reatto, Phys. Rev. A **66**, 043603 (2002)
44. L. Salasnich, A. Parola, L. Reatto, Phys. Rev. A **65**, 043614 (2002)
45. J. Cuevas, P.G. Kevrekidis, B.A. Malomed, P. Dyke, R.G. Hulet, New J. Phys. **15**, 063006 (2013)
46. K.E. Strecker, G.B. Partridge, A.G. Truscott, R.G. Hulet, Nature **417**, 150 (2002)
47. L. Khaykovich et al., Science **296**, 1290 (2002)
48. J.H.V. Nguyen, P. Dyke, D. Luo, B.A. Malomed, R.G. Hulet, Nat. Phys. **10**, 918 (2014)
49. A.L. Marchant, T.P. Billam, T.P. Wiles, M.M.H. Yu, S.A. Gardiner, S.L. Cornish, Nature Commun. **4**, 1865 (2013)
50. J.H.V. Nguyen, D. Luo, R.G. Hulet, Science **356**, 422 (2017)
51. L.D. Carr, J. Brand, Phys. Rev. Lett. **92**, 040401 (2004)
52. B. Eiermann et al., Phys. Rev. Lett. **92**, 230401 (2004)
53. A. Trombettoni, A. Smerzi, Phys. Rev. Lett. **86**, 2353 (2001)
54. B. Kibler, J. Fatome, C. Finot, G. Millot, F. Dias, G. Genty, N. Akhmediev, J.M. Dudley, Nat. Phys. **6**, 790 (2010)
55. M. Onorato, D. Proment, A. Toffoli, Phys. Rev. Lett. **107**, 184502 (2011)
56. J. Satsuma, N. Yajima, Prog. Theor. Phys. Suppl. **55**, 284 (1974)
57. A. Di Carli et al., Phys. Rev. Lett. **123**, 123602 (2019)
58. D. Luo et al., Phys. Rev. Lett. **125**, 183902 (2020)
59. R. Cruickshank et al., Phys. Rev. Lett. **135**, 263404 (2025)

Superfluid Helium

6

Abstract

In this chapter we consider the main properties of superfluid helium—the first quantum fluid which was discovered.

6.1 Helium 4 and Helium 3

The helium atom consists of two electrons, two protons and two neutrons: it is therefore a boson. There are only two stable isotopes of helium, the common isotope ^{4}He and the rare isotope ^{3}He, which lacks one neutron (hence the ^{3}He atom is a fermion).

The purity of helium is remarkable compared to other substances used in condensed matter physics experiments. Firstly, at the very low temperatures under consideration, any contaminant which may be present (e.g. air) freezes out and sticks to the walls of the containing vessel. The only possible impurity is ^{3}He. Secondly, the concentration of ^{3}He in natural helium samples from minerals or deposits of natural gas is only few atoms in 10^7; this is also the typical concentration of commercially available liquid helium. Using a heat flush technique, the ^{3}He concentration can be reduced [1] to few atoms in 10^{15}. A sample ^{4}He prepared in this way is thus the purest macroscopic sample of any element yet to have been prepared on Earth.

6.2 Helium I and Helium II

Soon after 1908 when helium was first liquified, physicists realized that, at sufficiently low temperature, liquid helium behaves differently from ordinary (classical) liquids. To appreciate these differences, it is instructive to compare the pressure-temperature phase diagram of helium to the phase diagram of water (the most common ordinary liquid), as shown in Fig. 6.1. In the phase diagram of water, notice the existence of a *triple point* where the solid, liquid and vapour (gas) phases coexist. The phase

© The Author(s), under exclusive license to Springer Nature Switzerland AG 2026

C. F. Barenghi et al., *Quantum Fluids, Solitons, and Vortices*, Lecture Notes in Physics 1050, https://doi.org/10.1007/978-3-032-20171-3_6

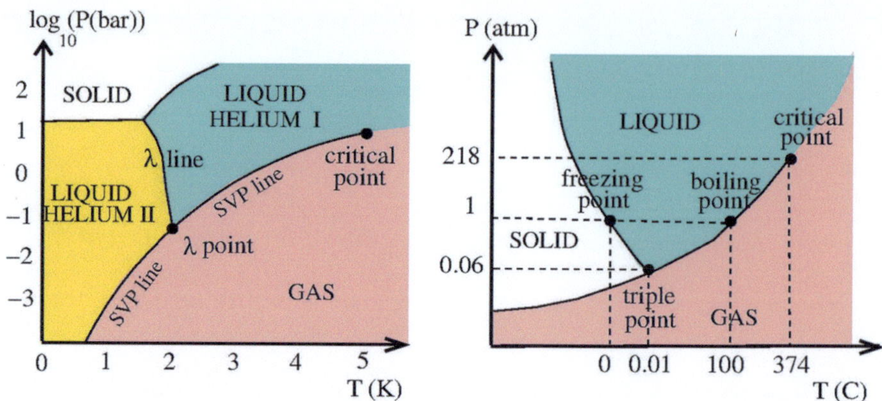

Fig. 6.1 Schematic phase diagrams of water (right) and helium (left) at varying pressure P and temperature T. Notice that the two figures use different temperature scales. For clarity, the normal freezing and boiling points of water (at $P = 1$ atm, $T = 0$ C and 100 C respectively) are also marked

diagram of helium lacks a triple point, and the liquid phase is divided in two regions called *helium I* and *helium II*, which are separated by the λ-*line*. This line marks the transition from a classical liquid (helium I) to a Bose-Einstein condensate (helium II). The point along the SVP (saturated vapour pressure) boundary which crosses the λ-line is called the *lambda-point*: here the temperature is $T_\lambda = 2.1768$ K, the critical temperature of Bose-Einstein condensation at SVP. The name lambda point arises from the shape of the specific heat vs temperature curve which recalls the Greek letter λ, shown in Fig. 6.2.

Another striking feature of the phase diagram of helium is that the liquid phase extends down to absolute zero, unlike ordinary liquids which become solid in this limit. To obtain solid helium, a considerable pressure must be applied–about 25 bar.

The distinction between helium I and helium II was made by the early researchers because the behaviour of liquid helium II differs greatly from liquid helium I. For example, if a sample of helium I is cooled by reducing the pressure of the vapour above the free surface using a pump, it is observed that for $T > T_\lambda$ the liquid boils violently, but, upon reaching T_λ, the boiling suddenly ceases. This is because helium II is a perfect heat conductor, as we shall see in Sect. 6.6, so it cools uniformly without local temperature gradients which would induce bubbles.

Other striking properties of helium II are the fountain effect, the spike of the specific heat vs temperature curve at T_λ, see Fig. 6.2, and the ability of the superfluid component to flow without viscous dissipation, a property called *superfluidity* by Kapitza in analogy with superconductivity (discovered by Heike Kamerlingh Onnes in 1911).

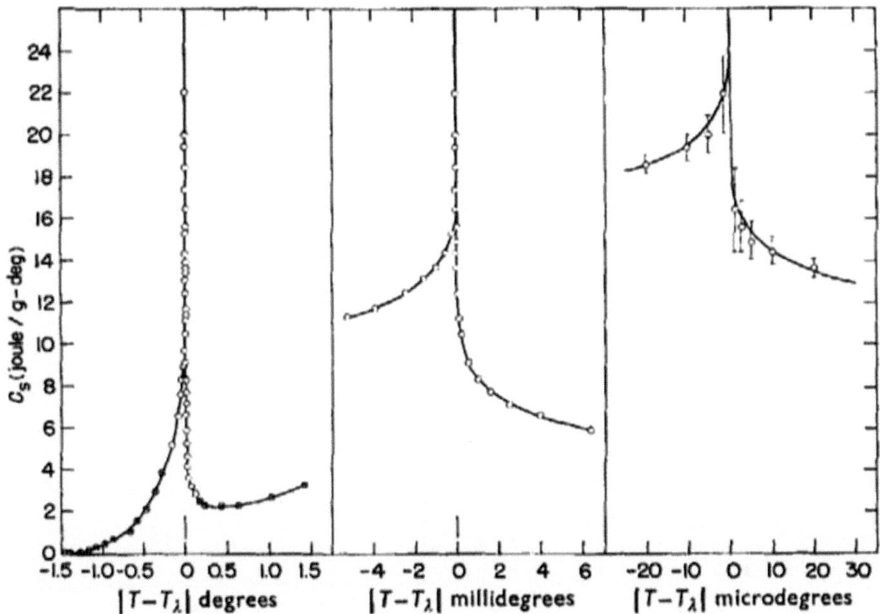

Fig. 6.2 Specific heat, C, of liquid helium vs temperature, T, shown at the Kelvin, the mK and the μK scales zooming into T_λ to demonstrate the scaling law of C at the phase transition. Reprinted with permission from [2]. Copyright 1961, North-Holland Publishing Co. All rights reserved. Compare this spike with the cusp of the C vs T curve for BEC in an ideal Bose gas shown in Fig. 2.11 (left)

6.3   Two-Fluid Model

To account for these effects, Lev Landau and, independently, László Tisza, introduced the *two-fluid model* [3]. This model describes helium II as the mixture of two inter-penetrating fluid components called the *normal fluid* and the *superfluid*. Each component has its own independent density field and velocity field, ρ_n and $\mathbf{v}_n$ for the normal fluid and ρ_s and $\mathbf{v}_s$ for the superfluid. The actual density of the liquid helium is the sum of normal fluid and superfluid densities,

$$\rho = \rho_n + \rho_s. \tag{6.1}$$

The model postulates that the superfluid component carries zero entropy and viscosity, whereas the normal fluid's specific entropy, S (entropy per unit mass), and viscosity, η, are nonzero. Table 6.1 summarises the properties of the two-fluid model.

By studying the period and the damping of the torsional oscillations of a pendulum consisting of a stack of closely-spaced disks, Andronikashvili determined the relative fractions ρ_n/ρ and ρ_s/ρ of normal fluid an superfluid as a function of temperature. The damping was observed to be the same in helium I and helium II,

Table 6.1 The two-fluid model of helium II. Numerical values of these and other helium parameters can be found at https://www.mas.ncl.ac.uk/helium/

Fluid component	Density	Velocity	Entropy	Viscosity
Normal fluid	ρ_n	$\boldsymbol{v}_n$	S	η
Superfluid	ρ_s	$\boldsymbol{v}_s$	0	0

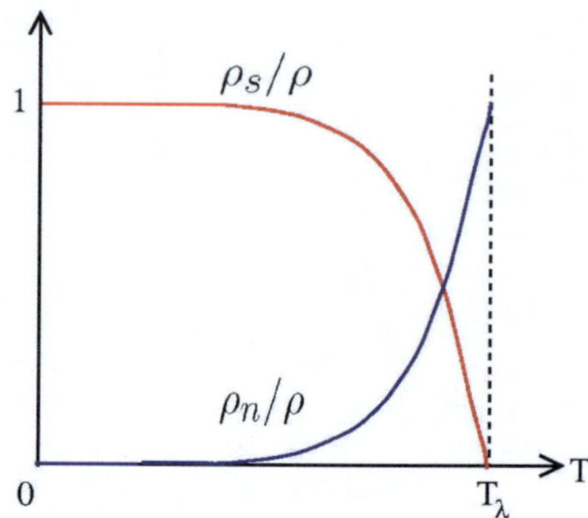

Fig. 6.3 Schematic normal fluid and superfluid fractions ρ_n/ρ and ρ_s/ρ versus temperature T

but the period (which depends on the inertia of the liquid trapped within the disks) is temperature-dependent, yielding the normal fluid and superfluid fractions showed in Fig. 6.3. Notice that $\rho_s \to 0$ for $T \to T_\lambda$ and $\rho_s \to \rho$ for $T \to 0$; vice versa. $\rho_n \to \rho$ for $T \to T_\lambda$ and $\rho_n \to 0$ for $T \to 0$. Whereas the density of helium, ρ, changes little with temperature, the normal fluid and superfluid densities change rapidly with temperature. For example the superfluid fraction, which is zero at T_λ, increases rapidly as the temperature is reduced. At $T = 1$ K it is $\rho_s/\rho = 99.3\%$, and we conclude that for $T < 1$ K liquid helium is effectively a pure superfluid.

Finally, notice the similarity between the temperature dependence of the superfluid fraction in helium II (Fig. 6.3) and the temperature dependence of the condensate fraction observed in atomic BECs (Fig. 2.12).

Mathematically, the two-fluid model consists of the following equations for normal fluid and superfluid velocities, $\boldsymbol{v}_n$ and $\boldsymbol{v}_s$, temperature, T, and pressure, P,

$$\frac{\partial \rho}{\partial t} + \nabla \cdot (\rho_n \boldsymbol{v}_n + \rho_s \boldsymbol{v}_s) = 0, \tag{6.2}$$

$$\frac{\partial (\rho S)}{\partial t} + \nabla \cdot (\rho S \boldsymbol{v}_n) = 0, \tag{6.3}$$

$$\rho_s \left(\frac{\partial \boldsymbol{v}_s}{\partial t} + (\boldsymbol{v}_s \cdot \nabla)\boldsymbol{v}_s \right) = -\frac{\rho_s}{\rho}\nabla P + \rho_s S \nabla T + \frac{\rho_n \rho_s}{2\rho}\nabla(\boldsymbol{v}_n - \boldsymbol{v}_s)^2, \tag{6.4}$$

$$\rho_n \left(\frac{\partial \boldsymbol{v}_n}{\partial t} + (\boldsymbol{v}_n \cdot \nabla) \boldsymbol{v}_n \right) = -\frac{\rho_n}{\rho} \nabla P - \rho_s S \nabla T - \frac{\rho_n \rho_s}{2\rho} \nabla (\boldsymbol{v}_n - \boldsymbol{v}_s)^2 + \eta \nabla^2 \boldsymbol{v}_n.$$

$$(6.5)$$

The first and second equations represent conservation of mass and entropy respectively, where $\rho_n \boldsymbol{v}_n + \rho_s \boldsymbol{v}_s$ is the mass flux and $\rho S \boldsymbol{v}_n$ is the entropy flux. The third and fourth equations are the superfluid and normal fluid equations of motion respectively.

If we set $\nabla T = 0$, then the superfluid and normal fluid equations have two interesting limits. If $T \to T_\lambda$, then $\rho_s \to 0$, $\rho_n \to \rho$, and the normal fluid equation reduces to the classical Navier-Stokes equation for a viscous fluid. Vice versa, if $T \to 0$, then $\rho_n \to 0$, $\rho_s \to \rho$, and the superfluid equation reduces to the classical Euler equation for an inviscid fluid. Indeed it is observed [4] that pressure waves (sound) propagate in helium II with the classical speed $u = (\partial p / \partial \rho)^{1/2}$ predicted by the Euler equation.

However, in the general case that $\nabla T \neq 0$, helium II displays phenomena such as the fountain effect and second sound which are not consistent with the Euler equation, and which are described in the next sections.

6.4   Fountain Effect

The fountain or thermo-mechanical effect, discovered by Allen and Jones in 1938 [5], is illustrated in Fig. 6.4. Two vessels A and B containing helium II at temperature T are separated by a *superleak* S, a narrow passage through which the viscous normal fluid cannot flow. A superleak consists either of a very narrow constriction less than a micron wide, or a small volume of very finely packed powder (e.g. jeweller's rouge). Initially the temperature and the level of the liquid are the same in both A and B. If power is dissipated in the resistor R, the temperature of the liquid in A increases from T to $T + \Delta T$, and the superfluid flows through S from B to A, creating a difference Δz between the levels of the liquid in A and B. This level difference corresponds to a pressure difference: $\Delta P = \rho g \Delta z$ where $g = 9.81$ m/s^2 is gravity's acceleration. It is found that $\Delta P / \rho = S \Delta T$, according to the superfluid Eq. (6.4). A large pressure difference generates a jet or fountain, hence the name of this effect.

The fountain effect is not consistent with classical Euler hydrodynamics. A temperature difference in a classical fluid can lead to a flow (convection), but the entropy law requires that the velocity of the fluid, $\boldsymbol{v}$, is in the same direction of the entropy flow, $\rho S \boldsymbol{v}$.

The inverse of the fountain effect was also observed. If the two vessels are initially at the same temperature T but the pressure in B is larger (for example because the level of the liquid in B is higher), the superfluid component flows through the superleak, and the liquid in A becomes colder (we can think of the superfluid component as a liquid at zero temperature).

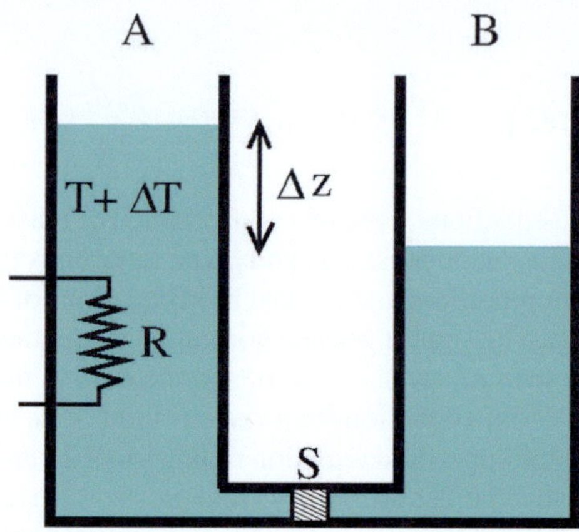

Fig. 6.4 Schematic fountain effect

6.5 Second Sound

Using the two-fluid equations, Lev Landau predicted the existence of two kinds of sound waves in helium II: *first sound* and *second sound*. The experimental verification of this prediction by Vasilii Peshkov in 1944 firmly established the two-fluid model. First sound, a density/pressure wave in which entropy/temperature are approximately constant and the two fluid components move in phase, is analog to ordinary sound in helium I and in classical fluids; second sound is an unusual temperature/entropy wave in which density/pressure are approximately constant. The first and second sound speeds u_1 and u_2 are,

$$u_1 = \sqrt{\frac{\partial p}{\partial \rho}}, \qquad u_2 = \sqrt{\frac{\rho_s T S^2}{\rho_n C}}, \tag{6.6}$$

where C is the specific heat. Note that u_2 vanishes at $T = T_\lambda$ because second sound does not exist in helium I (Fig. 6.5).

Second sound is important because it is used to detect the presence of quantised vortices in the flow and measure their length. The reason is that a second sound wave is scattered by the superfluid velocity field around one or more vortex lines which are perpendicular to the direction of sound propagation. This scattering process is temperature dependent and is parameterised by *mutual friction coefficients B* and B', or, equivalently, by the coefficients $\alpha = \rho_n B/(2\rho)$ and $\alpha' = \rho_n B'/(2\rho)$. The temperature dependence of the friction coefficients is shown in Fig. 6.6.

Consider a second sound transmitter/receiver pair set up across the helium channel. A particular resonance frequency f_0 (which depends on the channel's geometry) is chosen, and the peak's amplitude A_0 and width Δf_0 of this peak are measured when helium II is at rest. After vortex lines are introduced in the system for example

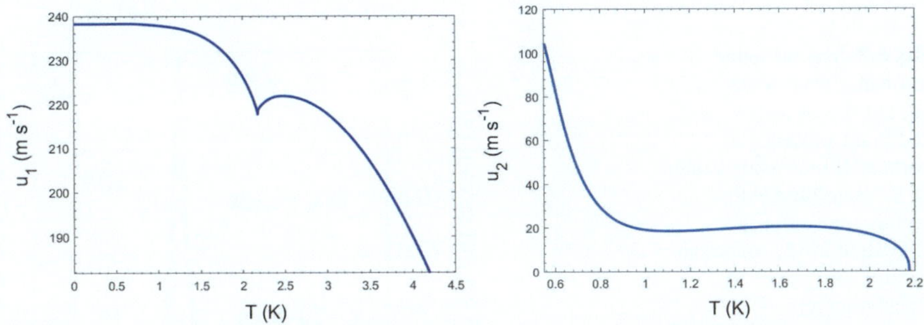

Fig. 6.5 First sound speed u_1 (left) and second sound speed u_2 (right) versus temperature T (available online at https://www.mas.ncl.ac.uk/helium/ together with other helium properties. Data reprinted with permission from [6]. Copyright 1998, American Institute of Physics. All rights reserved

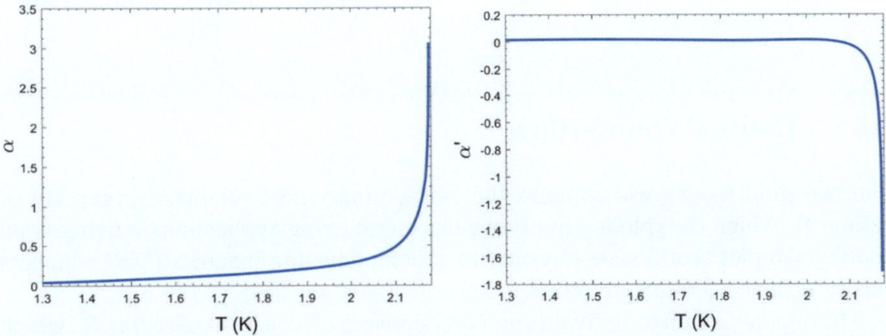

Fig. 6.6 Temperature dependence of the mutual friction coefficients α and α'. Data reprinted with permission from [6]. Copyright 1998, American Institute of Physics. All rights reserved. See also the website https://www.mas.ncl.ac.uk/helium/

by applying a longitudinal flow or counterflow of velocity V, the resonance peak's amplitude A_1 is measured again. Since second sound is attenuated by the vortex lines, the peak's amplitude is less ($A_1 < A_0$) as shown in Fig. 6.7. The *vortex line density* (defined as the length of quantised vortex lines per unit volume) can be estimated as,

$$L = \frac{6\pi \, \Delta f_0}{\kappa B} \left(\frac{A_0}{A_1} - 1 \right), \tag{6.7}$$

where κ is the quantum of circulation (see Chap. 9) and B is the first mutual friction coefficient.

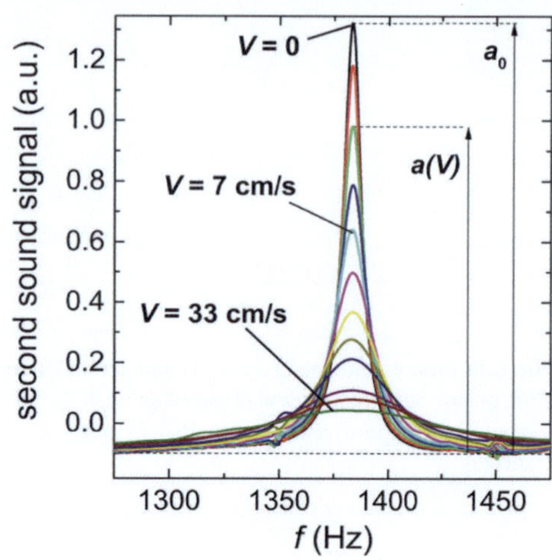

Fig. 6.7 Second sound resonance peak when helium II is at rest ($V = 0$) and in the presence of vortices at increasing values of V. Reprinted with permission from [7]. Copyright 2015, American Institute of Physics. All rights reserved

6.6 Thermal Counterflow

The two-fluid model also explains the extraordinary heat conducting property of helium II, which is exploited in cryogenics engineering applications ranging from cooling samples (solid state physics) to superconducting magnets (MRI scanners and particle accelerators) to infrared detectors (astrophysics).

The popular set-up to study this property consists of a channel of radius R_0 which is closed at one end and is open to the helium bath at the other end held at temperature T and pressure P, as shown in Fig. 6.8. At the closed end, a heater R (an electrical resistor) dissipates a known heat flux $W = \rho S T V_n$, defined as heat power $\dot{Q}$ per cross-sectional area of the channel, $W = \dot{Q}/(\pi R_0^2)$. This heat is carried away from the resistor by the normal fluid. The superfluid flows in the opposite direction to conserve mass, hence a *counterflow velocity* $V_{ns} = V_n - V_s$ is set up akin to an internal convection, where here for simplicity V_n, V_s and V_{ns} are velocities averaged over the cross section of the channel.

It can be shown (see Problem 6.2) that the counterflow velocity is proportional to applied heat flux,

$$V_{ns} = \frac{W}{\rho_s S T}.$$

(6.8)

If $\dot{Q}$ exceeds a critical value, however, we shall see that this perfect internal convection breaks down into turbulence.

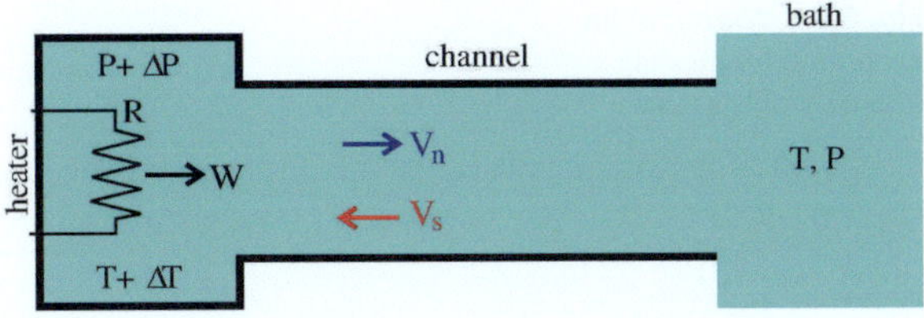

Fig. 6.8 Schematic counterflow channel

6.7 Superfluidity

The most striking effect exhibited by helium II is the ability to flow along narrow slits or channels without an applied pressure gradient, as if it had zero viscosity. The effect was first demonstrated by Kapitza [8], who determined the viscosity from the pressure drop when the liquid helium flows through the gap between two closely spaced discs. The result was confirmed by Allen and Misener [9] in an article which appeared on the next page of the same issue of Nature of Kapitza's paper.

However, if one attempts to determine the viscosity of helium II by measuring the damping of a vibrating wire or the torque between two widely-spaced concentric cylinders, one finds non-zero values. To solve this apparent paradox, one must remark that in the letter experiments what is measured is the viscosity of the normal fluid, whereas in Kapitza's experiment the normal fluid cannot move in between the closely spaced disks: what has zero viscosity is the superfluid component only.

6.8 Relation Between Helium II and BEC

The connection between the unusual properties of helium II and Bose-Einstein condensation was made as early as 1938 by London [10], but was largely ignored by many physicists until Bose-Einstein condensation was achieved in atomic gases in 1995 as originally envisaged by Bose and Einstein. Nowdays it is widely accepted that helium II is a condensate.

In his pioneering paper, London noticed the similarity between the cusp of the specific heat vs temperature curve predicted by Bose and Einstein for an ideal gas, see Fig. 2.11(left), and the observed spike of the specific heat of liquid helium, see Figs. 2.11(right) and 6.2. He also calculated the critical temperature for BEC at which an ideal gas with the same density of helium would condense, and found $T = 3.1$ K, which is numerically close to $T_\lambda = 2.1768$ K. Finally he noticed [11] the very large value of the zero point energy of liquid helium (hence the importance of quantum effects), which explains why helium does not become solid at low pressure.

It is important to notice that whereas Bose and Einstein's original idea referred to an ideal (non-interacting) Bose gas and the GPE describes a weakly-interacting Bose gas, helium II is a strongly interacting liquid, not a dilute gas. The GPE is therefore only a qualitative model of helium II: it captures some physical ingredients (notably the dynamics of vortex lines including their reconnections) but not all.

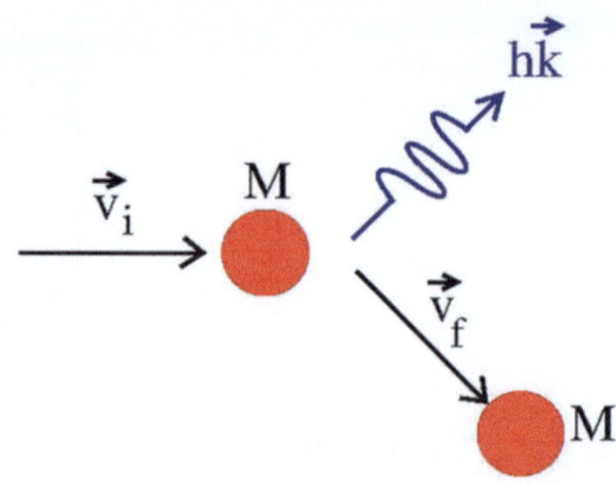

Fig. 6.9 Creation of an excitation by moving impurity of mass M

The condensate fraction should not be identified with the superfluid fraction. The condensate fraction of helium II can be measured experimentally by deep inelastic neutron scattering off the superfluid, yielding values ranging from 7.5% [12] to 14% [13] at low temperature.

6.9   Landau's Critical Velocity

Superfluidity means that the superfluid component can flow along a channel or past an object without losing energy, or, changing the reference frame, that an object can move with respect to the superfluid component without suffering a drag. However experiments of all kinds demonstrated that this property holds true only provided the velocity is relatively small. Landau [14] discovered a simple but powerful criterion to understand under which condition energy dissipation sets in. The argument, first developed for helium II, applies to a condensate described by the GPE too.

Consider a homogeneous ground state condensate, into which an impurity or an object of mass M enters with initial velocity $\boldsymbol{v}_i$, see Fig. 6.9. Let us imagine that the impurity imparts an excitation of the condensate with energy $\hbar\omega$ and momentum $\hbar\boldsymbol{k}$; the subsequent velocity of the impurity is $\boldsymbol{v}_f$. The initial energy (relative to the static background condensate) is just the initial kinetic energy of the impurity, $Mv_i^2/2$, while the final energy after generating the excitation is $Mv_f^2/2 + \hbar\omega$, where $v_i = |\boldsymbol{v}_i|$ and $v_f = |\boldsymbol{v}_f|$. Applying conservation of energy gives,

$$\frac{1}{2}Mv_i^2 = \frac{1}{2}Mv_f^2 + \hbar\omega. \tag{6.9}$$

Similarly, applying conservation of momentum before and after the event gives,

$$M\boldsymbol{v}_i = M\boldsymbol{v}_f + \hbar\boldsymbol{k}. \tag{6.10}$$

Inserting $\boldsymbol{v}_i$ from Eq. (6.10) into Eq. (6.9) and simplifying gives,

$$\hbar\omega = \hbar\boldsymbol{k} \cdot \boldsymbol{v}_i - \frac{\hbar^2 k^2}{2M}. \tag{6.11}$$

If M is sufficiently large, the second term at the right hand side can be neglected. For excitations to be energetically favoured, the initial velocity v_i then has to satisfy,

$$v_i \geq \left| \frac{\boldsymbol{k} \cdot \boldsymbol{v}_i}{k} \right| \geq \frac{\omega}{k}. \tag{6.12}$$

One can instead write this as,

$$v_i \geq v_{\rm L}, \tag{6.13}$$

where $v_{\rm L}$ is termed *Landau's critical velocity*,

$$v_{\rm L} = \min\left(\frac{\omega}{k}\right). \tag{6.14}$$

where the minimum is found by varying k. This is a defining property of superfluidity. For $v_i < v_{\rm L}$ the impurity propagates with no drag, i.e. moving in an inviscid fluid, while for $v_i \geq v_{\rm L}$ the energy of the motion is lost as energy and momentum are transferred from the impurity to waves in the fluid. Equation (6.14) marks the breakdown of superfluidity. We conclude that superfluidity is a consequence of the particular shape of the dispersion relation of the elementary excitations of a system– it is not a consequence of Bose-Einstein condensation. This result highlights the importance of the dispersion relation $\omega = \omega(k)$, which, since $E = \hbar\omega$ and $\boldsymbol{p} = \hbar\boldsymbol{k}$, is often expressed as an energy-momentum relation $E = E(p)$. In terms of energy and momentum, Landau's critical velocity becomes,

$$v_{\rm L} = \min\left(\frac{E}{p}\right). \tag{6.15}$$

The dispersion relation $\omega = \omega(k)$ is typically a non-trivial function of k. For a weakly-interacting homogeneous condensate, we derived the Bogoliubov dispersion relation, Eq. (4.17), from the GPE; using this result, if we minimize ω/k, we obtain,

$$v_{\rm L} = c. \tag{6.16}$$

where c is the speed of sound.

Therefore for $v < c$ the impurity moves with no drag or hindrance, a defining characteristic of superfluidity. For $v > c$ drag can occur through the creation of condensate's excitations. This result applies well to an atomic Bose-Einstein condensate.

6.10 Landau's Spectrum

When Landau developed this argument, the dispersion relation of the elementary excitations of helium II was not known. He guessed that it has the shape shown in Fig. 6.10, consisting of a *phonon* region at low k and a *roton* region at large k

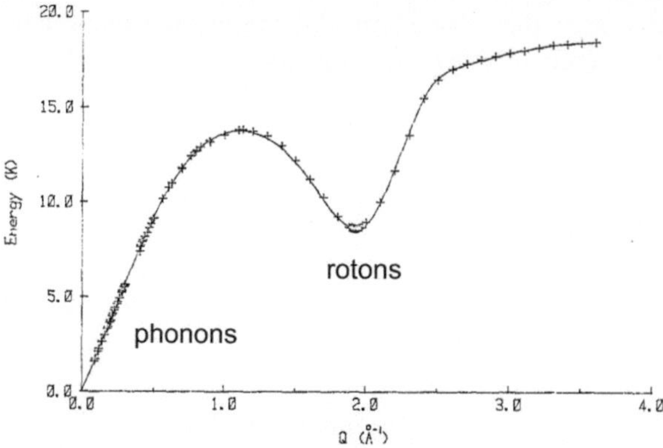

Fig.6.10 Landau's spectrum of helium II's elementary excitations as observed in neutron scattering experiments. Reprinted with permission from [15]. Copyright 1981, Plenum Publishing Corporation. All rights reserved. Here the energy is expressed in Kelvin degrees, and the wavenumber in inverse Angstrom

with a minimum at some $k = k_0$. Experiments in which neutrons were scattered by helium II verified Landau's conjecture [12,13].

A convenient approximation to Landau's spectrum expressed in the energy-momentum form $E = E(p)$,

$$E(p) = \begin{cases} u_1 p, & \text{in the phonon region} \\ \Delta_0 + \frac{(p-p_0)^2}{2\mu_0} & \text{in the roton region} \end{cases} \tag{6.17}$$

where, at very low pressure, $u_1 = 240$ m/s is the speed of first sound, $\Delta_0 = 1.20 \times 10^{-22}$ J is called the energy gap, $p_0 = \hbar k_0 = 2.02 \times 10^{-24}$ kg m/s is the momentum at the roton minimum, $\mu_0 = 0.161 m_4$ is the effective mass of a roton, and $m_4 = 6.65 \times 10^{-27}$ kg is the mass of one helium atom. At large pressure the shape of the Landau's spectrum changes slightly.

Using Eq. (6.17), we find that Landau's critical velocity is approximately $v_L = 59$ m/s in helium II at low pressure (see Problem 6.4).

Landau's critical velocity was verified by the ion experiments of Peter McClintock and collaborators [16]. The negative helium ion in helium II is a charged empty bubble of radius between 1.15 nm (near the melting pressure) and 1.7 nm (at SVP); its effective mass varies between 87 m_4 (at melting pressure) and 243 m_4 (at SVP). By applying suitable electric field, negative ions can be driven to a desired velocity which can be measured by a time of flight technique. In most experimental conditions, at SVP, the moving ion nucleates a quantum vortex at velocities which are much smaller than Landau's critical velocity; this vortex nucleation process therefore hides Landau's fundamental mechanism of roton creation.

McClintock and collaborators [16] found that, if a pressure is applied, Landau's critical velocity is less than the vortex nucleation velocity, and the fundamental mechanism envisaged by Landau can be observed. They performed experiments with low temperature helium I under pressure in the range $0.28 < T < 0.50$ K,

Fig. 6.11 Schematic measure of drag force on an ion vs ion velocity [16] at $T = 0.35$ K in helium II (red line), and at $T = 4.0$ K in helium I (blue line)

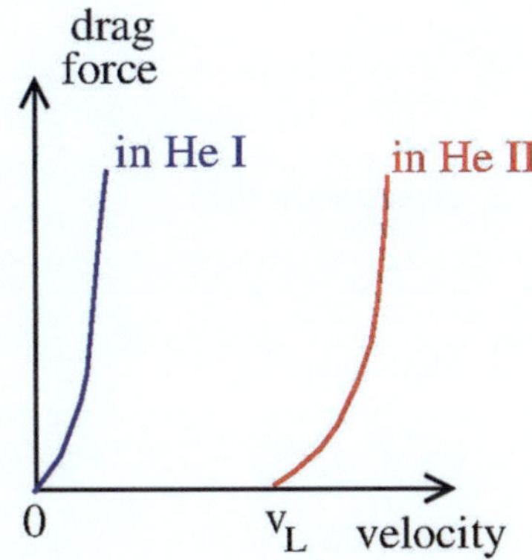

$20 \times 10^5 < P < 25 \times 10^5$ Pa. Figure 6.11 shows the main finding schematically: in helium II the drag force on the moving ion is zero for all velocities $v_i < v_L$, as predicted by Landau at the relevant pressure, whereas in helium I the classical drag force on the ion is nonzero at any velocity.

Finally it is interesting to notice that Landau's critical velocity for an ideal Bose gas is zero, in other words the ideal Bose gas is not a superfluid, see Problem 6.4.

Problems

6.1 *First sound and second sound*

Consider a uniform sample of helium II at rest with constant densities $\rho = \rho_0$, $\rho_n = \rho_{n0}$, $\rho_s = \rho_{s0}$, specific entropy $S = S_0$, pressure $p = p_0$, temperature $T = T_0$ and velocities $\boldsymbol{v}_n = \boldsymbol{v}_{n0} = \boldsymbol{0}$, $\boldsymbol{v}_s = \boldsymbol{v}_{s0} = \boldsymbol{0}$.

1. Introduce small perturbations (denoted by primes) $\rho = \rho_0 + \rho'$, $\rho_n = \rho_{n0} + \rho'_n$, $\rho_s = \rho_{s0} + \rho'_s$, $S = S_0 + S'$, $p = p_0 + p'$, $T = T_0 + T'$, $\boldsymbol{v}_n = \boldsymbol{v}_{n0} + \boldsymbol{v}'_n$ and $\boldsymbol{v}_s = \boldsymbol{v}_{s0} + \boldsymbol{v}'_s$. Neglecting the viscous term in the normal fluid equation (we know already that it would damp out any sound waves), show that the linearized equations for the perturbations are

$$\frac{\partial \rho'}{\partial t} + \nabla \cdot (\rho_{n0} \boldsymbol{v}'_n + \rho_{s0} \boldsymbol{v}'_s) = 0, \tag{6.18}$$

$$\rho_0 \frac{\partial S'}{\partial t} + S_0 \frac{\partial \rho'}{\partial t} + \rho_0 S_0 \nabla \cdot \boldsymbol{v}'_n = 0, \tag{6.19}$$

$$\rho_{s0} \frac{\partial \boldsymbol{v}'_s}{\partial t} = -\frac{\rho_{s0}}{\rho_0} \nabla P' + \rho_{s0} S_0 \nabla T', \tag{6.20}$$

$$\rho_{n0}\frac{\partial \boldsymbol{v}'_n}{\partial t} = -\frac{\rho_{n0}}{\rho_0}\nabla P' - \rho_{s0}S_0\nabla T'. \tag{6.21}$$

2. Hence show that

$$\frac{\partial^2 \rho'}{\partial t^2} = \nabla^2 P'. \tag{6.22}$$

3. Show that

$$\frac{\partial}{\partial t}(\boldsymbol{v}'_n - \boldsymbol{v}'_s) = -\frac{S_0\rho_0}{\rho_{n0}}\nabla T'. \tag{6.23}$$

4. Show that

$$\frac{\partial^2 S}{\partial t^2} = \frac{\rho_{s0}S_0^2}{\rho_{n0}}\nabla T' = 0. \tag{6.24}$$

5. Derive the following equations for P' and T':

$$\left(\frac{\partial \rho}{\partial P}\right)_T \frac{\partial^2 P'}{\partial t^2} + \left(\frac{\partial \rho}{\partial T}\right)_P \frac{\partial^2 T'}{\partial t^2} = \nabla^2 P', \tag{6.25}$$

$$\left(\frac{\partial S}{\partial P}\right)_T \frac{\partial^2 P'}{\partial t^2} + \left(\frac{\partial S}{\partial T}\right)_P \frac{\partial^2 T'}{\partial t^2} = \frac{S_0^2\rho_{s0}}{\rho_{n0}}\nabla^2 T'. \tag{6.26}$$

6. Assume that P' and T' depend on x and t as $e^{i\omega t - ikx} = e^{i\omega(t-x/u)}$ where ω is the angular velocity, k is the wavenumber, and $u = \omega/k$ is the phase speed. Derive the following linear system for P' and T',

$$\left[-\omega^2\left(\frac{\partial \rho}{\partial P}\right)_T + k^2\right]P' - \omega^2\left(\frac{\partial \rho}{\partial T}\right)_P T' = 0, \tag{6.27}$$

$$-\omega^2\left(\frac{\partial S}{\partial P}\right)_T P' + \left[-\omega^2\left(\frac{\partial S}{\partial T}\right)_P + k^2\frac{S_0^2\rho_{s0}}{\rho_{n0}}\right]T' = 0. \tag{6.28}$$

7. The linear system has nontrivial solutions only if the determinant of the matrix is zero, hence, after introducing the Jacobian

$$\frac{\partial(S,\rho)}{\partial(T,P)} = \left(\frac{\partial S}{\partial T}\right)_P\left(\frac{\partial \rho}{\partial P}\right)_T - \left(\frac{\partial S}{\partial P}\right)_T\left(\frac{\partial \rho}{\partial T}\right)_P, \tag{6.29}$$

derive the condition

$$\frac{\partial(S,\rho)}{\partial(T,P)}u^4 - u^2\left(\frac{\rho_{s0}S_0^2}{\rho_{n0}}\left(\frac{\partial \rho}{\partial P}\right)_T + \left(\frac{\partial S}{\partial T}\right)_P\right) + \frac{\rho_{s0}S_0^2}{\rho_{n0}} = 0. \tag{6.30}$$

From thermodynamics we know that

$$\frac{\partial(S, \rho)}{\partial(T, P)} = \frac{C_V}{T}\left(\frac{\partial \rho}{\partial P}\right)_T, \qquad \left(\frac{\partial S}{\partial T}\right)_P \frac{T}{C_V}\left(\frac{\partial P}{\partial \rho}\right)_T = \left(\frac{\partial P}{\partial \rho}\right)_S, \qquad (6.31)$$

hence, by approximating

$$C_V \approx C_P = C, \qquad \text{and} \qquad \left(\frac{\partial P}{\partial \rho}\right)_S \approx \left(\frac{\partial P}{\partial \rho}\right)_T = \frac{\partial P}{\partial \rho}, \qquad (6.32)$$

show that u satisfies the quartic equation

$$u^4 - \left[\frac{\partial P}{\partial \rho} + \frac{T_0 S_0^2 \rho_{s0}}{\rho_{n0} C}\right] u^2 + \left(\frac{T_0 S_0^2 \rho_{s0}}{\rho_{n0} C}\right)\frac{\partial P}{\partial \rho} = 0. \qquad (6.33)$$

8. Solve the equation for u and show that the roots are

$$u_1 = \left(\frac{\partial P}{\partial \rho}\right)^{1/2}, \qquad u_2 = \left(\frac{\rho_{s0} T_0 S_0^2}{\rho_{n0} C}\right)^{1/2},$$

representing respectively the speed of first sound and the speed of second sound.

6.2 *Thermal counterflow I*

A heater dissipates a known heat flux W at the resistor's end of a counterflow channel, see Fig. 6.8. Assuming uniform velocities, show that both the average normal fluid velocity, V_n and the average counterflow velocity, V_{ns} are proportional to the applied heat flux W.

6.3 *Thermal counterflow II*

Consider the previous problem but this time in more detail, taking into account the flow profile in the channel.

1. Assuming a steady state and that the velocities are small (i.e. neglecting quadratic terms in the velocities), show that the normal fluid and superfluid Eqs. 6.5 and 6.4 reduce to

$$\eta \nabla^2 \boldsymbol{v}_n = \nabla P, \qquad \nabla P = \rho S \nabla T.$$

2. Assume that the counterflow channel is a cylinder of radius R and length Δz. Using cylindrical coordinates r, θ, z where z is the axis of the cylinder, assume that $\mathbf{v}_n = v_n \widehat{\mathbf{z}}$ where $\widehat{\mathbf{z}}$ is the unit vector along the z-direction, and $v_n = v_n(r)$ depends on r only. Imposing viscous no-slip boundary conditions on the normal fluid, show that

$$v_n(r) = \frac{\Delta P R^2}{4\eta \Delta z} \left(1 - \frac{r^2}{R^2} \right).$$

Here the pressure gradient along z (which is negative for v_n to be positive) is approximated as $dP/dz \approx -\Delta P/\Delta z$ where Δz is the length of the channel and $\Delta P > 0$ is the pressure difference between the heater and the bath.

3. Show that the average normal fluid velocity is

$$V_n = \frac{\Delta P R^2}{8\eta \Delta z}.$$

4. Show that ΔP and ΔT, are proportional to the heat flux W.

6.4 *Landau's critical velocity*

Consider an object of mass M moving at velocity $\mathbf{v}_i$ which creates an excitation of energy E and momentum $\mathbf{p} = \hbar \mathbf{k}$.

1. Show that Landau's critical velocity, $v_L = \min(E/p)$, is equivalent to the condition $dE/dp = E/p$.
2. Using the approximate expression for Landau's spectrum given by Eq. (6.17), show that in liquid helium II Landau's critical velocity is $v_L \approx 59$ m/s at low pressure.
3. Show that an ideal Bose gas, which has the free-particle dispersion relation $E = p^2/(2M)$, is not a superfluid (its Landau's critical velocity is zero).

6.5 *Fountain effect*

Consider the arrangement of Fig. 6.4. Initially, the level of helium is the same in A and B. If the heater is switched on, the superfluid flows from B to A through the superleak. If the temperature difference between A and B is $\Delta T = 1$ mK, what is the height difference Δz ? Assume that the temperature is $T = 2$ K.

References

1. P.V.E. McClintock, Cryogenics **18**, 201 (1978)
2. M.J. Buckingham, W.M. Fairbank, Prog. Low Temp. Phys., Chapter III, **3**, 80–112, North-Holland (1961)
3. S. Balibar, J. Low Temp. Phys. **146**, 665 (2007)

4. J.F. Allen, A.D. Misener, Proc. Roy. Soc. A **172**, 467 (1939)
5. J.F. Allen, H. Jones, Nature **141**, 243 (1938)
6. R.J. Donnelly, C.F. Barenghi, J. Phys. Chem. Ref. Data **27**, 1217 (1998)
7. E. Varga, S. Babuin, L. Skrbek, Phys. Fluids **27**, 065101 (2015)
8. P. Kapitza, Nature **141**, 74 (1938)
9. J.F. Allen, A.D. Misener, Nature **141**, 75 (1938)
10. F. London, Nature **141**, 643 (1938)
11. F. London, Proc. Roy. Soc. A **153**, 576 (1936)
12. H.R. Glyde, R.T. Azuah, W.G. Stirling, Phys. Rev. B **62**, 14337 (2000)
13. V.F. Sears, E.C. Svensson, P. Martel, A.D.B. Woods, Phys. Rev. Lett. **49**, 279 (1982)
14. P. Nozieres, D. Pines, *The Theory of Quantum Liquids* (Perseus Books, Cambridge, 1999)
15. R.J. Donnelly, J.A. Donnelly, R.N. Hills, J. Low Temp. Phys. **44**, 471 (1981)
16. D.R. Allum, P.V.E. McClintock, A. Phillips, R.M. Bowley, Phil. Trans. Roy Soc. A **284**, 79 (1977)

Rotation and Vortices

7

Abstract

As well as being free from viscosity, Bose-Einstein condensates have another striking property–they are constrained to circulate only through the presence of whirlpools of fixed size and quantised circulation. In contrast, in conventional fluids, the eddies can have arbitrary size and circulation. Here we establish the form of these quantum vortices, their key properties and how they are formed and modelled.

7.1 Phase Defects

The condensate wavefunction is a complex quantity. We have seen that it can be written as $\Psi(\mathbf{r}, t) = R(\mathbf{r}, t)e^{iS(\mathbf{r},t)}$ (Madelung transform), where $R(\mathbf{r}, t)$ and $S(\mathbf{r}, t)$ are respectively the phase and amplitude distributions at time t. Consider following a closed path C of arbitrary shape through a region of the condensate. As we go around the path, the integrated change in the phase is,

$$\Delta S = \oint_C \nabla S \cdot d\boldsymbol{\ell}, \tag{7.1}$$

where the vector $d\boldsymbol{\ell}$ is the line element of integration. Let the wavefunction be Ψ_0 and Ψ_1 respectively at the starting point and at the final point of C. Since the two points are the same and Ψ must be single-valued, the condition $\Psi_1 = \Psi_0$ means that,

$$\Delta S = 2\pi q, \qquad q = 0, \pm 1, \pm 2, \cdots \tag{7.2}$$

If the integer number $q \neq 0$ then, somewhere within the region enclosed by C, there must be a *phase defect*, a point where the phase wraps by the amount $2\pi q$. At this point the phase of the wavefunction takes on every value, and the only way that Ψ can remain single-valued here is if Ψ is exactly zero.

© The Author(s), under exclusive license to Springer Nature Switzerland AG 2026   123
C. F. Barenghi et al., *Quantum Fluids, Solitons, and Vortices*, Lecture Notes in Physics 1050, https://doi.org/10.1007/978-3-032-20171-3_7

7.2   Quantised Vortices

What does the presence of a phase defect mean for the condensate as a fluid? Recalling that the phase distribution defines the fluid's velocity via $\mathbf{v} = (\hbar/m)\nabla S$, Eq. (7.2) implies that the *circulation* Γ around the path C is either zero or a multiple of the *quantum of circulation* κ,

$$\Gamma = \oint_C \mathbf{v} \cdot \mathrm{d}\boldsymbol{\ell} = q\kappa, \qquad \kappa = \frac{h}{m}. \tag{7.3}$$

This important result (the *quantization of the circulation*) tells us that the condensate flows very differently from ordinary fluids, where the circulation takes arbitrary values.

Assume that $q \neq 0$, and that the path C is a circle of radius r centred at the singularity. Consider the simple case of two-dimensional flow in the xy plane. Using polar coordinates (r, θ), the line element is $\mathrm{d}\boldsymbol{\ell} = r\mathrm{d}\theta\,\widehat{\boldsymbol{e}}_\theta$, where $\widehat{\boldsymbol{e}}_\theta$ is the unit vector in the azimuthal direction θ. Then the circulation becomes,

$$\Gamma = \oint_C \mathbf{v} \cdot \mathrm{d}\boldsymbol{\ell} = \int_0^{2\pi} r\mathbf{v} \cdot \widehat{\boldsymbol{e}}_\theta\, \mathrm{d}\theta = 2\pi r v_\theta. \tag{7.4}$$

Comparison with Eq. (7.3) shows that the fluid's azimuthal speed around the singularity is,

$$v_\theta = \frac{q\hbar}{mr} = \frac{q\kappa}{2\pi r}. \tag{7.5}$$

Since the condensate is a fluid without viscosity, this flow around the singularity should go on forever, at least in principle.

For $q \neq 0$, Eq. (7.5) tells us that the velocity around the singularity decreases to zero at infinity ($v_\theta \to 0$ as $r \to \infty$), and that, as we approach the axis, the flow becomes faster and faster, and diverges ($v_\theta \to \infty$ as $r \to 0$). If we increase q, the flow speed increases discontinuously, because q takes only discrete values. The sign of q determines the direction of the flow (clockwise or anticlockwise) around the singularity.

We now have a better picture of the nature of the singularity: it is a *quantised vortex line*, a whirlpool in the fluid. The quantity q is called the *charge* of the vortex. Figure 7.1(left) represents a straight vortex line through the origin, parallel to the z axis. Since the flow is the same on all planes perpendicular to the z axis, the flow of the (three-dimensional) straight vortex can be more simply described as the flow due to a two-dimensional *vortex point* on the xy plane, as in Fig. 7.1(middle). If these conditions are not met, such as the curved vortex line shown in Fig. 7.1(right), then the flow is fully three-dimensional and cannot be represented by a vortex point.

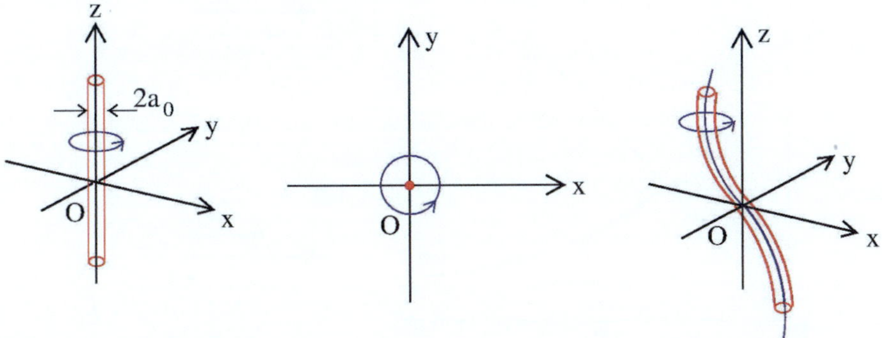

Fig. 7.1 Left: Schematic (three-dimensional) straight vortex line through the origin and parallel to the z axis. The red tube around the vortex axis of radius a_0 represents the vortex core. Middle: Since the vortex line is straight, it suffices to consider the two-dimensional flow of a vortex point on the xy plane (the flow on other planes parallel to the xy plane will be the same). Right: For a more general bent vortex line the flow is fully three-dimensional

7.3 Classical Versus Quantum Vortices

The flow of the condensate is different from the flow of an ordinary fluid in two respects. Firstly, and as we showed in Sect. 3.3, it is inviscid (there is no viscosity to slow down the flow and bring it to a stop). Secondly, the circulation is quantised, as we showed above. To appreciate the second difference we recall the *vorticity* field (the local rotation), defined as,

$$\boldsymbol{\omega} = \nabla \times \mathbf{v}. \tag{7.6}$$

The following examples illustrate velocity fields with the associated vorticity fields:

(i) Consider water inside a bucket rotating at constant angular velocity Ω. We use cylindrical coordinates (r, θ, z) where z is the axis of rotation.[1] The velocity field is $\mathbf{v} = v_\theta \widehat{\boldsymbol{e}}_\theta = \Omega r \widehat{\boldsymbol{e}}_\theta$ and the vorticity is $\boldsymbol{\omega} = 2\Omega \widehat{\boldsymbol{e}}_z$ (where $\widehat{\boldsymbol{e}}_\theta$ and $\widehat{\boldsymbol{e}}_z$ are the unit vectors along θ and z). The azimuthal speed v_θ of this flow as a function of r is shown by case (i) of Fig. 7.2a. This flow is called *solid body rotation*.

(ii) As derived above, the velocity field around a vortex line in a condensate is $v_\theta = q\hbar/(mr)$, shown by case (ii) in Fig. 7.2a. It is easy to verify that its vorticity is zero: we say that this flow is *irrotational*. Physically, a parcel of fluid which goes around the vortex axis does not 'turn' (as it does in solid

---

[1] We recall that in cylindrical coordinates, the curl of the vector $\mathbf{A} = (A_r, A_\theta, A_z)$ is

$$\nabla \times \mathbf{A} = \left(\frac{1}{r}\frac{\partial A_z}{\partial \theta} - \frac{\partial A_\theta}{\partial z}, \ \frac{\partial A_r}{\partial z} - \frac{\partial A_z}{\partial r}, \ \frac{1}{r}\frac{\partial (rA_\theta)}{\partial r} - \frac{1}{r}\frac{\partial A_r}{\partial \theta} \right).$$

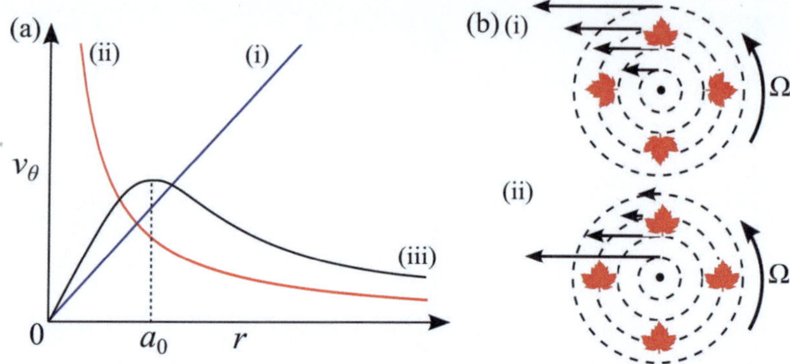

Fig. 7.2 a Examples of rotation curves. (i) solid body rotation, (ii) vortex line in a condensate (irrotational flow), and (iii) flow around a hurricane or a bathtub vortex, which combines solid body rotation in the inner region $r \ll a_0$ and irrotational flow in the outer region $r \gg a_0$. **b** Schematic of the two-dimensional flow for cases (i) and (ii), showing the orientation of an object, here a maple leaf, in the flow. Notice that in case (i) the leaf rotates about its own axis, whereas in case (ii) it does

body rotation), but retains its orientation (like a gondola of a Ferris wheel); this flow is depicted in case (ii) of Fig. 7.2b. It also follows mathematically: the condensate's velocity is proportional to the gradient of the quantum mechanical phase, and the curl of a gradient is always zero. However, the singularity itself contributes vorticity according to,

$$\boldsymbol{\omega} = \kappa \delta^{(2)}(\mathbf{r})\widehat{\boldsymbol{e}}_z, \tag{7.7}$$

where $\delta^{(2)}(\mathbf{r})$ is the two-dimensional delta function satisfying $\delta^{(2)}(\mathbf{r}=0)=1$ and $\delta^{(2)}(\mathbf{r}\neq 0)=0$. At first it may surprise that a quantum vortex has zero vorticity, but the result is expected - the key point is that motion in the condensate is irrotational, but isolated vortex line singularities are allowed.

(iii)   The velocity of the wind around the centre of a hurricane, case (iii) of Fig. 7.2a, combines solid body rotation in the inner region ($r \ll a_0$) with irrotational motion in the outer region ($r \gg a_0$) where a_0 is called the *vortex core radius*.

In ordinary fluids the vorticity $\boldsymbol{\omega}$ is arbitrary, and therefore vortices can be weak or strong, big or small. In a condensate, Eq. (7.3) is a strict quantum mechanical constraint: motion around a singularity has fixed form and intensity.

7.4   The Nature of the Vortex Core

A natural question is: what is the structure of the vortex, particularly towards the axis of the vortex ($r \to 0$), where, according to Eq. (7.5), the velocity becomes infinite? Using cylindrical coordinates (r, θ, z) again, we consider a straight vortex line aligned in the z direction in a homogeneous condensate ($V = 0$). Assuming $\Psi(r, \theta, z) = A(r)e^{iq\theta}$ and substituting into the GPE of Eq. (3.17) we obtain the following differential equation,[2] for the function $A(r)$,

$$\mu A = -\frac{\hbar^2}{2m}\frac{1}{r}\frac{\mathrm{d}}{\mathrm{d}r}\left(r\frac{\mathrm{d}A}{\mathrm{d}r}\right) + \frac{\hbar^2 q^2}{2mr^2}A + gA^3. \tag{7.9}$$

The terms on the right-hand side arise from the quantum kinetic energy, the kinetic energy of the circulating flow and the interaction energy, respectively. The boundary conditions are that $A(r) \to 0$ for $r \to 0$ and $A(r) \to \psi_0$ for $r \to \infty$. The equation has no exact solution and must be solved numerically for $A(r)$; the corresponding density profile $n(r) = A^2$ is shown in Fig. 7.3a. It is apparent that the axis of the vortex is surrounded by a region of depleted density, essentially a tube of radius $a_0 \approx 5\xi$, called the *vortex core radius*. For small r, the density scales as $r^{|q|}$. We see that although the velocity diverges for $r \to 0$, the density vanishes–no atom moves at infinite speed! We can therefore interpret a vortex as a 'hole' surrounded by (quantised) circulation. Recall from Sect. 3.4.2 that if a static and otherwise homogeneous condensate is pinned to zero density, then the density 'heals' back to the background density with a characteristic profile $\tanh^2(x/\xi)$. The vortex density profile is slightly wider than this profile and relaxes more slowly to the background density, as seen in Fig. 7.3a. This is due to the kinetic energy of the circulating flow, which gives rise to an outwards centrifugal force on the fluid.

While there is no exact analytic form for the vortex density profile, a useful approximation for a single-charged vortex is,

$$n(r) = n_0\left(1 - \frac{1}{1 + r'^2}\right), \tag{7.10}$$

where $r' = r/\xi$.

This result (a vortex line is a 'hole' surrounded by circulating flow) has an interesting mathematical consequence: a condensate with vortices is a multiply-connected region, and the classical Stokes Theorem[3] does not apply.

In a trapped condensate the vortex creates a similar tube surrounded by quantised circulation; the only difference is that the density of the condensate is not uniform (as in a homogeneous condensate). In typical 2D column-integrated images of the

---

[2] We have expressed the Laplacian in its cylindrically symmetric form,

$$\nabla^2 = \frac{1}{r}\frac{\partial}{\partial r}\left(r\frac{\partial}{\partial r}\right) + \frac{1}{r^2}\frac{\partial^2}{\partial\theta^2} + \frac{\partial^2}{\partial z^2}. \tag{7.8}$$

[3] Stokes Theorem states that

$$\oint_C \mathbf{A} \cdot \mathrm{d}\boldsymbol{\ell} = \int_S (\nabla \times \mathbf{A}) \cdot \mathrm{d}\mathbf{S},$$

where the surface S enclosed by the oriented curve C is simply-connected, i.e. any closed curve on S can be shrunk continuously to a point within S.

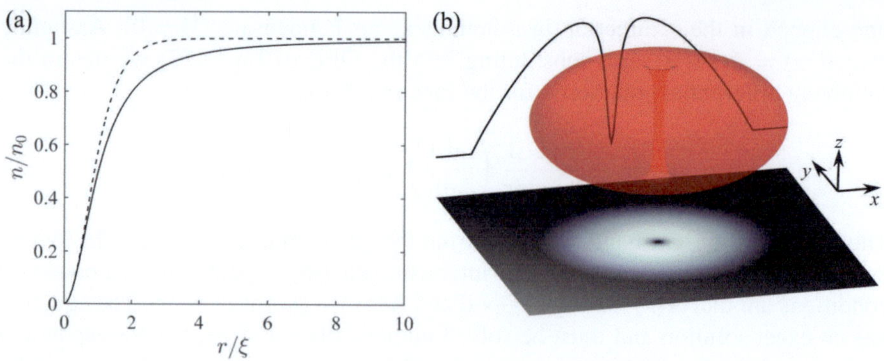

Fig. 7.3 Left: The radial density profile $n(r)$ of a $q = 1$ vortex in a homogeneous condensate (solid line). Shown for comparison is the 'healing' profile for a static condensate whose density is pinned to zero. Right: Appearance of a vortex lying along the axis of a trapped condensate. Shown is an isosurface of the 3D density (with the vortex appearing as a central tube), a 2D density profile column integrated along z (with the vortex appearing as a black dot), and a 1D density profile column-integrated along y and z

condensate, the vortex appears as a low density dot. Since the healing length depends on the local density, in a trapped condensate the thickness of the vortex core depends on the position. If the condensate is in the Thomas-Fermi regime and the vortex along the z axis, then an approximation for the density profile can be constructed as the product of the static Thomas-Fermi profile, Eq. (3.41), and the vortex density, Eq. (7.10), i.e.,

$$n(x, y, z) = n_0 \left(1 - \frac{x^2}{R_x^2} - \frac{y^2}{R_y^2} - \frac{z^2}{R_z^2} \right) \left(1 - \frac{1}{1 + r'^2} \right), \qquad (7.11)$$

where $r' = r/\xi$ is defined is terms of the healing length evaluated at the condensate centre.

7.5   Vortex Energy and Angular Momentum

We now evaluate some useful properties associated with a quantum vortex: its energy and angular momentum. For simplicity, we still consider the case of a single straight vortex lying along the z-axis of a cylindrically-symmetric condensate of constant density; assuming that the condensate size is much larger than the healing length, the density depletion at the axis of the vortex and near the walls can be neglected. A cylindrical bucket of height H_0 and radius R_0 containing superfluid liquid helium would be a realistic example. For trapped atomic condensates, where the vortex size is significant relative to the system size and the condensate density varies in space, these ideas can be generalized by, for example, taking the density profile to be of the form of Eq. (7.11), or by estimating the necessary integrals numerically.

The kinetic energy E_{kin} of the swirling fluid is obtained from summing the contributions of the atoms, each carrying kinetic energy $mv_\theta^2/2$ where $\mathbf{v} = v_\theta \widehat{\boldsymbol{e}}_\theta = (q\hbar/mr)\,\widehat{\boldsymbol{e}}_\theta$ is the velocity. Summing over all atoms we have,

$$E_{\text{kin}} = \int \frac{1}{2} mn(\mathbf{r}) v_\theta^2(\mathbf{r})\, \mathrm{d}^3\mathbf{r}, \tag{7.12}$$

where the integral is performed over the bucket volume. Using cylindrical coordinates,

$$E_{\text{kin}} = \int_0^{H_0} \mathrm{d}z \int_0^{2\pi} \mathrm{d}\theta \int_0^{R_0} \frac{mn_0}{2} \left(\frac{q\hbar}{mr}\right)^2 r\, \mathrm{d}r = \pi H_0 \frac{n_0 q^2 \hbar^2}{m} \int_0^{R_0} \frac{\mathrm{d}r}{r}. \tag{7.13}$$

To prevent the integral from diverging at $r \to 0$ we introduce a cutoff length a_0,[4] the vortex core radius; in doing so, we recognize that the density vanishes at the axis of the vortex, but simplify the core structure, assuming that the core is hollow up to the distance $r = a_0$. Notice that without the outer limit of integration (the size of the container R) the integral would also diverge at $r \to \infty$. We then obtain,

$$E_{\text{kin}} = \pi H_0 \frac{n_0 q^2 \hbar^2}{m} \int_{a_0}^{R_0} \frac{\mathrm{d}r}{r} = \pi H_0 \frac{n_0 q^2 \hbar^2}{m} \ln\left(\frac{R_0}{a_0}\right). \tag{7.14}$$

We conclude that the kinetic energy per unit length of the vortex, $E_{\text{kin}}/H_0 = \pi n_0 (q^2 \hbar^2/m) \ln(R_0/a_0)$, is constant.

Each atom swirling around the axis of the vortex carries angular momentum $L_z = mv_\theta r$. The total angular momentum of the flow is therefore,

$$L_z = \int mn(\mathbf{r}) v_\theta(\mathbf{r}) r\, \mathrm{d}^3\mathbf{r}. \tag{7.15}$$

Proceeding as for the kinetic energy, we find,

$$L_z = 2\pi H_0 n_0 q \hbar \left(\frac{R_0^2}{2} - \frac{a_0^2}{2}\right) \approx \pi H_0 n_0 q \hbar R_0^2. \tag{7.16}$$

Consider a condensate in a state with an arbitrary high angular momentum L_z. We can construct this state as either (i) one vortex with large q or (ii) many vortices with $q = 1$. Which situation is preferred? Since E_{kin} scales as q^2, a state with many singly-charged vortices has less energy than a state with a single multi-charged vortex. Experiments confirm that this is indeed the case: in Ref. [1] a $q = 2$ vortex was seen to quickly decay into two singly-charged vortices. Hereafter we assume that all vortices are singly-charged, with $q = \pm 1$.

---

[4] Often this cutoff is taken instead as the healing length ξ.

7.6   Vortex Lattice in Bucket

Vortices can be created by rotating the condensate [2,3]. Consider again a cylindrical condensate of height H_0, radius R_0 and uniform density. A vortex appears only if the system, by creating a vortex, lowers its energy. In a rotating system at very low temperature, it is not the energy E which must be minimized, but rather the free energy $F = E - \Omega L_z$ where Ω is the angular velocity of rotation. A state without any vortex, hence without angular momentum, has free energy $F_1 = E_0$ where E_0 is the internal energy. A state with a vortex has free energy $F_2 = E_0 + E_{\text{kin}} - \Omega L_z$. The free energy difference is thus,

$$\Delta F = F_2 - F_1 = E_{\text{kin}} - \Omega L_z = 2\pi H_0 \frac{\hbar^2}{m^2} \ln\left(\frac{R_0}{a_0}\right) - \Omega 2\pi H_0 n_0 \hbar R_0^2. \quad (7.17)$$

Therefore $\Delta F < 0$ (the free energy is reduced by creating a vortex) provided that the rotational velocity is larger than a critical value Ω_{c1},

$$\Omega > \Omega_{c1} = \frac{\hbar}{m R_0^2} \ln\left(\frac{R_0}{a_0}\right). \quad (7.18)$$

For superfluid helium ($m = 6.7 \times 10^{-27}$ kg, $\kappa = 9.97 \times 10^{-8}$ m^2/s, $a_0 \approx 10^{-10}$ m) inside a container of radius $R_0 = 10^{-2}$ m, the critical angular velocity is $\Omega_{c1} = 3 \times 10^{-3}$ s^{-1}. States with two, three and more vortices onset at higher critical velocities Ω_{c2}, Ω_{c3} etc, as shown in Fig. 7.4 for superfluid helium and in Fig. 7.7 for atomic condensates. Note that the vortices are parallel to the rotation axis and arrange themselves in a *vortex lattice* like atoms in a crystal with triangular symmetry. The vortex lattice is therefore a steady configuration in the frame of reference rotating at angular velocity Ω.

Vortices are topological defects which can only be created at a boundary or spontaneously with an oppositely-charged vortex. Where then do the vortices in a vortex lattice originate from?

For a rotating container of helium, with even a relatively small rotation frequency, the roughness of the container surface is expected to seed vortices, providing a constant source of vortices from which to develop a vortex lattice in the bulk if the critical rotation frequency is exceeded.

According to *Feynman's rule*, the density of vortices (number of vortices per unit area) is,

$$n_v = \frac{2\Omega}{\kappa}. \quad (7.19)$$

Since each vortex contributes vorticity according to Eq. (7.7), the average vorticity per unit area is,

$$\bar{\omega} = \kappa n_v \widehat{e}_z = 2\Omega \widehat{e}_z. \quad (7.20)$$

This tells us that the averaged vorticity (averaged over distance larger than the inter-vortex spacing) reproduces the vorticity 2Ω of an ordinary fluid in rotation. Similarly,

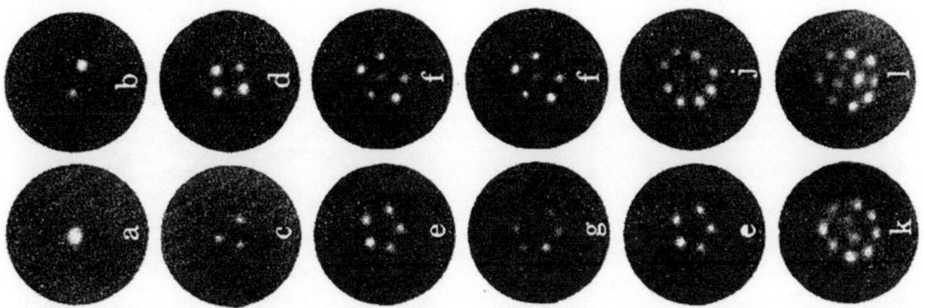

Fig. 7.4 Experimental images of vortex lattices at increasing angular velocities Ω in superfluid helium. Reprinted figure with permission from [4]. Copyright 1979 by the American Physical Society

the large-scale azimuthal flow is $\mathbf{v} \approx \Omega r \widehat{\boldsymbol{e}}_\theta$. Remarkably, the many quantised vortices mimic classical solid body rotational flow. Note that the local velocity field around vortices can remain rather complicated.

In the frame rotating at angular frequency Ω about the z-axis, the GPE of Eq. (3.54) is,

$$i\hbar\frac{\partial \psi}{\partial t} = -\frac{\hbar^2}{2m}\nabla^2\psi + g|\psi|^2\psi + V\psi + \Omega L_z\psi - \mu\psi, \qquad (7.21)$$

where,

$$L_z = i\hbar\left(y\frac{\partial}{\partial x} - x\frac{\partial}{\partial y}\right), \qquad (7.22)$$

is the angular momentum operator in the z direction. The vortex lattices are the ground-state stationary solutions of this equation (providing Ω is large enough). Figure 7.5 shows such a vortex lattice solution for a condensate being rotated in a bucket. The above bucket scenario is modelled through the *bucket potential*,

$$V(r) = \begin{cases} 0 & \text{if } r \le R_0, \\ \infty & \text{if } r > R_0. \end{cases} \qquad (7.23)$$

The lattice in Fig. 7.5 features $N_{\mathrm{v}} = 56$ vortices. Note the appearance of the phase "dislocations" in the phase profile at each vortex position. At the boundary there are as many 2π phase slips as there are vortices. The average flow speed around the edge of the bucket can then be approximated by evaluating the magnitude of $\mathbf{v} = (\hbar/m)\nabla S$ around the boundary, i. e.,

$$v_r(r = R_0) = \frac{\hbar}{m}\frac{2\pi N_{\mathrm{v}}}{2\pi R_0} = \frac{\hbar}{m}\frac{56}{29\xi} \approx 1.93c. \qquad (7.24)$$

This is close to what one would expect for solid body rotation, $v_r(r = R_0) = \Omega R_0 = 2.3c$.

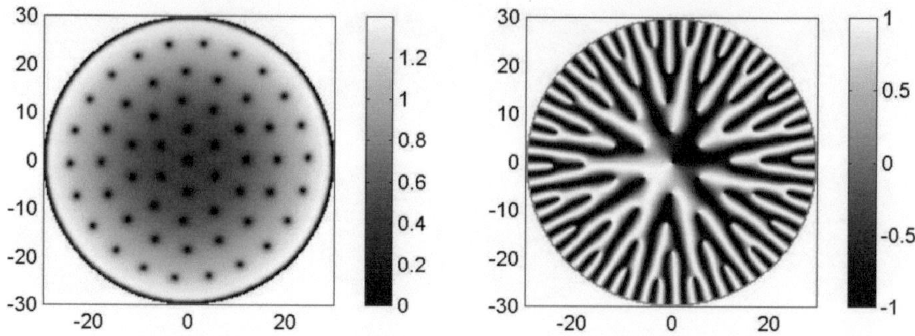

Fig. 7.5 Vortex lattice formed in a bucket potential rotating about the z-axis. Shown are the **a** density (in arbitrary units) and **b** phase (in units of π) in the xy-plane (position presented in units of the healing length ξ), corresponding to the stationary solution of the rotating-frame GPE of Eq. (7.21). The bucket has radius $R = 29\xi$ and the rotation frequency is $\Omega = 0.08\,c/\xi$. Reprinted with permission from [5]. Copyright 2001, The Author(s). All rights reserved

In a small system, at the same value of Ω one often observes vortex configurations which are slightly different from each other. This is because there is a very small energy difference between these slightly rearranged states. For example, Fig. 7.4 shows two states with six vortices each (in one case the six vortices are distributed around a circle, in the other case there are five vortices around a circle and one vortex in the middle).

Notice how the background density for the rotating bucket solution in Fig. 7.5 features a *meniscus*, that is, it is raised towards the edge of the bucket. Let us determine this background density profile. We denote the rotation vector $= \Omega \widehat{\boldsymbol{e}}_z$. Recall the fluid interpretation of the GPE. Using the Madelung transformation $\psi = \sqrt{n}e^{iS}$ and the fluid velocity definition $\mathbf{v} = (\hbar/m)\nabla S$, the rotating-frame GPE of Eq. (7.21) is equivalent to the modified fluid equations,

$$\frac{\partial n}{\partial t} = -\nabla \cdot [n\,(\mathbf{v} - \times \mathbf{r})]\,, \tag{7.25}$$

$$m\frac{\partial \mathbf{v}}{\partial t} = -\nabla \cdot \left(\frac{1}{2}mv^2 + V + gn - \frac{\hbar^2}{2m}\frac{\nabla^2\sqrt{n}}{\sqrt{n}} - m\mathbf{v}\cdot[\times \mathbf{r}]\right)\,, \tag{7.26}$$

where the $\boldsymbol{\omega} \times \mathbf{r}$ terms account for frame rotation and $\mathbf{v}$ is the velocity field in the laboratory frame (expressed in the coordinates of the rotating frame). We assume the Thomas-Fermi approximation by neglecting the quantum pressure term in Eq. (7.26), and seek the stationary velocity profile. Setting $\partial \mathbf{v}/\partial t = 0$ and integrating gives,

$$\frac{1}{2}mv^2 + V + gn - m\mathbf{v}\cdot[\times \mathbf{r}] = \mu\,, \tag{7.27}$$

where the chemical potential μ is the integration constant.

We consider a coarse-grained scale, ignoring the structure of the individual vortices and for which the velocity field approximates the solid body form $\mathbf{v}(r) = \Omega r \widehat{\boldsymbol{e}}_\theta$. We then obtain,

$$gn + V - \frac{1}{2}m\Omega^2 r^2 = \mu, \tag{7.28}$$

where we have used $\widehat{\boldsymbol{e}}_z \times \widehat{\boldsymbol{e}}_r = \widehat{\boldsymbol{e}}_\theta$. Rearranging for the density,

$$n(r) = \frac{1}{g}\left(\mu - V + \frac{1}{2}m\Omega^2 r^2\right), \tag{7.29}$$

which is valid for $n(r) > 0$; otherwise $n(r) = 0$. We conclude that rotation causes a parabolic increase in the coarse-grained density, consistent with the behaviour visible in Fig. 7.5. The is due to centrifugal effects, and is observed in rotating classical fluids. Note that μ can be determined by normalizing the profile.

7.7 Vortex Lattice in a Trapped Condensate

To predict the critical rotation frequency for vortices to become favoured in a harmonically-trapped condensate, one can repeat the above approach but the inhomogeneous density profile must be accounted for (i.e. replacing n_0 above with $n(\mathbf{r})$). One way to approximate this is by the Thomas-Fermi density profile. For a trap which is symmetric in the plane of rotation, with frequency $\omega_\perp$, the critical rotation frequency is then,

$$\Omega_{\mathrm{c}1} = \frac{5}{2}\frac{\hbar}{mR^2}\ln\left(\frac{0.67R_\perp}{\xi}\right), \tag{7.30}$$

where $R_\perp$ is the Thomas-Fermi radius in the plane of rotation. For typical atomic condensates, $\Omega_{\mathrm{c}1} \sim 0.3\omega_\perp$.

Rotating an axi-symmetric harmonic trap applies no torque to the condensate, and so in practice the trap is made slightly anisotropic in the plane of rotation in order to form a vortex lattice. Surprisingly, experiments observed vortices at rotation frequencies $\Omega \sim 0.7\omega_\perp$, considerably higher than the frequency at which they become energetically favourable. The traps are so smooth that vortex nucleation is very different to that of helium.

We can examine this by considering the planar potential to be weakly elliptical, with frequencies $\omega_x = \sqrt{1-\epsilon}\,\omega_\perp$ and $\omega_y = \sqrt{1+\epsilon}\,\omega_\perp$, where ϵ is the *trap ellipticity*. We follow the approaches of Refs. [6,7]. We seek the stationary solutions of the trapped vortex-free condensate under rotation about z. Under the Thomas-Fermi approximation, the solutions must satisfy Eq. (7.27). Furthermore, we look for solutions with the phase profile, and corresponding velocity profile, given by,

$$S(x, y) = \beta xy, \quad \mathbf{v}(x, y) = \frac{\hbar}{m}\nabla S = \frac{\beta\hbar}{m}(y\widehat{\boldsymbol{e}}_x + x\widehat{\boldsymbol{e}}_y), \tag{7.31}$$

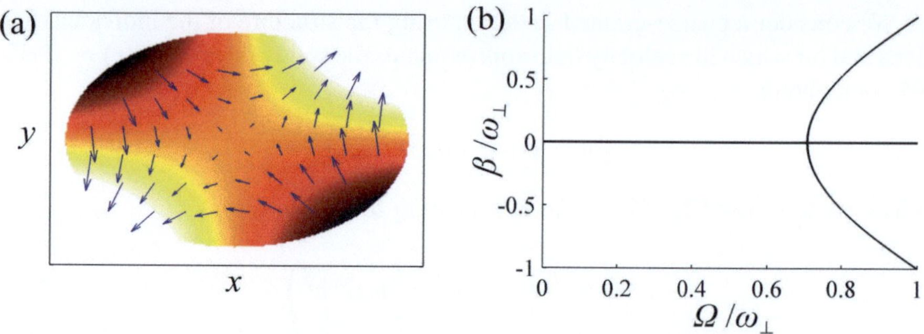

Fig. 7.6 a Illustration of the irrotational flow pattern of a rotating elliptically-trapped condensate, according to Eqs. (7.31). The color indicates the phase $S(x, y)$ while the velocity field is shown by arrows. **b** The velocity field amplitude β as a function of rotation frequency Ω for an axi-symmetric trap ($\epsilon = 0$). At $\Omega = \omega_\perp/\sqrt{2}$ the solutions trifurcate. In this region, these solutions become unstable

where β is a parameter to be determined below. Inserting into Eq. (7.27), and noting that $\times \mathbf{r} = \Omega(x\widehat{\boldsymbol{e}}_y - y\widehat{\boldsymbol{e}}_x)$, leads to the density profile,

$$n = \frac{1}{g}\left(\mu - \frac{1}{2}m\tilde{\omega}_x^2 x^2 + \tilde{\omega}_y^2 y^2 + \omega_z^2 z^2)\right), \tag{7.32}$$

where the effect of the rotation is to introduce *effective trap frequencies* in the xy-plane,

$$\tilde{\omega}_x^2 = (1 - \epsilon)\omega_\perp^2 + \beta^2 - 2\beta\Omega, \tag{7.33}$$
$$\tilde{\omega}_y^2 = (1 + \epsilon)\omega_\perp^2 + \beta^2 + 2\beta\Omega. \tag{7.34}$$

Plugging this density profile into the rotating-frame continuity equation, Eq. (7.25), and setting $\partial n/\partial t = 0$, leads to an expression for β,

$$\beta^3 + \beta(\omega_\perp^2 - 2\Omega^2) - \epsilon\Omega\omega_\perp^2 = 0. \tag{7.35}$$

Hence the stationary solution of the condensate in the rotating frame has been completely specified. In the laboratory frame, this solution has an elliptical density profile which rotates about z. However, the fluid remains irrotational, thanks to the special velocity field which distorts the density is such a way as to mimic rotation, as depicted in Fig. 7.6a.

Analysing the case of $\epsilon = 0$ for simplicity, there exists one solution, with $\beta = 0$, for $\Omega \leq \omega_\perp/\sqrt{2}$; this represents a motion-less and axisymmetric condensate. However, for $\Omega > \omega_\perp/\sqrt{2}$ the solutions trifurcate, with two new branches with $\beta \neq 0$ and corresponding to non-axisymmetric solutions of the form shown in Fig. 7.6. This trifurcation leads to an instability of the condensate (as can be confirmed via linearizing about these solutions [7]) in which perturbations grow at the condensate surface and develop into vortices. Experiments [8] and simulations [9] of the GPE show that this instability then allows the condensate to evolve into a vortex lattice, the lowest energy state.

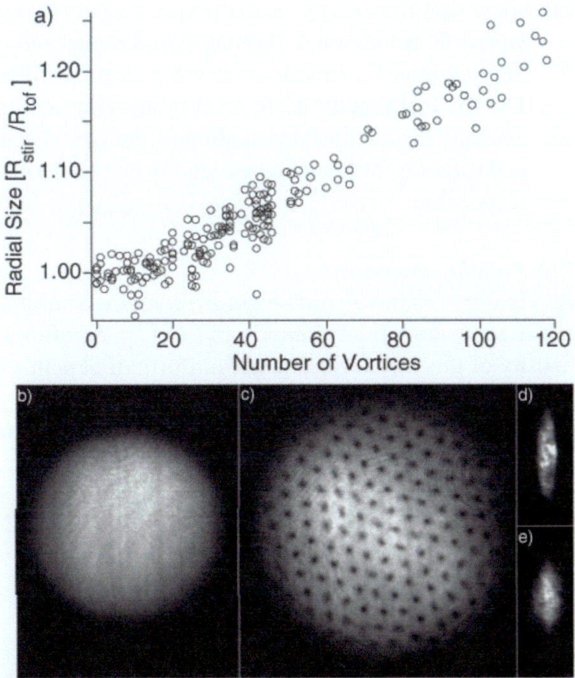

Fig. 7.7 An experimental vortex lattice (**c**) formed in a flattened trapped rotating atomic condensate (image represents the condensate density). **b** shows the profile of the corresponding non-rotating condensate. **d** and **e** show the side views of the non-rotating and rotating condensates, respectively. The condensate grew in radius with the number of vortices, as shown in (**a**). Reprinted figure with permission from [10]. Copyright 2001 by the American Physical Society

Figure 7.7 shows a vortex lattice produced in a rotating trapped atomic condensate. Note the regularity and density of the vortex lattice. Note also that the rotating condensate is significantly broader than the non-rotating condensate. In the presence of the vortex lattice, we can predict the coarse-grained density profile of the condensate. Considering an axi-symmetric trap ($\omega_x = \omega_y \equiv \omega_\perp$), then the coarse-grained density profile of Eq. (7.29) gives,

$$n(r) = \frac{1}{g}\left(\mu - \frac{1}{2}m(\omega_r^2 r^2 - \Omega^2)r^2\right). \tag{7.36}$$

There is a competition between the quadratic trapping potential, which pushes atoms inwards, and the quadratic centrifugal potential, which pushes atoms outwards. The net potential is quadratic with effective harmonic potential $\omega_r^2 - \Omega^2$. As Ω is increased, the condensate expands, and when $\Omega \geq \omega_r$ it becomes untrapped.

Problems

7.1 *Critical rotation*

Replace the bucket of Sect. 7.5 with a harmonic potential $V(r) = \frac{1}{2}m\omega_r^2 r^2$ where r is the radial direction perpendicular to the axis of the cylinder. Take the condensate to adopt the Thomas-Fermi profile.

(a) Show that the energy of the vortex-free condensate is $E_0 = \pi m n_0 \omega_r^2 H_0 R_r^4 / 6$, where R_r is the radial Thomas-Fermi radius and n_0 is the density along the axis.

(b) Now estimate the kinetic energy E_{kin} due to a vortex along the axis via Eq. (7.12). Use the fact that $a_0 \ll R_r$ to simplify your final result.

(c) Estimate the angular momentum of the vortex state, and hence estimate the critical rotation frequency at which the presence of a vortex becomes energetically favourable.

7.2 *Angular momentum*

A cylinder of radius R and height H_0 contains a number of parallel vortex lines which are aligned along the cylinder's axis (which herefter we take as the z-direction). The density of the vortex lines is not uniform; it depends on the radial distance r, and it is described by the average vorticity $\bar{\omega}_z(r)$.

Show that the z-component of the angular momentum is

$$L_z = \pi \rho_s R^2 H_0 \int_0^R dr\, r \bar{\omega}_z(r) W(r),$$

hence determine the weight function $W(r)$.

7.3 *Pseudo-vorticity*

Let the wavefunction of the condensate be $\psi = \sqrt{n} e^{i\phi}$ where $n = |\psi|^2$ is the number density, $\rho = mn$ is the mass density, ϕ is the phase and the velocity is $u = (\hbar/m)\nabla\phi$.

1. Show that the current density $j = \rho u$ can also be written as

$$j = \frac{\hbar}{2i}(\psi^* \nabla \psi - \psi \nabla \psi^*).$$

2. The vorticity $\nabla \times u$ is zero because the curl of a gradient is always zero, but, using the current density, we can define the *pseudo-vorticity* as

$$\omega_p = \frac{1}{2}\nabla \times j.$$

   Show that

$$\omega_p = \frac{\hbar}{2i}\nabla\psi^* \times \nabla\psi.$$

3. Separate real and imaginary parts of ψ by writing $\psi = \psi_r + i\psi_i$, hence show that the pseudo-vorticity can be written as

$$\omega_p = \hbar \nabla\psi_r \times \nabla\psi_i.$$

4. Consider a vortex located at $x = 0$, $y = 0$ and aligned along the z-direction. Its wavefunction is $\psi = \sqrt{n}e^{i\theta}$ where

$$n \approx n_0 \left(1 - \frac{1}{(1 + r^2/\xi^2)} \right).$$

Show that the pseudo-vorticity is

$$\omega_p = \frac{\hbar}{2r} \frac{\partial n}{\partial r} \widehat{z},$$

where $\widehat{z}$ is the unit vector along the z-direction. Finally show that ω_p is constant in the core region of the vortex but vanishes at large r.

References

1. Y. Shin, M. Saba, M. Vengalattore, T.A. Pasquini, C. Sanner, A.E. Leanhardt, M. Prentiss, D.E. Pritchard, W. Ketterle, Phys. Rev. Lett. **93**, 160406 (2004)
2. A.L. Fetter, Rev. Mod. Phys. **81**, 647 (2009)
3. M. Tsubota, K. Kasamatsu, M. Ueda, Phys. Rev. A **65**, 023603 (2002)
4. E.J. Yarmchuck, M.J.V. Gordon, R.E. Packard, Phys. Rev. Lett. **43**, 214 (1979)
5. T. Winiecki, *Numerical Studies of Superfluids and Superconductors*, Ph.D thesis, University of Durham (2001)
6. A. Recati, F. Zambelli, S. Stringari, Phys. Rev. Lett. **86**, 377 (2001)
7. S. Sinha, Y. Castin, Phys. Rev. Lett. **87**, 190402 (2001)
8. K.W. Madison, F. Chevy, V. Bretin, J. Dalibard, Phys. Rev. Lett. **86**, 4443 (2001)
9. N.G. Parker, R.M.W. van Bijnen, A.M. Martin, Phys. Rev. A **73**, 061603(R) (2006)
10. C. Raman, J.R. Abo-Shaeer, J.M. Vogels, K. Xu, W. Ketterle, Phys. Rev. Lett. **87**, 210402 (2001)

Hydrodynamic Methods

8

Abstract

This chapter reviews some mathematical methods which are useful to solve simple but instructive problems of superfluid hydrodynamics which are concerned with sound waves and vortices. Besides the GPE, which is the major model used to study quantum fluids, these methods include the Vortex Point Model (VPM) for 2D flows, the Vortex Filament Model for 3D flows, their finite-temperature generalization (the Schwarz equation), and, finally, the Hall-Vinen-Bekharevic-Khalatnikov (HVBK) equations, which generalise Landau's two-fluid equations presented in Chap. 6 to the presence of vortex lines in the system. A basic understanding of these methods will be required to make sense of the rich variety of vortex phenomena which will be presented in Chap. 9.

8.1 The Generalized Helmholtz Theorem According to the GPE

In classical fluid dynamics, Helmholtz Theorem states that vortices move with the fluid; in other words, although vorticity (i.e. local rotation) is a dynamically active variable, it is transported by the flow as if it were a passive (vectorial) tracer. A generalized form of Helmholtz Theorem which also accounts for density gradients can be derived from the GPE in the following way.

For simplicity, consider a vortex in a 2D condensate. Let $\boldsymbol{r}_0 = (x_0, y_0)$ be the initial position of the vortex at time $t = 0$. Following Ref. [1], it is convenient to use the complex variable $z = x + iy$ to represent a point of the xy-plane, hence we say that the vortex is initially at $z_0 = x_0 + iy_0$. The initial wavefunction is therefore,

$$\psi_0 = (z - z_0)\rho e^{i\phi}, \tag{8.1}$$

where $\rho(\boldsymbol{r}, t)$ and $\phi(\boldsymbol{r}, t)$ are smoothly varying functions which describe the background density and phase in the absence of the vortex, and the factor $z - z_0$ accounts for density and phase near the vortex. Note that $\psi_0 = 0$ at the vortex centre, $z = z_0$.

© The Author(s), under exclusive license to Springer Nature Switzerland AG 2026

C. F. Barenghi et al., *Quantum Fluids, Solitons, and Vortices*, Lecture Notes in Physics 1050, https://doi.org/10.1007/978-3-032-20171-3_8

According to the GPE, at time $t = \delta t$ the wavefunction becomes,

$$\psi_1 = \exp\left(-\frac{i\mathcal{H}\delta t}{\hbar}\right)\psi_0 \approx \left(1 - \frac{i\delta t}{\hbar}\mathcal{H} + \cdots\right)\psi_0, \qquad (8.2)$$

where $\mathcal{H} = -(\hbar^2/(2m))\nabla^2 + \mathcal{U}$ and $\mathcal{U} = g|\psi|^2 + V_{trap}$. Hence,

$$\psi_1 \approx (z - z_0)\rho e^{i\phi} - \frac{i}{\hbar}\delta t\left(-\frac{\hbar^2}{2m}\nabla^2 + \mathcal{U}\right)(z - z_0)\rho e^{i\phi}. \qquad (8.3)$$

We compute $\nabla^2(z - z_0)\rho e^{i\phi}$, and evaluate ψ_1 at the point $z = z_0 + \delta z$. Clearly $\psi_1 = 0$ at $z = z_0 + \delta z$ because this point is the location of the vortex at time $t = \delta t$. Taking the limit $\delta z \to 0$ and $\delta t \to 0$, we find that the vortex velocity $\boldsymbol{v} = (v_x, v_y)$ is,

$$v_x + i v_y = \frac{\delta z}{\delta t} = \frac{\hbar}{m}\left((1, i) \cdot \nabla\phi + (-i, 1) \cdot \nabla\ln\rho\right), \qquad (8.4)$$

which is,

$$\boldsymbol{v} = \frac{\hbar}{m}\nabla\phi - \frac{\hbar}{m}\widehat{\boldsymbol{\kappa}} \times \nabla\ln\rho = \boldsymbol{v}_s + \boldsymbol{v}_\rho, \qquad (8.5)$$

where $\widehat{\boldsymbol{\kappa}}$ is the unit vector along the vortex (the direction perpendicular to the xy-plane) and $\kappa = h/m$ is the quantum of circulation.

We conclude that the vortex velocity consists of two parts. The first part is the background superfluid velocity, $\boldsymbol{v}_s = (\hbar/m)\nabla\phi$, which is due to the ambient phase, for example to other vortices; this term is consistent with the classical Helmholtz Theorem. The second term is the contribution of the spatially-dependent density background, $\boldsymbol{v}_\rho = (\hbar/m)\widehat{\boldsymbol{\kappa}} \times \nabla\ln\rho$, which is particularly important in harmonically-trapped condensates where the density decreases near the edge.

Let us consider three simple but important consequences of Helmholtz Theorem in superfluids. Firstly, a straight vortex in an infinite system will remain straight without moving. Secondly, a vortex which is bent in the shape of a ring will move in the direction perpendicular to the plane of the ring; this is because each point P along the ring is advected by all the other points P' $\neq$ P around the ring - an effect called *self-induced motion*. This effect applies not only to vortex ring (a closed curve of constant curvature) but to any curved vortex line in general. Thirdly, two straight parallel vortex lines will advect each other, executing either translational or rotational motion, depending on the whether they are parallel (forming a *vortex-vortex pair*) or anti-parallel (forming a *vortex-antivortex pair*), as shown in Figs. 8.1 and 8.2.

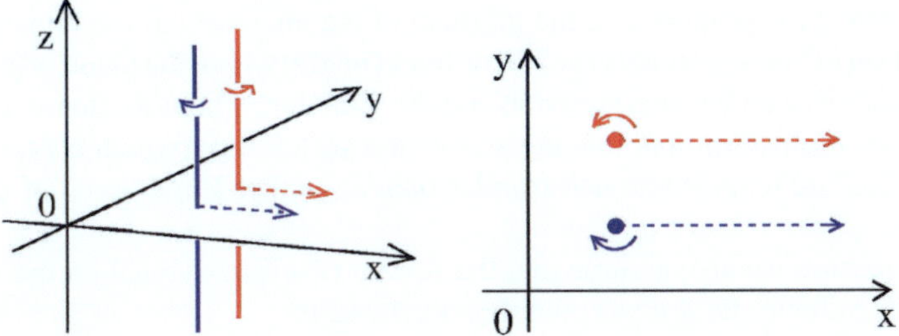

Fig. 8.1 Left: Schematic vortex-antivortex pair in the xyz-space. The red solid line and the blue solid line denote respectively the positive (anti-clockwise) vortex and the negative (clockwise) vortex. The pair moves on the xy-plane as shown by the dashed red and blue trajectories. Right: simpler representation of the vortex-antivortex pair in the xy-plane

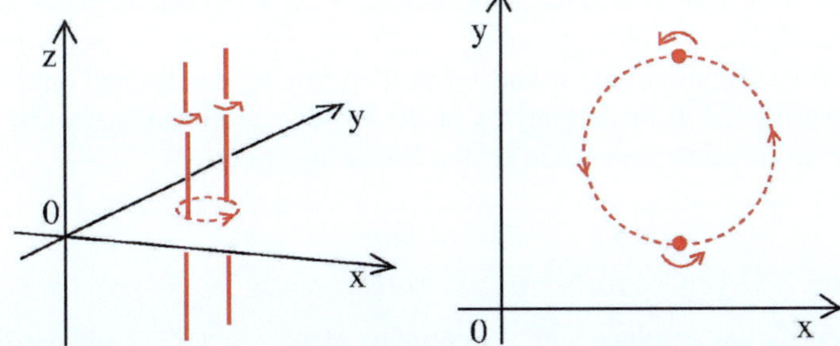

Fig. 8.2 Left: Schematic vortex-vortex pair in the xyz-space. The two red solid lines denote the two positive (anti-clockwise) vortices. The pair rotates on the xy plane as shown by the dashed red trajectory. Right: simpler representation of the vortex-vortex pair on the xy-plane

8.2 The 2D Vortex Point Model (VPM)

The other models that we shall describe (the PVM, the VFM, the Schwarz equation and the HVBK equations) do not suffer the main limitation of the GPE, which is that it only describes a condensate at $T = 0$ (or at least at sufficiently low temperatures), but account realistically for finite-temperature effects.

We have seen that, at length scales larger than the healing length ξ, the Gross-Pitaevskii equation reduces to the classical continuity equation and the compressible Euler equation for an inviscid fluid. In the further limit of velocity smaller than the speed of sound (i.e. for small Mach numbers), density variations can be neglected; in this limit, the compressible Euler equation reduces to the incompressible Euler equation,

$$\frac{\partial \boldsymbol{v}}{\partial t} + (\boldsymbol{v} \cdot \nabla)\boldsymbol{v} = -\frac{1}{\rho}\nabla p, \tag{8.6}$$

where ρ is constant, and the continuity equation becomes the solenoidal condition $\nabla \cdot \boldsymbol{v} = 0$. The irrotational condition $\nabla \times \boldsymbol{v} = \boldsymbol{0}$ is satisfied because in a quantum fluid the velocity is proportional the gradient of the phase of the wavefunction, with the exception of singular points (in 2D) or lines (in 3D) where the phase is not defined and the vorticity can be interpreted as a delta-function. The next step is to assume the *thin-core approximation*, that the size of the vortex core (which is typically few times the healing length) is much smaller than any other length scale of interest in the flow.

In this section we also assume that the flow is two-dimensional on the xy-plane, and that the velocity field v has components u and v,

$$v(x, y) = (u(x, y), v(x, y)). \tag{8.7}$$

If the flow is irrotational ($\nabla \times v = 0$) then (since the curl of a gradient is always zero) there exists a *scalar potential* Φ such that $v = \nabla\Phi$, that is $u = \partial\Phi/\partial x$ and $v = \partial\Phi/\partial y$. If the flow is also incompressible ($\nabla \cdot v = 0$) then there exists a *stream function* Ψ such that $u = \partial\Psi/\partial y$ and $v = -\partial\Psi/\partial x$ so that the velocity is divergence-free. The velocity components u and v can therefore be determined either by computing derivatives of Φ or derivatives of Ψ. We conclude that the scalar potential and the stream function are related by the following equations,

$$\frac{\partial\Phi}{\partial x} = \frac{\partial\Psi}{\partial y}, \qquad \frac{\partial\Phi}{\partial y} = -\frac{\partial\Psi}{\partial x}. \tag{8.8}$$

Now we identify the xy-plane with the complex plane $z = x + iy$ and use the theory of complex functions. Consider the complex function,

$$\Omega(z) = \Phi(x, y) + i\Psi(x, y), \tag{8.9}$$

of the complex variable z, where Φ and Ψ are the real and the imaginary parts of Ω. The complex function $\Omega(z)$ is said *analytic* if the derivative with respect to z exists and is the same irrespective of the direction of differentiation, in which case one can show that Eqs. (8.8) relating the real and the imaginary parts of $\Omega(z)$ hold true. Eqs. (8.8) are called the *Cauchy-Riemann relations*. Therefore any analytic function represents a 2D incompressible irrotational flow, and, viceversa, any incompressible irrotational flow is represented by an analytic function. Such analytic function $\Omega(z)$ is called the *complex potential*; its real part is the velocity potential and its imaginary part is the stream function. It is easy to verify that u and v can be obtained directly from the complex potential by computing its derivative with respect to z (which for simplicity we denote by a prime),

$$\Omega'(z) = \frac{d}{dz}\Omega(z) = u - iv. \tag{8.10}$$

Note the minus sign at the right hand side.

An example of complex potential is $\Omega(z) = U_0 z$, which describes a uniform flow of speed U_0 in the x-direction (in fact $d\Omega/dz = U_0$, hence $u = U_0$ and $v = 0$). Another example is,

$$\Omega(z) = -\frac{i\kappa_0}{2\pi} \log(z - z_0), \tag{8.11}$$

which describes a vortex of circulation $\kappa_0 > 0$ (anticlockwise motion) at position $z_0 = x_0 + iy_0$. The corresponding velocity components are,

$$u = -\frac{1}{2\pi} \frac{\kappa_0(y - y_0)}{[(x - x_0)^2 + (y - y_0)^2]}, \quad v = \frac{1}{2\pi} \frac{\kappa_0(x - x_0)}{[(x - x_0)^2 + (y - y_0)^2]}. \tag{8.12}$$

Since, as it can easily verified, both Φ and Ψ obey Laplace's equation, which is a linear equation, the sum of two complex potentials (each describing a flow) describes a more complicated flow (*superposition principle*). By applying the superposition principle, we conclude that the complex potential of a system of N vortices of circulation κ_n and instantaneous position $z_n = x_n + iy_n$ ($n = 1, \cdots N$) is simply,

$$\Omega(z) = -\frac{i}{2\pi} \sum_{n=1}^{n=N} \kappa_n \log(z - z_n), \tag{8.13}$$

resulting in the velocity components,

$$u(x, y) = -\frac{1}{2\pi} \sum_{n=1}^{n=N} \frac{\kappa_n(y - y_n)}{[(x - x_n)^2 + (y - y_n)^2]}, \tag{8.14}$$

$$v(x, y) = \frac{1}{2\pi} \sum_{n=1}^{n=N} \frac{\kappa_n(x - x_n)}{[(x - x_n)^2 + (y - y_n)^2]}. \tag{8.15}$$

Notice that Eqs. (8.14) and (8.15) give the velocity components at all points of the plane but at the locations of the vortices, $x = x_n$ and $y = y_n$, where the velocity is not defined (and, in the GPE, the quantum phase is not defined).

Since each vortex j at position $z_j = x_j + iy_j$ ($j = 1, \cdots N$) is advected by all the other $N - 1$ vortices but not by itself, the velocity of vortex j is

$$\frac{dx_j}{dt} = u(x_j, y_j) = -\frac{1}{2\pi} \sum_{n=1, n \neq j}^{n=N} \frac{\kappa_n(y_j - y_n)}{[(x_j - x_n)^2 + (y_j - y_n)^2]}, \tag{8.16}$$

$$\frac{dy_j}{dt} = v(x_j, y_j) = \frac{1}{2\pi} \sum_{n=1, n \neq j}^{n=N} \frac{\kappa_n(x_j - x_n)}{[(x_j - x_n)^2 + (y_j - y_n)^2]}, \tag{8.17}$$

where, when computing the velocity of the vortex j, we skip the contribution of vortex j itself (a vortex does not advect itself). Equations (8.16) and (8.17) can also be derived directly from the complex potential of N vortices.

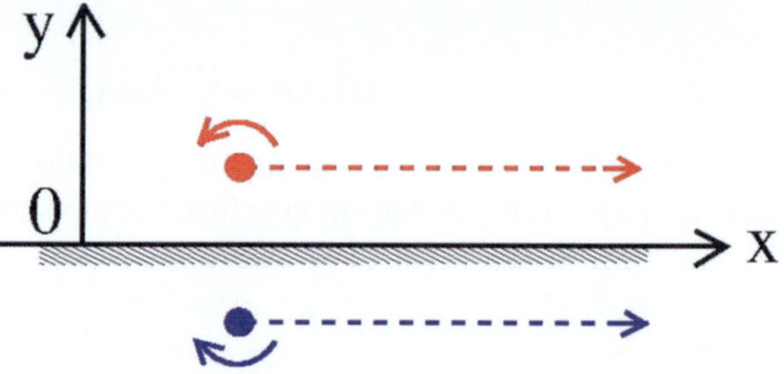

Fig. 8.3 Schematic positive vortex (red) the boundary at $y = 0$ (shadowed) and its image (a negative vortex, blue) on the other side of the boundary

A simple application of Eqs. (8.16) and (8.17) is the case $N = 2$ (vortex-vortex pair and vortex-antivortex pair), which can be easily solved by hand. Problem 8.2 shows how to use the complex potential to prove that if the two vortices have the same circulation (i.e. they are parallel) then they rotate around each others at constant angular velocity which is inversely proportional to their separation; if they have opposite circulation (i.e. they are antiparallel) then their motion is translational with speed inversally proportional to the separation. These effects are shown schematically in Figs. 8.1 and 8.2. The case of three vortices ($N = 3$) is more difficult but can still be solved, but the motion of four or more vortices is chaotic [2].

By using the *method of images*, the VPM can be adapted to find the motion of vortices in the presence of boundaries. For example, suppose that the fluid is confined to the semi-infinite plane $-\infty < x < \infty$, $0 \leq y < \infty$ by a boundary along the x-axis ($y = 0$), and that a vortex is at position $z_0 = x_0 + i y_0$ at distance y_0 from the boundary. To solve the problem we introduce an image vortex *outside* the system, so that the correct boundary condition is satisfied, see Fig. 8.3. In this way, the complex potential of the flow, $\Omega(z)$, is the sum of two complex potentials, $\Omega_1(z)$ and $\Omega_2(z)$; the first is the complex potential of the vortex at z_0, while the second is the complex potential of an image vortex of opposite circulation placed symmetrically across the boundary at $z_0^* = x_0 - i y_0$,

$$\Omega(z) = \Omega_1(z) + \Omega_2(z) = -\frac{i\kappa}{2\pi} \log(z - z_0) + \frac{i\kappa}{2\pi} \log(z - z_0^*). \qquad (8.18)$$

In fact, using Eq. 8.10, we find that the velocity $\boldsymbol{u} = (u, v)$ at any point $x \neq x_0$, $y \neq y_0$ of the semi-infinite plane is,

$$u(x, y) - i v(x, y) = \Omega'(z) = -\frac{i\kappa}{2\pi} \left(\frac{1}{(z - z_0)} - \frac{1}{(z - z_0^*)} \right). \qquad (8.19)$$

Setting $y = 0$, the velocity field at the boundary is,

$$u(x, 0) = \frac{\kappa}{\pi} \frac{y_0}{(x - x_0)^2 + y_0^2}, \qquad v(x, 0) = 0, \qquad (8.20)$$

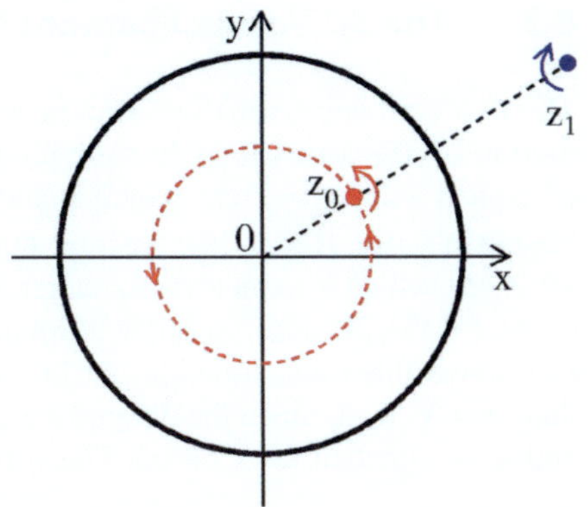

Fig. 8.4 Schematic positive vortex (red) at position $z_0 = Re^{i\omega t}$ inside a disk of radius a. The image vortex of opposite circulation (blue) is at position $z_1 = (a^2/R^2)z_0$. The vortex travels along a circular orbit (red dashed line)

hence it satisfies the boundary condition of the Euler equation that, at the boundary, there is no flow *across* the boundary ($v = 0$ at $y = 0$). The existence of a nonzero flow *along* the boundary ($u \neq 0$ at $y = 0$) is not an issue since the fluid lacks viscosity (it is a superfluid).

The actual velocity (u_0, v_0) of the vortex at z_0 near the boundary is found simply from $u_0 - iv_0 = \Omega_2'(z_0)$ (we do not use the complex potential Ω_1 because the vortex does not advect itself); we have,

$$u_0 = \frac{\kappa}{4\pi\, y_0}, \qquad v_0 = 0, \tag{8.21}$$

hence the vortex moves to the right driven by its image (the closer it is to the boundary, the faster it travels). In conclusion, the problem of one vortex close to a straight boundary is mathematically equivalent to the problem of a vortex-antivortex pair in the infinite plane.

The case of a vortex at $z_0 = x_0 + iy_0 = Re^{i\omega t}$ inside a circular boundary of radius a (a disk) is relevant to many experiments with atomic condensates, and is dealt in a similar way [3] by adding an image vortex of the opposite circulation at $z_1 = a^2/z_0^* = (a^2/R^2)z_0$ outside the disk. The total complex potential is thus,

$$\Omega(z) = -\frac{i\kappa}{2\pi} \log(z - z_0) + \frac{i\kappa}{2\pi} \log(z - z_1). \tag{8.22}$$

On can easily verify that the circular boundary $z = ae^{i\theta}$ is indeed a streamline, so the boundary condition is satisfied (the flow cannot cross the circular boundary), and that the vortex moves inside the disk along a circular orbit with azimuthal velocity,

$$V = \frac{\kappa R}{2\pi(a^2 - R^2)}, \tag{8.23}$$

and angular frequency $\omega = V/R$, see Fig. 8.4. This means that if the vortex is at the centre it does not move; if it is near the boundary, the vortex and its image move at a speed of a vortex-antivortex pair of half-separation $\delta = a - R \ll a$.

8.3   The 3D Vortex Filament Model (VFM)

The thin-core approximation used in the last section is particularly relevant to the motion of 3D vortex lines in superfluid helium. Infact, consider a helium sample of typical size $D = 1$ cm which rotates at the typical frequency of 1 Hz. Using Feynman's rule (Eq. 7.19), we find that the inter-vortex distance is $\ell = n_v^{-1/2} \approx 10^{-2}$ cm, which is indeed vastly larger than the vortex core radius, $a_0 = 10^{-8}$ cm.

As for the 2D case, in order to model a superfluid at $T = 0$, we start from the incompressible Euler equation. We introduce [4] the vector potential A defined such that $v = \nabla \times A$. Since the divergence of a curl is always zero, we have $\nabla \cdot \mathbf{A} = 0$, and $\mathbf{A} \to$ constant for $\mathbf{x} \to \infty$. The vorticity $\boldsymbol{\omega}$ can be written as,

$$\boldsymbol{\omega} = \nabla \times \mathbf{v} = \nabla \times (\nabla \times \mathbf{A}) = \nabla(\nabla \cdot \mathbf{A}) - \nabla^2 \mathbf{A} = -\nabla^2 \mathbf{A}. \tag{8.24}$$

Given the vorticity distribution $\boldsymbol{\omega}(\mathbf{r}, t)$ at the time t, the vector potential $\mathbf{A}(\mathbf{r}, t)$ is obtained by solving Poisson's equation,

$$\nabla^2 \mathbf{A} = -\boldsymbol{\omega}. \tag{8.25}$$

The solution of Eq. (8.25) at the point $\mathbf{s}$ is,

$$\mathbf{A}(\mathbf{s}, t) = \frac{1}{4\pi} \int_{\mathcal{V}} \frac{\boldsymbol{\omega}(\mathbf{r}, t)}{|\mathbf{s} - \mathbf{r}|} \mathrm{d}^3 \mathbf{r}, \tag{8.26}$$

where $\mathbf{r}$ is the variable of integration and $\mathcal{V}$ is volume. Taking the curl (with respect to $\mathbf{s}$), we obtain the *Biot-Savart law*,

$$\mathbf{v}(\mathbf{s}, t) = \frac{1}{4\pi} \int_{V} \frac{\boldsymbol{\omega}(\mathbf{r}, t) \times (\mathbf{s} - \mathbf{r})}{|\mathbf{s} - \mathbf{r}|^3} \mathrm{d}^3 \mathbf{r}. \tag{8.27}$$

In electromagnetism, the Biot-Savart law determines the magnetic field as a function of the distribution of currents. Here, the Biot-Savart law determines the velocity as a function of the distribution of vorticity $\boldsymbol{\omega}$.

The next step is to assume that the vorticity $\boldsymbol{\omega}$ is concentrated on filaments of infinitesimal thickness with circulation κ. We formally replace $\boldsymbol{\omega}(\mathbf{r}, t)\mathrm{d}^3\mathbf{r}$ with $\kappa \mathrm{d}\mathbf{r}$, the volume integral, Eq. (8.27), becomes a line integral over the vortex line configuration $\mathcal{L}$, and the Biot-Savart law reduces to [5],

$$\mathbf{v}(\mathbf{s}, t) = -\frac{\kappa}{4\pi} \oint_{\mathcal{L}} \frac{(\mathbf{s} - \mathbf{r})}{|\mathbf{s} - \mathbf{r}|^3} \times \mathrm{d}\mathbf{r}. \tag{8.28}$$

Equation (8.28) is the cornerstone of the 3D vortex filament model (VFM).

The VFM describes a configuration of vortex lines at $T = 0$ as a collection of oriented 3D space curves $\mathbf{s}(\zeta, t)$ of circulation κ, where ζ is the *arc length* along each vortex line measured from some arbitrary position. Since, according to Helmholtz's

Theorem, a vortex line moves with the flow, the time evolution of the vortex configuration at the point s is given by,

$$\frac{d\mathbf{s}}{dt} = \boldsymbol{v}_{si},$$
(8.29)

where,

$$\boldsymbol{v}_{si}(\boldsymbol{s}) = -\frac{\kappa}{4\pi} \oint_{\mathcal{L}} \frac{(\mathbf{s} - \mathbf{r})}{|\mathbf{s} - \mathbf{r}|^3} \times d\mathbf{r},$$
(8.30)

is the superfluid velocity which is induced at the point s by all the vortex lines in the vortex configuration $\mathcal{L}$.

Notice that the integrand of Eq. (8.30) diverges as $\mathbf{r} \to \mathbf{s}$, so it must be desingularized. A physically sensible cutoff length scale for the desingularization is the vortex core radius a_0. The induced superfluid velocity is therefore divided into *local* and *nonlocal* contributions in the following way [5],

$$\boldsymbol{v}_{si}(\boldsymbol{s}) = \beta \mathbf{s}' \times \mathbf{s}'' - \frac{\kappa}{4\pi} \oint_{\mathcal{L}'} \frac{(\mathbf{s} - \mathbf{r})}{|\mathbf{s} - \mathbf{r}|^3} \times d\mathbf{r}.$$
(8.31)

where $\mathcal{L}'$ is the vortex configuration which excludes the vicinity of the point s, and,

$$\beta = \frac{\kappa}{4\pi} \ln\left(\frac{R}{a_0}\right).$$
(8.32)

Denoting with a prime the derivative with respect to arc length, we have defined $\mathbf{s}' = d\mathbf{s}/d\zeta = \widehat{\boldsymbol{t}}$ the *unit tangent vector* at $\mathbf{s}$, $\mathbf{s}'' = d^2\mathbf{s}/d\zeta^2$, $\kappa_0 = |\mathbf{s}''|$ the *curvature*, and $R = 1/|\mathbf{s}''|$ the *local radius of curvature* (do not confuse the curvature and the radius of curvature, they are the inverse of each other). Besides $\widehat{\boldsymbol{t}}$, two more orthogonal unit vectors can be defined at the point s to create a *Frenet-Serret frame*: the *normal* unit vector $\widehat{\boldsymbol{n}} = \mathbf{s}''/|\mathbf{s}''|$ and the *binormal* unit vector $\widehat{\boldsymbol{b}} = \widehat{\boldsymbol{t}} \times \widehat{\boldsymbol{n}}$. The three orthogonal unit vectors obey the following *Frenet-Serret formulae*,

$$\frac{d\widehat{\boldsymbol{t}}}{d\zeta} = \kappa_0 \widehat{\boldsymbol{n}},$$
(8.33)

$$\frac{d\widehat{\boldsymbol{n}}}{d\zeta} = -\kappa_0 \widehat{\boldsymbol{t}} + \tau_0 \widehat{\boldsymbol{b}},$$
(8.34)

$$\frac{d\widehat{\boldsymbol{b}}}{d\zeta} = -\tau_0 \widehat{\boldsymbol{n}},$$
(8.35)

where τ_0 is the *torsion*.

To implement the VFM, vortex lines are divided into a large number of discretization points $\mathbf{s}_j$ ($j = 1, 2, \cdots, N_p$) separated by a distance $\delta \ll \ell$. Each discretization point evolves in time according to Eq. (8.31). When two vortex lines collide, vortex reconnections (discussed in Sect. 9.9) are performed algorithmically. Notice that the chosen discretization δ determines the numerical estimation of s'', hence of R, and what exactly is meant by the 'vicinity' of the point s (i.e. the difference between $\mathcal{L}$ and $\mathcal{L}'$).

Since the CPU time required to evolve a vortex configuration from time t to time $t + \Delta t$ scales with N_p^2, in his pioneering work, Schwarz [5] neglected the computationally-expensive nonlocal term of Eq. (8.31), and used simply the local term,

$$\boldsymbol{v}_{si}(s) \approx \beta \boldsymbol{s}' \times \boldsymbol{s}''. \tag{8.36}$$

Equation (8.36) is called the *Local Induction Approximation* or LIA.

The meaning of Eq. (8.36) is thus the following: at each point s along a vortex line, the vortex moves in the binormal direction with speed which is inversely proportional to the local radius of curvature. Note that a straight vortex line does not move, as its radius of curvature is infinite.

To illustrate the LIA, we compute the velocity of a vortex ring of radius R located on the $z = 0$ plane at $t = 0$. The ring is described by the space curve $\mathbf{s} = (R\cos(\theta), R\sin(\theta), 0)$, where θ is the angle and $\zeta = R\theta$ is the arc length. Taking derivatives with respect to ζ we have $\mathbf{s}' = (-\sin(\zeta/R), \cos(\zeta/R), 0)$ and $\mathbf{s}'' = (-1/R)(\cos(\zeta/R), \sin(\zeta/R), 0)$. Using Eq. (8.36), we conclude that the vortex ring moves in the z direction with velocity,

$$\boldsymbol{v}_{si} = \frac{\kappa}{4\pi R} \ln(R/a_0)\widehat{z}. \tag{8.37}$$

This result is in good agreement with a more precise solution of the Euler equation based on a hollow core at constant volume, which is,

$$\boldsymbol{v}_{si} = \frac{\kappa}{4\pi R} \left(\ln\left(\frac{8R}{a_0} \right) - \frac{1}{2} \right) \widehat{z}. \tag{8.38}$$

Using the GPE, Roberts and Grant [6] found that a vortex ring of radius much larger than the healing length moves with velocity,

$$\boldsymbol{v}_{si} = \frac{\kappa}{4\pi R} \left(\ln\left(\frac{8R}{a_0} \right) - 0.615 \right) \widehat{z}. \tag{8.39}$$

8.4   The Schwarz Equation

The 2D vortex point model and the 3D vortex filament model can be generalized to helium at non-zero temperatures. This is very important as it allows to model actual experiments. Consider a vortex line around which the superfluid circulation is κ. Let s be the position of a point along the vortex line, $\boldsymbol{s}' = d\boldsymbol{s}/d\zeta$ be the tangent unit vector at s, and $\boldsymbol{v}_L = d\boldsymbol{s}/dt$ be the velocity of the vortex line at s. Let also $\boldsymbol{v}_s$ and $\boldsymbol{v}_n$ be any normal fluid and superfluid velocites which are externally imposed (for example by bellows driving $\boldsymbol{v}_n$ and $\boldsymbol{v}_s$ in the same direction, or by thermal counterflow driving $\boldsymbol{v}_n$ and $\boldsymbol{v}_s$ in opposite directions). Let $\boldsymbol{v}_{si}$ be the superfluid velocity which is induced by the vortex configuration itself (according to Eq. (8.31)). The *Magnus force* per unit length on the vortex line is,

$$\boldsymbol{f}_M = \rho_s \kappa \boldsymbol{s}' \times (\boldsymbol{v}_L - \boldsymbol{v}_s - \boldsymbol{v}_{si}), \tag{8.40}$$

where $\boldsymbol{v}_s + \boldsymbol{v}_{si}$ is the total superfluid velocity. At nonzero temperatures, the velocity field around the vortex lines scatters the thermal excitations which make up the normal fluid, creating a *mutual friction force* (per unit length) on the vortex line [7] which is usually written as,

$$\boldsymbol{f}_D = -\alpha \rho_s \kappa \boldsymbol{s}' \times [\boldsymbol{s}' \times (\boldsymbol{v}_n - \boldsymbol{v}_s - \boldsymbol{v}_{si})] - \alpha' \rho_s \kappa \boldsymbol{s}' \times (\boldsymbol{v}_n - \boldsymbol{v}_s - \boldsymbol{v}_{si}), \tag{8.41}$$

where α and α' are dimensionless temperature-dependent friction coefficients which are determined in the experiments. Notice that whereas the α term is dissipative, the α' term is not. Since the radius of the vortex core is very small, we neglect the inertia of the vortex core, and assume that the sum of all forces on the vortex line (per unit length) is zero,

$$\boldsymbol{f}_M + \boldsymbol{f}_D = \boldsymbol{0}. \tag{8.42}$$

Solving for $\boldsymbol{v}_L = d\boldsymbol{s}/dt$, we obtain the *Schwarz equation* [5]

$$\frac{d\boldsymbol{s}}{dt} = \boldsymbol{v}_s + \boldsymbol{v}_{si} + \alpha \boldsymbol{s}' \times (\boldsymbol{v}_n - \boldsymbol{v}_s - \boldsymbol{v}_{si}) - \alpha' \boldsymbol{s}' \times (\boldsymbol{s}' \times (\boldsymbol{v}_n - \boldsymbol{v}_s - \boldsymbol{v}_{si})). \tag{8.43}$$

Notice that the Schwarz equation can be implemented in both 2D and 3D, thus generalizing the PVM and the VFM to nonzero temperatures.

There are two interesting limiting cases to consider. If $T = 0$, then the normal fluid is absent (hence $\boldsymbol{v}_n = \boldsymbol{0}$) and the friction vanishes (hence $\alpha = \alpha' = 0$); the Schwarz equation reduces to $d\boldsymbol{s}/dt = \boldsymbol{v}_s + \boldsymbol{v}_{si}$, consistently with Helmholtz Theorem.

The second interesting case is the case of a vortex line (without any imposed superflow, hence $\boldsymbol{v}_s = \boldsymbol{0}$) at $T \neq 0$ moving in a normal fluid background which is at rest (hence $\boldsymbol{v}_n = \boldsymbol{0}$). In this case the Schwarz equation reduces to

$$\frac{d\boldsymbol{s}}{dt} = \boldsymbol{v}_{si} - \alpha \boldsymbol{s}' \times \boldsymbol{v}_{si} + \alpha' \boldsymbol{s}' \times (\boldsymbol{s}' \times \boldsymbol{v}_{si}). \tag{8.44}$$

The main limitation of the Schwarz equation, hence of both PVM/VFM at nonzero temperatures. is that the normal fluid velocity, $\boldsymbol{v}_n$, is externally imposed rather than computed self-consistently. In other words, the back-reaction of the vortex lines on the normal fluid is neglected. Modern variants of the VFM address this limitation by coupling the Schwarz equation with a Navier-Stokes equation for $\boldsymbol{v}_n$ which has been suitably modified by the introduction of the mutual friction force [8–11].

8.5   The HVBK Equations

In the original two-fluid equations the superfluid is irrotational: vortex lines were not included. The Hall-Vinen-Bekharevich-Kalatnikov (HVBK) equations [12,13] generalise the original two-fluid equations to the presence of vortex lines by allowing non zero superfluid vorticity $\boldsymbol{\omega}_s = \nabla \times \boldsymbol{v}_s$, but only in a coarse-grained sense. Essentially, the HVBK equations describe the motion of a *continuum* of vortex lines rather than individual vortex lines as described by the Schwarz equation.

In the HVBK equations, normal fluid and superfluid are dynamically coupled by the mutual friction $\boldsymbol{F}$ (unlike the Schwarz equation). A vortex tension force $\boldsymbol{T}$ arises because the vortices have energy per unit length. The HVBK equations are the following,

$$\frac{\partial \boldsymbol{v}_n}{\partial t} + (\boldsymbol{v}_n \cdot \nabla)\boldsymbol{v}_n = -\frac{1}{\rho}\nabla P - \frac{\rho_s}{\rho}S\nabla T + \nu_n \nabla^2 \boldsymbol{v}_n + \frac{\rho_s}{\rho}\boldsymbol{F}, \tag{8.45}$$

$$\frac{\partial \boldsymbol{v}_s}{\partial t} + (\boldsymbol{v}_s \cdot \nabla)\boldsymbol{v}_s = -\frac{1}{\rho}\nabla P + S\nabla T - \frac{\rho_n}{\rho}\boldsymbol{F} + \boldsymbol{T}, \tag{8.46}$$

where $\widehat{\boldsymbol{\omega}}_s = \boldsymbol{\omega}_s/|\boldsymbol{\omega}_s|$ is the unit vector in the direction of the superfluid vorticity, and $\nu_n = \eta/\rho_n$ is the normal fluid's kinematic viscosity. The mutual friction force is,

$$\boldsymbol{F} = \frac{B}{2}\widehat{\boldsymbol{\omega}}_s \times (\boldsymbol{\omega}_s \times \boldsymbol{c}) + \frac{B'}{2}\boldsymbol{\omega}_s \times \boldsymbol{c}, \tag{8.47}$$

where B and B' are related to the friction coefficients α and α' by,

$$\alpha = \frac{B\rho_n}{2\rho}, \qquad \alpha' = \frac{B'\rho_n}{2\rho}, \tag{8.48}$$

and,

$$\boldsymbol{c} = \boldsymbol{v}_n - \boldsymbol{v}_s - \nu_s \nabla \times \widehat{\boldsymbol{\omega}}_s. \tag{8.49}$$

Finally the vortex tension force is,

$$\boldsymbol{T} = -\nu_s \boldsymbol{\omega}_s \times (\nabla \times \widehat{\boldsymbol{\omega}}_s), \tag{8.50}$$

where the *vortex tension parameter* is,

$$\nu_s = \frac{\kappa}{4\pi} \ln R/a_0. \tag{8.51}$$

The tension force can also be written as,

$$\boldsymbol{T} = -\nu_s (\boldsymbol{\omega}_s \cdot \nabla)\widehat{\boldsymbol{\omega}}_s. \tag{8.52}$$

It is an exercise (Exercise 8.6) to demonstrate this.

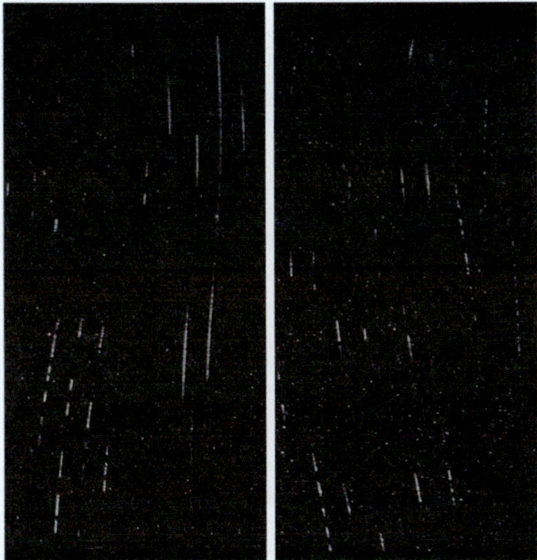

Fig. 8.5 Visualisation of vortex lines in rotating superfluid helium using frozen hydrogen particles [18,19]. The two images (downloaded from the repository [21] https://doi.org/10.57745/AHTGVK on 4 Dec 2025) display a side view of the vortex lattice (the axis of rotation being along the vertical direction) at two different times. A large number of tracer particles are clearly trapped along the vortices, making them visible. Notice also that the vortices are not straight: the visible time-dependent long-wavelength ondulations of the vortices is a collective Kelvin wave (a heater has just been switched on, driving an axial flow and triggering the Glaberson-Donnelly instability

8.6 Particle Dynamics in Superfluid Helium

In classical physics and engineering the study of fluids is assisted by flow visualisation via small tracer particles which are suspended in the fluid and then illuminated by a laser sheet focused on a thin plane. Successive images of the particles taken by a fast camera are used to determine the trajectories of individual particles (*Particle Tracking Velocimetry*, or PTV), or instantaneous streamlines (*Particle Image Velocimetry*, or PIV).

Pioneering work led by Steven Van Sciver implemented these experimental techniques to the low temperature environment of liquid helium using suitable small tracers made of glass, plastic, or, more frequently, solid hydrogen (sometimes a mixture of hydrogen and deuterium to improve the buoyancy). Visualisation via tracer particles has allowed detailed investigations of flow patterns and velocity statistics [14–16]. Since the particles can become trapped in the vortex lines, this visualisation technique has also been used to study vortex lattices [17], Kelvin waves [18,19] and vortex reconnections [20]. An example is shown in Fig. 8.5.

To interpret the results obtained by this technique, it is necessary to understand the dynamics of particles moving in superfluid helium. In fluid dynamics, a fluid particle is an idealised concept: it is an infinitesimal volume of fluid (a *material point*) identifiable in its history while moving with the fluid. The fluid's velocity $\boldsymbol{v}$ at location $\boldsymbol{r}$ is simply the rate of change of the position $\boldsymbol{r}$ of a material particle,

$$\boldsymbol{v}(\boldsymbol{r}, t) = \frac{d\boldsymbol{r}}{dt}. \tag{8.53}$$

On the contrary, a tracer particle is a physical object of a certain mass and size which obeys Newtons's law. Let the tracer particle be a small sphere of radius a_p and density ρ_p. If the particle is not trapped inside a vortex, its position $\boldsymbol{r}_p$ and velocity $\boldsymbol{v}_p$ are related not by the single Eq. (8.53) but by two equations [22],

$$\frac{d\boldsymbol{r}_p}{dt} = \boldsymbol{v}_p, \tag{8.54}$$

$$\frac{d\boldsymbol{v}_p}{dt} = \frac{1}{\tau}(\boldsymbol{v}_n - \boldsymbol{v}_p) + \frac{(\rho_p - \rho)}{\rho_0}\mathbf{g} + \frac{3\rho_n}{2\rho_0}\frac{D\boldsymbol{v}_n}{Dt} + \frac{3\rho_s}{2\rho_0}\frac{D\boldsymbol{v}_s}{Dt}. \tag{8.55}$$

Here $\mathbf{g}$ is the acceleration due to gravity, $\boldsymbol{v}_n$ and $\boldsymbol{v}_s$ are respectively the normal fluid and superfluid velocity,

$$\rho_0 = \rho_p + \frac{1}{2}\rho, \qquad \tau = \frac{2\rho_0 a_p^2}{9\eta}. \tag{8.56}$$

τ is called the *Stokes relaxation time*, η and ρ are the viscosity and the density of liquid helium, and,

$$\frac{D\boldsymbol{v}_n}{Dt} = \frac{\partial \boldsymbol{v}_n}{\partial t} + (\boldsymbol{v}_n \cdot \nabla)\boldsymbol{v}_n, \qquad \frac{D\boldsymbol{v}_s}{Dt} = \frac{\partial \boldsymbol{v}_s}{\partial t} + (\boldsymbol{v}_s \cdot \nabla)\boldsymbol{v}_s. \tag{8.57}$$

If the particle is buoyant ($\rho_p = \rho$) the equations simplify slightly. The density of helium II is $\rho \approx 145$ kg/m^{-3}), the viscosity ranges from $\eta = 1.3 \times 10^{-6}$ Ps s at $T = 1.6$ K to 2.2×10^{-6} Pas s at $T = 2.15$ K, and the typical particle radius in the experiments is $a_p \approx 10^{-6}$ m.

The first and second term at the right hand side of Eq. (8.55) is the viscous Stokes drag with respect to the normal fluid and the buoyancy acceleration; the third and fourth term are inertial effects arising respectively from spatial and/or temporal changes of the normal fluid and the superfluid.

If a particle becomes close to a vortex line, it may become trapped into it as shown in Fig. 8.6, because the energy of the system is reduced. The precise dynamics is quite complex [23]: at different times, almost all terms of Eq. (6.5) become relevant. At first the superfllow generated by the particle pushes away the vortex, but, as the particle becomes closer, the vortex interacts with its image in the particle. This effect makes the vortex to rotate around the particle and to bend locally, creating superfluid inertial effects. Eventually the vortex reconnects with its image and the particle is trapped, as excess energy in the form of Kelvin waves is either carried away or damped by friction. If the initial velocity of the particle is sufficiently large, the particle may become initially trapped, but then it frees itself of the vortex and escapes.

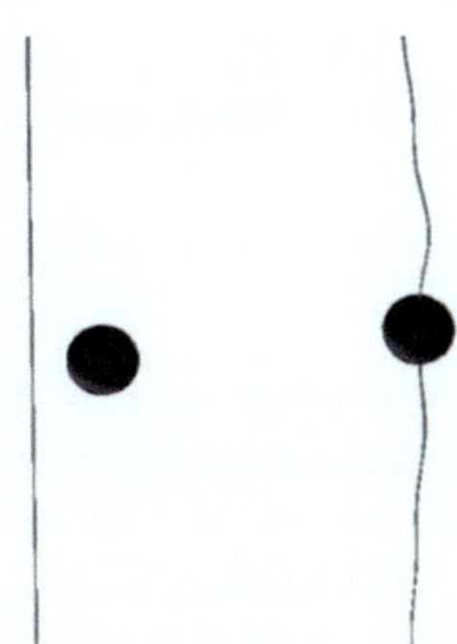

Fig. 8.6 Left: a particle approaches a vortex (left) and becomes trapped in it (right). Calculation performed with the VFM. Reprinted figure with permission from Ref. [23]. Copyrights (2007) of the American Physical Society

Problems

8.1 *Generalized Helmholtz Theorem for the GPE*

Derive Eq. (8.5), that in 2D condensates the velocity of a vortex is $\boldsymbol{v} = \boldsymbol{v}_s + \boldsymbol{v}_\rho$ where $\boldsymbol{v}_s = (\hbar/m)\nabla\phi$ and $\boldsymbol{v}_\rho = (\hbar/m)\widehat{\boldsymbol{\kappa}} \times \nabla \ln \rho$.

8.2 *Vortex-antivortex pair*

Using the VPM (vortex point method) and the complex potential, determine the translational speed of a vortex-antivortex pair (each vortex having circulation κ) separated by the distance $2D$.

8.3 *Vortex-vortex pair*

Using the VPM (vortex point method) and the complex potential, determine the period of rotation of a vortex-vortex pair (each vortex having circulation κ) separated by the distance $2D$.

8.4 *Kelvin wave frequency*

Use the LIA (Eq. 8.36) to determine the angular frequency of rotation of a Kelvin wave of wavelength $\lambda = 2\pi/k$ (where k is the wavenumber) on a vortex line with circulation κ.

8.5 *Attenuation of second sound*

Consider helium II at rest at temperature T_0, pressure p_0, entropy S_0, and densities ρ_0, ρ_{s0} and ρ_{n0} respectively. Introduce small perturbations T', p', S', ρ', ρ'_s and ρ'_n. Assume also small velocity perturbations $\boldsymbol{v}'_n$ and $\boldsymbol{v}'_s$. We want to find the effect that vortex lines have on second sound.

1. Using Eq. (6.2) (conservation of mass) and Eq. (6.3) (conservation of entropy), show that temperature and velocity perturbations satisfy,

$$\frac{\partial T'}{\partial t} = -\frac{\rho_{s0} S_0 T_0}{\rho_0 C} \nabla \cdot \boldsymbol{q}',$$

   where $\boldsymbol{q}' = \boldsymbol{v}'_n - \boldsymbol{v}'_s$. For simplicity, we assume that the specific heats at constant volume and pressure are the same: $C = T(\partial S/\partial T)$.

2. Consider the HVBK Eqs. (8.45) and (8.46) for the velocity perturbations. Neglect the nonlinear terms at the left hand sides (the velocity perturbations are small) and the viscosity of the normal fluid (we know that it would simply damp any second sound wave). Assume that the friction $\boldsymbol{F}$ is caused by an array of vortex lines rotating at angular velocity $\boldsymbol{\Omega} = \Omega\widehat{\boldsymbol{z}}$, corresponding to the superfluid vorticity $\boldsymbol{\omega}_s = 2\boldsymbol{\Omega}$. Do not perturb the vortex lines, which remain straight (i.e. the velocity perturbations $\boldsymbol{v}'_s$ are irrotational). Show that,

$$\frac{\partial \boldsymbol{q}'}{\partial t} = -\frac{\rho_0 S_0}{\rho_{n0}} \nabla T' + \boldsymbol{F}.$$

3. In order to describe the perturbations in the rotating frame of reference, add the Coriolis acceleration $-2\boldsymbol{\Omega} \times \boldsymbol{q}'$ to the right hand side of the last equation, hence derive the following second sound wave equation in the rotating frame,

$$\frac{\partial^2 \boldsymbol{q}'}{\partial t^2} = c_2^2 \nabla(\nabla \cdot \boldsymbol{q}') - (2 - B')(\boldsymbol{\Omega} \times \frac{\partial \boldsymbol{q}'}{\partial t}) + B\widehat{\boldsymbol{\Omega}} \times (\boldsymbol{\Omega} \times \frac{\partial \boldsymbol{q}'}{\partial t}),$$

   where

$$c_2 = \sqrt{\frac{\rho_{s0} T_0 S_0^2}{\rho_{n0}}},$$

   is the speed of second sound.

4. Assume that the perturbation is a wave propagating in the x-direction: $\boldsymbol{q}' = (q_x, q_y, 0)e^{i\omega t - ikx}$ where ω is the angular velocity and k is the wavenumber. Derive a linear system of equations for q_x and q_y, hence show that non-trivial solutions exists only if, in the high frequency limit that $\Omega/\omega \ll 1$, the dispersion relation is,

$$k = \frac{\omega}{c_2}\left(1 - \frac{i\Omega B}{2\omega}\right).$$

5. Use this result to show that the amplitude of a second sound wave decreases as e^{-x/λ_0} where the attenuation coefficient λ_0 depends on the vortex line density.

8.6 *Vortex tension*

Show that the two expressions Eqs. (8.50) and (8.52) for the vortex tension are equivalent.

8.7 *Vortex in a disk*

Consider a vortex of circulation κ at position $z_0 = x_0 + iy_0$ inside a disk of radius a. For simplicity assume that $y_0 = 0$. The complex potential is given by Eq. (8.22),

$$\Omega(z) = -\frac{i\kappa}{2\pi} \log(z - z_0) + \frac{i\kappa}{2\pi} \log(a^2/z - z_0^*).$$

1. Show that the boundary of the disk, $z = ae^{i\theta}$ where θ is the angle, is a streamline, that is the velocity is along the boundary, with no component across it.
2. Determine the velocity of the vortex (u_0, v_0).

8.8 *Particle trapped in a vortex line*

If a particle of radius a_p is trapped in a vortex line, the energy of the helium is reduced by an amount ΔE equal to the kinetic energy of the displaced superfluid. Show that in a typical experiment ΔE is larger than the thermal energy $k_B T$ where k_B is Boltzmann's constant, hence the trapping is stable.

References

1. A.J. Groszek, D.M. Paganin, K. Helmerson, T.P. Simula, Phys. Rev. A **97**, 023617 (2018)
2. H. Aref, N. Pomphrey, Proc. Roy. Soc. Lond. A **380**, 359 (1982)
3. P.K. Newton, *The N-vortex problem*, Springer (2001)
4. P.G. Saffman *Vortex Dynamics*, Cambridge (1993)
5. K.W. Schwarz, Phys. Rev. B **38**, 2398 (1988)
6. P.H. Roberts, J. Grant, J. Phys. A: Gen. Phys. **4**, 55 (1971)
7. C.F. Barenghi, W.F. Vinen, R.J. Donnelly, J. Low Temp. Phys. **52**, 189 (1982)
8. D. Kivotides, C.F. Barenghi, D.C. Samuels, Science **290**, 777 (2000)
9. S. Yui, M. Tsubota, H. Kobayashi, Phys. Rev. Lett. **120**, 155301 (2018)
10. L. Galantucci, A.W. Baggaley, C.F. Barenghi, G. Krstulovic, Eur. Phys. J. Plus **135**, 547 (2020)
11. L. Galantucci, G. Krstulovic, C.F. Barenghi, Phys. Rev. Fluids **8**, 014702 (2023)
12. R.N. Hills, P.H. Roberts, Arch. Rat. Mech. Anal. **66**, 43 (1977)
13. C.F. Barenghi, C.A. Jones, J. Fluid Mech. **197**, 551 (1988)
14. M.S. Paoletti, M.E. Fisher, K.R. Sreenivasan, D.P. Lathrop, Phys. Rev. Lett. **101**, 154501 (2008)
15. B. Mastracci, W. Guo, Phys. Rev. Fluids **3**, 063304 (2018)
16. M. La Mantia, L. Skrbek, Europhys. Lett. **105**, 46002 (2014)
17. G.P. Bewley, D.P. Lathrop, K.R. Sreenivasan, Nature **441**, 588 (2006)
18. C. Peretti, J. Vessaire, E. Durozoy, M. Gibert, Sci. Adv. **9**, eadh2899 (2023)
19. J. Vassaire, C. Peretti, F. Lorin, E. Durozoy, G. Garde, P. Spathis, B. Chabaud, M. Gibert, Rev. Sci. Instrum. **97**, 025206 (2026). arXiv:2510.00026v1
20. G.P. Bewley, M.S. Paoletti, K.R. Sreenivasan, D.P. Lathrop, Proc. Nat. Acad. Sci. USA **105**, 3707 (2008)
21. M. Gibert, C. Peretti, F. Lorin, J. Vessaire, Direct observations of quantum vortices in helium II: particle decoration and counterflow-induced waves. Recherche Data Gouv (2025). https://doi.org/10.57745/AHTGVK
22. D.R. Poole, C.F. Barenghi, Y.A. Sergeev, W.F. Vinen, Phys. Rev. B **71**, 064514 (2005)
23. D. Kivotides, C.F. Barenghi, Y.A. Sergeev, Phys. Rev. B **75**, 212502 (2007)

Vortex Dynamics

9

Abstract

This chapter is concerned with the rich dynamics of quantum vortices which has been observed, from vortex rings to vortex knots, from vortex reconnections to vortex-sound interactions.

9.1   Vortex Dynamics According to the GPE

Consider a single straight vortex placed off-centre in a cylindrical box-trap containing an atomic condensate of constant density (i.e. a set-up similar to a bucket of superfluid helium). The radial component of the fluid velocity must be zero at the cylindrical boundary. We have seen in the previous chapter that, for this condition to be achieved, an image vortex with opposite circulation must exist on the other side of the boundary, outside the fluid. In this way, by forming a virtual vortex-antivortex pair with its image, the off-centre vortex precesses around the cylindrical container.

In a harmonically-trapped condensate, an off-centre vortex still precesses about the trap centre. However, the slow variation of the density towards the edge complicates the image interpretation. Instead, we can interpret the precession in terms of a Magnus force. Imagine the vortex line as a rotating cylinder, shown in Fig. 9.1(left). The vortex line feels a radial force due to its position in the condensate, and this gives rise to a motion of the vortex line with velocity $\mathbf{v}_L$ which is perpendicular to the force, an effect well-known in classical hydrodynamics. The force can be deduced from the free energy of the system. The energy decreases with the vortex position, r_0, as shown in Fig. 9.1(right). The radial force, which follows as $-\partial E/\partial r_0$, acts outwards and has contributions from the "buoyancy" of the vortex, which behaves like a bubble, as well as its kinetic energy. This force balances the Magnus force $-mn\boldsymbol{\kappa} \times \boldsymbol{v}_L$, leading to the expression,

$$\frac{\partial E}{\partial r_0} = mn\boldsymbol{\kappa} \times \boldsymbol{v}_L, \tag{9.1}$$

C. F. Barenghi et al., *Quantum Fluids, Solitons, and Vortices*, Lecture Notes in Physics 1050, https://doi.org/10.1007/978-3-032-20171-3_9

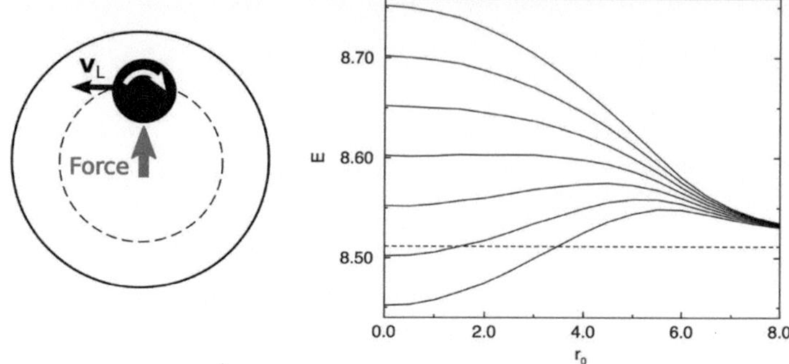

Fig. 9.1 Left: Schematic of the Magnus effect which causes an off-centre vortex to precess in a trapped condensate. Right: Free energy E of a trapped condensate versus the radial position of a vortex, r_0. The top line is for a non-rotating system, while the lower lines have increases rotation frequencies. Reprinted figure with permission from [2]. Copyright 1999 by the American Physical Society

where κ is the circulation vector. The net effect is a precession of the vortex about the trap centre. More generally, the vortex follows a path of constant free energy; for example, it will trace out a circular path in an axi-symmetric harmonic trap and an elliptical path in a non-axisymmetric harmonic trap. The experiment of Ref. [1] pioneered the real-time imaging of vortices in condensates and was able to directly monitor the precession of a vortex, finding it to agree well with theoretical predictions.

At the trap centre, the slope of the curve $E(r_0)$ becomes flat, hence the vortex ceases to precess; in fact, the trapped condensate with a central vortex line is a stationary state. For a non-rotating condensate, this state is energetically unstable ($E(r_0)$ is a maximum at the origin). Under sufficiently fast rotation, however, $E(r_0)$ changes shape such that this state becomes a minimum and thus energetically stable.

This analysis assumes the vortex line to be straight. This is also valid in flattened, quasi-2D geometries, but in 3D geometries the vortex line can bend and support excitations, and the motion of a vortex becomes more complicated.

9.2   Vortex Pairs

In Chap. 8 we have already seen the simplest vortex configurations: the vortex-antivortex pair and the vortex-vortex pair. They move respectively in a translational and in a rotational way, see Figs. 8.1 and 8.2. We have seen that these motions are direct consequences of the property that a vortex moves with the local fluid velocity, which in these cases is the velocity induced by the other vortex of the pair. The rotational motion of the vortex-vortex pair gives insight as to why many vortices of the same circulation rotate together, forming a vortex lattice.

It is instructive to calculate the energy of the vortex pairs in a cylindrical condensate (radius R_0, height H_0) by assuming a uniform density and integrating the kinetic energy, as we did to calculate the energy of a single vortex line in Eq. (7.14). The vortices have circulation q_1 and q_2, and individual velocity fields $\mathbf{v}_1$ and $\mathbf{v}_2$, respectively. The net velocity field of the two vortices is $\mathbf{v}_1 + \mathbf{v}_2$. Assuming $\xi \ll d \ll R$ then the (kinetic) energy of the pair is,

$$E_{\text{kin}} = \int mn_0|\mathbf{v}_1 + \mathbf{v}_2|^2 \, d\mathbf{r} = \frac{\pi n_0 H_0 \hbar^2}{m}\left[q_1^2 \ln \frac{R_0}{a_0} + q_2^2 \ln \frac{R_0}{a_0} + 2q_1 q_2 \ln \frac{R_0}{d}\right].$$
(9.2)

The first two terms are the energies of the individual vortices if they were isolated. The second term is the *interaction energy*, the change in energy arising from the interaction between the vortices. For a vortex-antivortex pair ($q_1 = -q_2$) the interaction energy is negative. This is because the flow fields tend cancel out in the bulk, reducing the total kinetic energy. Indeed, in the limit $d \to a_0$, the flow fields completely cancel and the total energy tends to zero; in reality the vortices annihilate with each other in this limit. For a corotating pair ($q_1 = q_2$), the interaction energy is positive; in the bulk the flow fields tend to reinforce, increasing the total kinetic energy.

In the presence of dissipation on the vortices, this result also informs us that vortex-antivortex pairs will shrink (ultimately annihilating when their cores begin to overlap) and corotating pairs will expand, until they reach the boundary of the condensate and ultimately annihilate against their images. Interestingly, at finite temperature and in 2D condensates, vortex-antivortex pairs can be created spontaneously [3].

9.3 Vortex Rings

A vortex line either terminates at a boundary (e.g. the vortex in the cylindrical container discussed in the previous section) or is a closed loop. A circular vortex loop is called a *vortex ring*. It is the three-dimensional analog of the (two-dimensional) vortex-antivortex pair: each element of the ring moves due to the flow induced by the rest of the ring, resulting in the ring travelling in a straight line at a constant speed which is inversely proportional to its radius. Figure 9.2 shows a vortex ring travelling towards, and interacting with, a straight vortex line.

In nature, vortex rings are easily created when a fluid is pushed through an orifice. The same strategy to create vortex rings is followed by cigarette smokers, volcanoes, dolphins, and low temperature physicists working with liquid helium [4], as shown in Fig. 9.3.

Another technique to generate vortex rings in helium consists of injecting electrons, as we shall see in the next section on vortex nucleation. Numerical simulations reveal that, after the nucleation, the electron becomes trapped in the core of the vortex ring. Using grids held at desired electrical potentials, one can perform a controlled time-of-flight experiment on the charge-carrier (the electron-ring complex) over a known travel distance, the energy of the charge-carrier being known from the applied potentials which control it. Such experiments [5] confirmed that charge-carriers behave indeed as vortex rings of circulation $\kappa = h/m$ which obey the

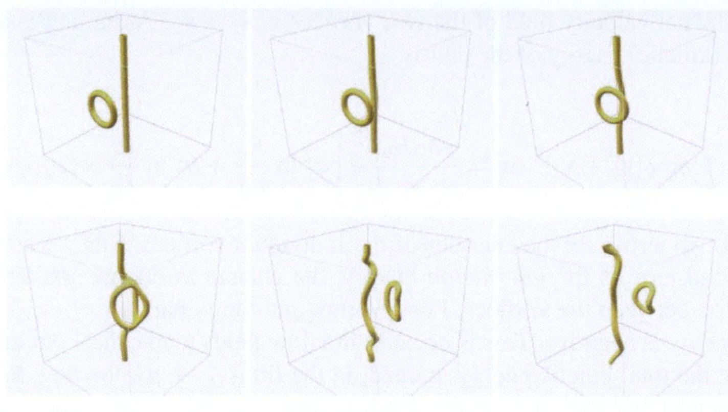

Fig. 9.2 Time sequence showing a vortex ring travelling towards a vortex line computed by numerically solving the GPE (courtesy of A.J. Youd). Notice the vortex reconnection and the Kelvin waves it produces

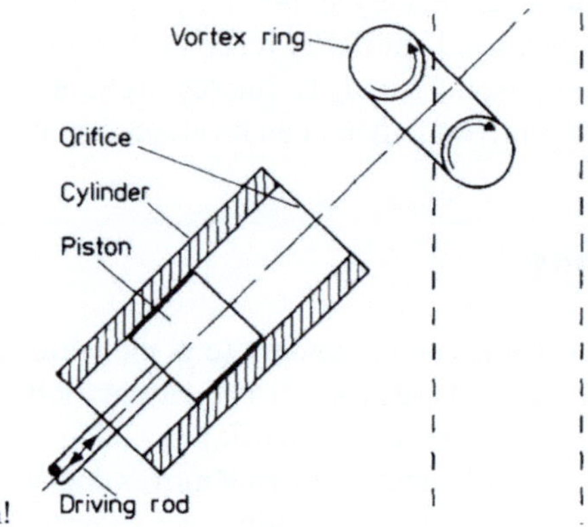

Fig. 9.3 Piston-cylinder arrangement used to generate vortex rings in liquid helium [4]. The fluid which exists the cylinder rolls up on the sharp edge of the circular orifice forming a large centimetre-size vortex ring. The position of the helium vortex ring as it travels is then detected acoustically using first sound and second sound. Reprinted with permission from [4]. Copyright 1983, American Institute of Physics. All rights reserved

classical equations (9.12), (9.13) and (9.14) for the energy, velocity and momentum of a vortex ring given in Exercise 7.9.

9.4   Vortex Bundles and Vortex Leapfrogging

The leapfrogging of two co-axial vortex-antivortex pairs (in 2D) or vortex rings (in 3D) when sufficiently close to each other is a striking periodic motion which was reported by von Helmholtz [6]. Figure 9.4 illustrates this motion. The pair (or ring)

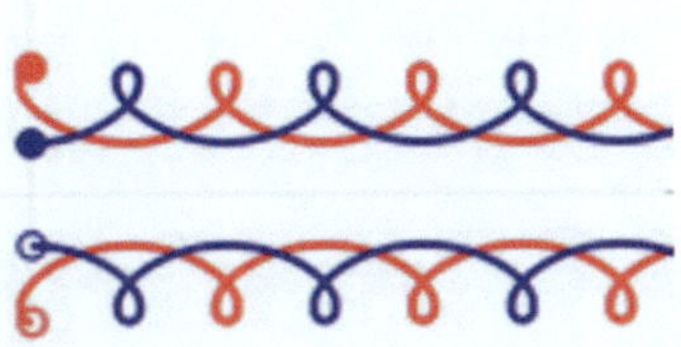

Fig. 9.4 Trajectories on the xy-plane of two vortex-antivortex pairs which leapfrog each other travelling from left to right, computed using the VPM [9]. The first pair is marked in red, the second pair in blue. Filled (open) symbols refer to positive (negative) vortices. Reprinted under CC-BY-4.0 license from [9]. Copyright 2021, The Author(s)

which is initially in front becomes larger, slows down, and moves to the back, while the pair (or ring) which is initially behind becomes smaller, speeds up, and moves to the front, and so on periodically. In viscous fluids, leapfrogging does not last forever because vorticity diffuses. In quantum fluids at low temperatures, leapfrogging is limited only by slow energy losses due to sound radiation and by dynamical instabilities (still poorly understood).

A toroidal *vortex bundle* is a configuration consisting of N coaxial vortex ring. It is apparent that, as they move, the rings leapfrog each other. The cylinder-piston apparatus shown in Fig. 9.3 which is used to create a single vortex ring generates a vortex bundle if the stroke of the piston is long enough. Borner and collaborators [4] were able to create large toroidal structures containing up to $N = 1000$ quanta (i.e. $N = 1000$ singly-quantised vortex rings) which were observed to travel a significant distance. Numerical simulations of the experiment using the VFM [7] revealed that vortex bundles are unstable and become disorganized, but as long as they are recognizably toroidal, the translational speed is not affected. Compare the motion of an $N = 3$ vortex bundle in Fig. 9.5, which is stable, with the motion of an $N = 7$ vortex bundle in Fig. 9.6 which is unstable (but still travels coherently as a compact structure). In both figures, each vortex loop is identified by a different colour. For example, in Fig. 9.5, notice the position of the red ring in the $N = 3$ vortex bundle. It is initially at the front of the bundle, then becomes larger, slides backwards over the other two rings, then becomes smaller, overtakes the other two rings from the inside, and finally, at the end of the cycle, recovers the initial position at the front of the bundle. On the contrary the $N = 7$ bundle quickly loses its axial symmetry, see Fig. 9.6. Larger and more disorganized vortex bundles but which still travel coherently have been studied more recently in Ref. [8].

Generating vortex rings (or, in general, vortex lines) in atomic Bose-Einstein condensates requires specialized techniques such as the *snake instability* [10], phase imprinting [11], laser spoons [12] and optical chopsticks [13].

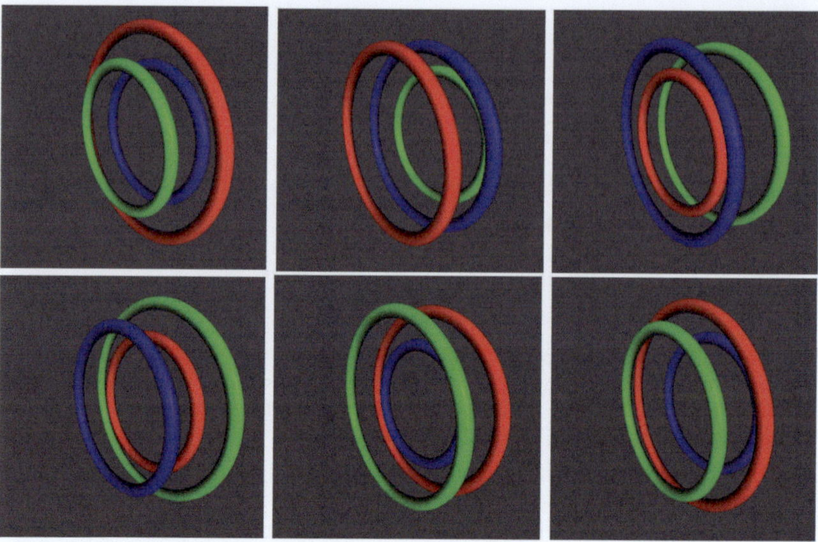

Fig. 9.5 Bundle of $N = 3$ vortex rings which leapfrog over each other at $T = 0$, computed using the VFM [7]. The snapshots were taken at $t = 0.0005$ s (top left), $t = 0.8$ s (top middle), $t = 1.6$ s (top right), $t = 2.0$ s (bottom left), $t = 2.8$ s (bottom middle), and $t = 3.2$ s (bottom right). Reprinted with permission from [7]. Copyright 2014, American Institute of Physics. All rights reserved

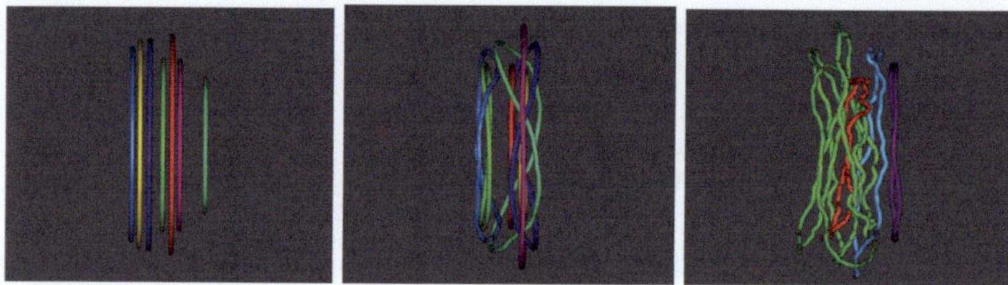

Fig. 9.6 Bundle of $N = 7$ vortex rings which leapfrog over each other at $T = 0$ and become unstable, computed using the VFM [7]. Snapshots taken at $t = 56$ s (left), $t = 60$ s (middle), and $t = 75$ s (right). Reprinted with permission from [7]. Copyright 2014, American Institute of Physics. All rights reserved

Both vortex rings and vortex-antivortex pairs are forms of solitary waves, since they propagate without spreading. Moreover, like dark solitons, they are stationary (excited) solutions of the homogeneous condensate in the frame moving with the ring/pair.

9.5 Vortex Nucleation

In Sect. 7.7 we have seen how vortices appear in a rotating condensate. In this section we deal with the more general process of *vortex nucleation*, particularly nucleation near moving objects and hard boundaries (the latter being still poorly understood).

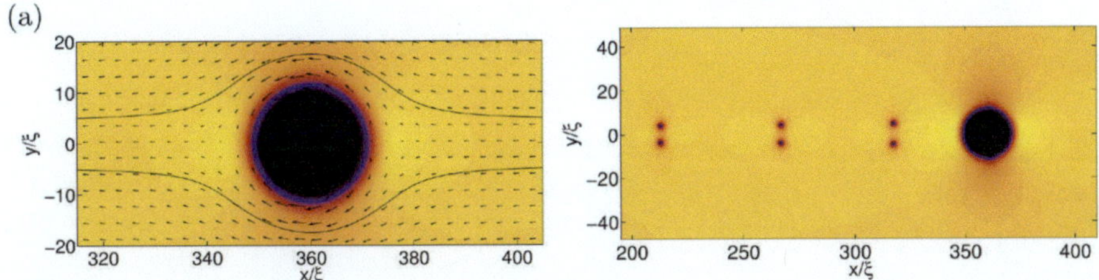

Fig. 9.7 Flow (right to left) of a homogeneous condensate past a cylindrical obstacle at velocity below (left) and above (right) the critical velocity. Shown is the condensate density, and the arrows (left) show the velocity field. Note that the obstacle punches a large hole in the condensate (the large black circle) which is much larger than the cores of the vortices (the small black circles). Notice also (right panel) that the vortices are created in regular pairs at the poles of the obstacle. At larger velocities symmetric and then asymmetric clusters of vortices are nucleated. Results are based on simulations of the 2D GPE in the moving frame. Reprinted under CC-BY-3.0 license from [15]. Copyright 2015, The Author(s)

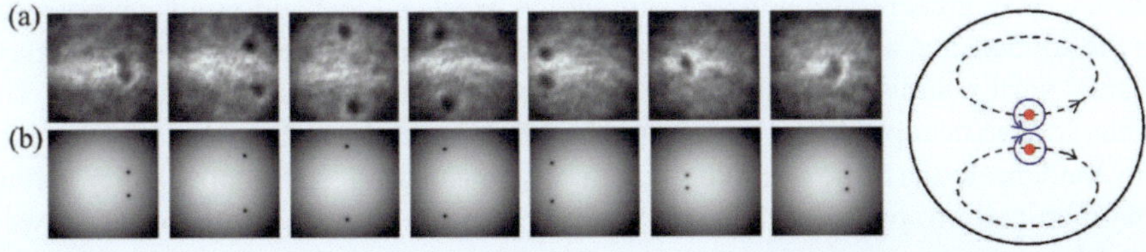

Fig. 9.8 **a** Experimental images of vortex-antivortex pair moving within a trapped condensate. **b** The vortex-antivortex pair's trajectory is reproduced by numerically solving the GPE. Note that the vortex core appears larger in the experimental images since the condensate is first expanded to aid in resolving the cores. Figure adapted with permission from Ref. [16]. Copyrighted by the American Physical Society. A schematic of the trajectory of a vortex-antivortex pair in a trapped condensate is shown on the right

To understand the nucleation mechanism, recall Landau's criterion for the generation of excitations in the condensate (Sect. 6.9). In the hydrodynamic picture, the speed of the atom/impurity is replaced by the local fluid velocity. Consider the scenario of a homogeneous condensate flowing with bulk speed v_∞ past a cylindrical obstacle (this is equivalent to the cylindrical obstacle moving at speed v_∞ through a static condensate but more convenient to simulate numerically). For low values of v_∞, the condensate undergoes undisturbed laminar flow around the obstacle, as shown in Fig. 9.7(left). At the poles of the obstacle the local flow speed is approximately twice as large as away at infinity, v_∞ (indeed, for an inviscid Euler fluid, the velocity at the poles is exactly $2v_\infty$). Therefore, when $v_\infty \approx 0.5c$, the local flow velocity at the poles exceeds the speed of sound, becomes critical, and, as per Landau's prediction, excitations are created. The excitations take the form of pairs of opposite circulation vortices [14], which periodically peal off from the poles of the obstacle and travel downstream to the left, as seen in Fig. 9.7(right). If instead the obstacle moves to the right in a stationary fluid, the nucleated vortex pairs travel to the right, but slightly slower than the obstacle, falling behind it.

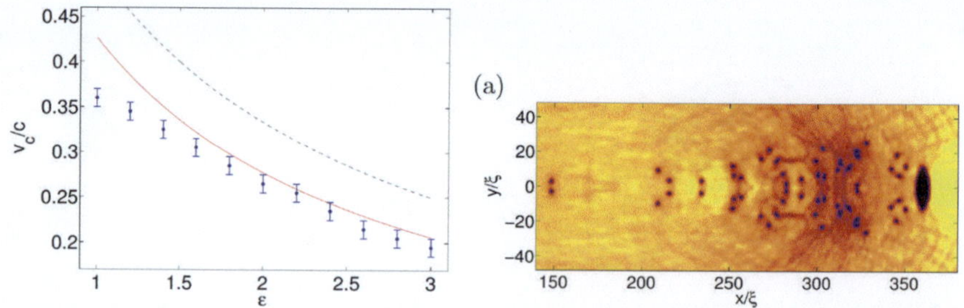

Fig. 9.9 Left: Dimensionless critical velocity v_c/c as function of obstacle's ellipticity, ϵ (blue circles with error bars) as computed by the GPE for a cylindrical obstacle of diameter 10ξ. The Euler prediction $v_c = c/(1 + \epsilon)$ and a corrected Euler prediction [17] are represented respectively by the black dashed line and the solid red line. Right: flow of a homogeneous condensate past an elliptical obstacle ($d = 5\xi$, $\epsilon = 3$) at $v = 0.53c$. Reproduced from Ref. [15] under CC-BY licence

This process has been studied experimentally in atomic condensates [16,21]. The obstacle is engineered by a laser beam which exerts a localized repulsive potential on the condensate, and moved relative to the condensate. Figure 9.8 shows an experimental vortex-antivortex pair which moves within a trapped condensate (left). The dynamics can be reproduced by solving the GPE (right). Note that whereas in an infinite condensate the vortex-antivortex pair has constant translational velocity, within a harmonically-trapped condensate the motion of each vortex of the pair follows a curved trajectory.

We have remarked that, for a homogeneous perfect Euler fluid moving around a circular obstacle, the velocity at the poles is exactly twice the velocity at infinity. If the obstacle is elliptical rather than circular, the velocity at the poles is enhanced by the obstacle's ellipticity, ϵ (defined as the ratio of the major and minor semi-axes), hence the critical velocity for vortex nucleation can be achieved at smaller value of v_∞, as shown in Fig. 9.9 (left). Moreover, at the same v_∞, more vortices are nucleated by an elliptical obstacle than a circular obstacle, hence it is easier to nucleate vortex clusters.

To illustrate this point, Fig. 9.10 shows a numerical simulation [15] of an experiment of Shin and collaborators [20] in which a circular obstacle, consisting of a laser beam, moves at velocity $v = 1.4\,\mathrm{mm/s}$ and nucleates vortices. Figure 9.11 simulates the same system with an elliptical obstacle nucleating as many vortices by moving at slower speed ($v = 0.8\,\mathrm{mm/s}$).

Similarly, 3D vortex rings arise when a spherical obstacle exceeds a critical speed relative to the condensate. Besides pushing liquid helium through an orifice, vortex rings can be created by injecting electrons with a sharp high-voltage tip; the electron's zero point motion carves a small, charged spherical bubble in the liquid of radius approximately 16×10^{-10} m which can be accelerated by a suitably applied electric field. Upon exceeding a critical velocity, a vortex ring peels off at the bubble's equator as described above; subsequently the electron falls into the vortex core, leaving a vortex ring with an electron bubble attached; the last part of the sequence is shown in Fig. 9.12.

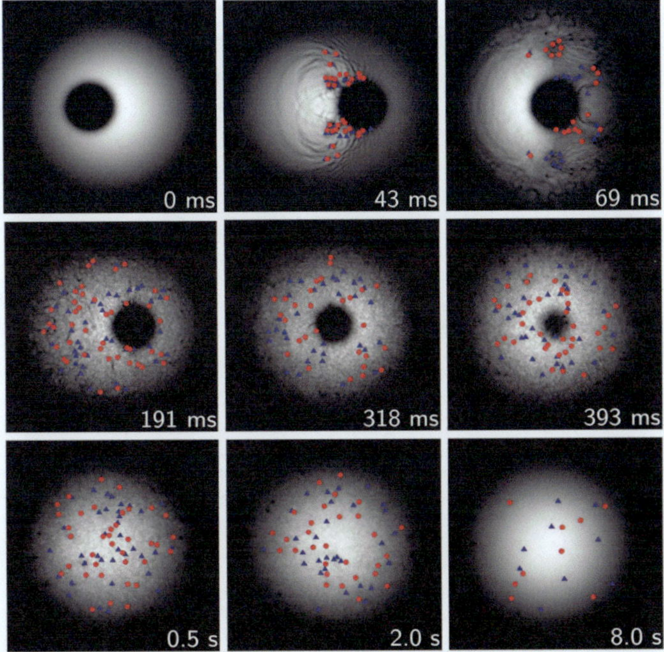

Fig. 9.10 GPE simulation [15] of the experiment of Ref. [20]: after creating positive and negative vortices (marked in red and blue respectively) by moving the obstacle at velocity up to $v = 1.4\,\mathrm{mm/s}$, the amplitude of the potential which represents the obstacle is gently ramped down to zero, leaving behind a circular condensate containing 2D quantum turbulence, see Sect. 10.9. Reprinted under CC-BY-3.0 license from [15]. Copyright 2015, The Author(s)

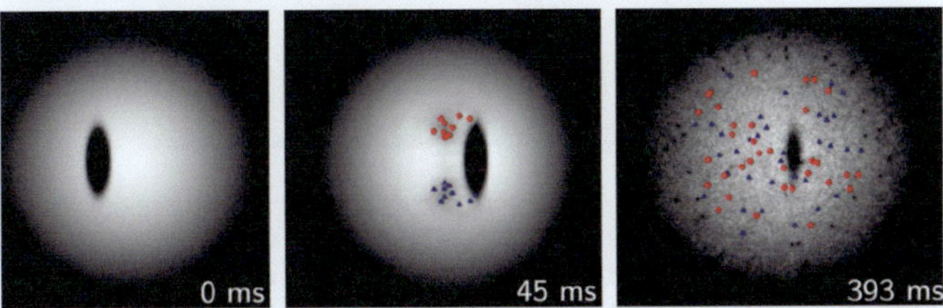

Fig. 9.11 As in Fig. 9.10 but the obstacle is elliptical ($\epsilon = 3$); although the velocity is lower ($v = 0.8\,\mathrm{mm/s}$), the number of vortices which are nucleated is larger. Reprinted under CC-BY-3.0 license from [15]. Copyright 2015, The Author(s)

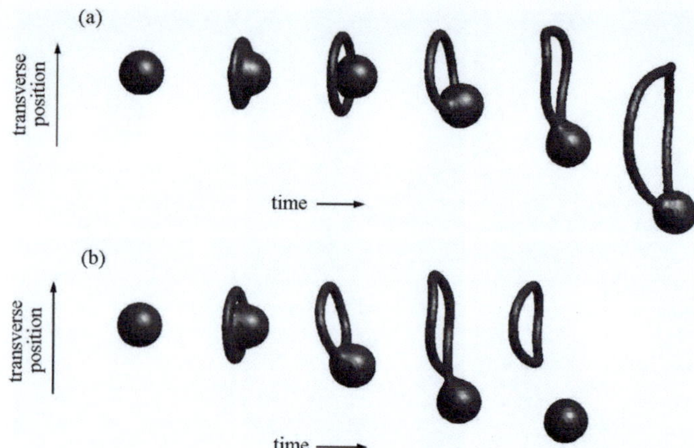

Fig. 9.12 Vortex rings nucleated by moving bubbles, computed by numerically solving the GPE. Reprinted with permission from [26]. Copyright 2000, EDP Sciences. All rights reserved

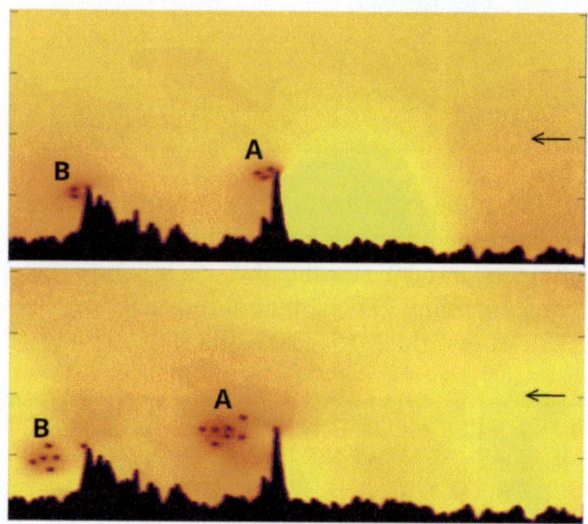

Fig. 9.13 GPE simulation of the constant flow (right to left) past a rough boundary. Shown is the condensate density in the domain $-150 \leq x \leq 150, 0 \leq y \leq 100$ (the actual computation extends to $y = 120$) at two successive times. The boundary conditions at $x = \pm 150$ are periodic, whereas a step potential $V \ll \mu$ confines the condensate in the y-direction (the bottom boundary condition is described in the main text). Note the vortex clusters A and B which are nucleated near the tallest "mountains" and then are advected to the left.

We now consider what happens at a boundary. It is reasonable to assume that the same nucleation mechanism which we have described for an obstacle takes place at the boundary of any vessel which contains a sample of superfluid liquid helium (^{4}He), provided that this boundary is not perfectly smooth. Indeed it would be difficult to build a metal boundary perfectly smooth at the length scale of the superfluid healing length, $\xi \approx 10^{-10}$m. Unfortunately, there is no technique to visualize the nucleation of an individual vortex near the boundary, but we can shed some light into the problem of nucleation by solving the GPE for a flow pasta rough boundary.

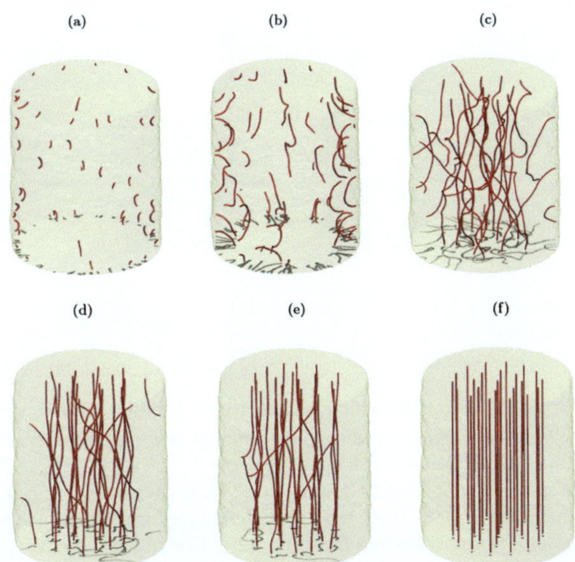

Fig. 9.14 Formation of a vortex lattice [18]: snapshots of the vortex lines at **a** $t = 100$, **b** 200, **c** 500, **d** 1000, **e** 1500 and **f** 3000. Reproduced with permission from Ref. [18]. Copyrights (2020) of the American Physical Society

A 2D example is shown in Fig. 9.13 where the rough boundary is the vertically-stretched 1D profile of the image of a small area of a vibrating wire used to stir superfluid helium at Lancaster University - image being taken by an atom force microscope. In the presence of a steady uniform flow (from right to left), the local velocity near the summits of the tallest "mountains" of the profile exceeds the critical velocity, thus clusters A and B of (anti-clockwise) vortices are quickly nucleated and then advected to the left. When a cluster contains a sufficient number of vortices, the (left to right) rotational flow which they induce on the side of the boundary opposes the (right to left) flow at infinity, shutting down the nucleation process briefly, until that cluster has moved away, and the local velocity at the summit exceeds the critical velocity again.

The related problem of nucleation in the flow over a sinusoidal boundary was tackled analytically in Ref. [19]. A 3D application of this nucleation idea allowed Keepfer et al. [18] to model the formation of a vortex lattice in rotating helium inside a rigid metal container. They solved the GPE inside a cylinder of radius $R = 50\xi$ and height $H = 100\xi$ where ξ is the healing length. The key feature of this simulation is that the later boundary of the cylinder is rough on the length scale of the healing length, and rotates at angular velocity Ω. The roughness of the lateral boundary is modelled by a step potential which is defined by fractal Brownian motion. The irregular surface of the wall extends radially approximately from 45ξ to 0ξ, corresponding to irregular "surface bumps" of height up to five healing lengths (which is of the order of the vortex core size). The top and bottom boundaries are smooth. As in the 2D calculation, the motion of the sharpest small bumps on the boundary with respect to the condensate (which is initially at rest) nucleates small vortex loops (U-vortices attached to the boundary), which interact, stretch, become

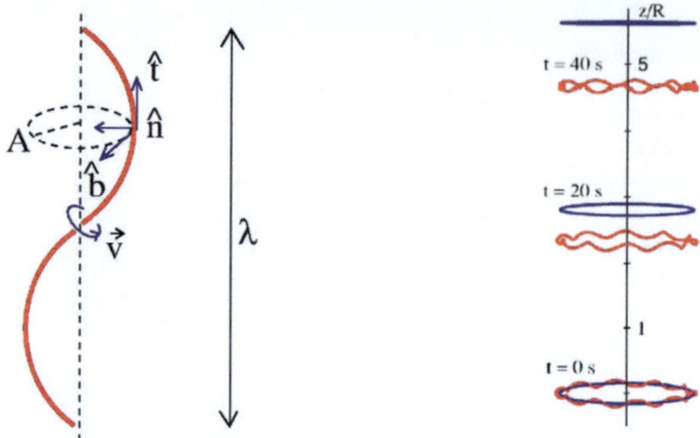

Fig. 9.15 Left: Schematic Kelvin wave of amplitude A and wavelength λ. The three unit vectors $\hat{t}$, $\hat{n}$, $\hat{b}$ in the tangent, normal and binormal directions are shown. The wave rotates along the binormal direction, in the direction opposite to the direction of the flow. Right: Comparison between motion of a vortex ring (radius $R = 0.1$ cm, blue) and vortex ring perturbed by Kelvin waves (relative amplitude $A/R = 0.05$, red). Calculation performed with the vortex filament model [29]. Figure adapted with permission from Ref. [29]. Copyrighted by the American Physical Society

longer and straighten by moving to the centre of the condensate, until, with the help of a small amount of dissipation ($\gamma = 0.05$), settle down as a vortex lattice, as shown in Fig. 9.14.

9.6   Kelvin Waves

A sinusoidal or helical perturbation of the vortex core away from its rest position is called a *Kelvin wave* [22–25]. Figure 9.15 (left) shows a Kelvin wave of amplitude A and wavelength λ. A Kelvin wave of infinitesimal amplitude A and wavelength $\lambda \gg a_0$ (where a_0 is the vortex core radius) rotates with angular velocity,

$$\omega_0 \approx \frac{\kappa k^2}{4\pi} \left(\ln \left(\frac{1}{ka_0} \right) - 0.116 \right), \tag{9.3}$$

in the direction opposite to that of the fluid's circulation, where $k = 2\pi/\lambda$ is the wavenumber; in other words, the shorter the wave, the faster it rotates.

Kelvin waves are induced on a vortex line either by the interaction with other vortex lines of by collisions (vortex reconnections). The time sequence shown in Fig. 9.2 shows a vortex ring which hits a straight vortex: notice how after the collision both the originally straight vortex and the vortex ring are perturbed by Kelvin waves.

Circular vortex rings can also be perturbed by Kelvin waves, see Fig. 9.15 (right). The vortex ring with waves travels slower than the unperturbed circular ring.

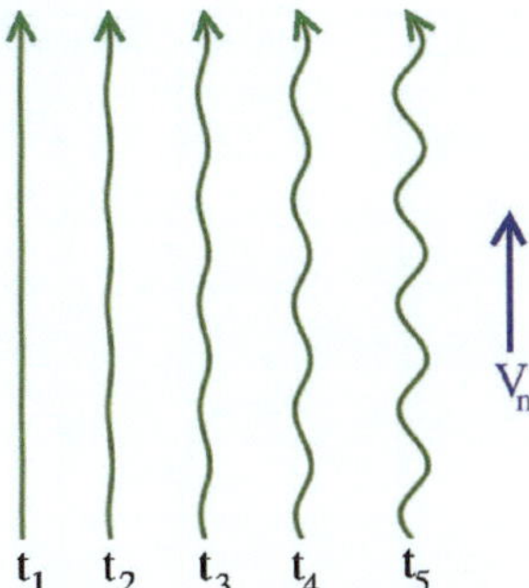

Fig. 9.16 Donnelly-Glaberson instability. Schematic time sequence in which a vortex line (green, the arrow denoting the direction of the vorticity) becomes unstable in the presence of the normal fluid velocity V_n (blue) in the same direction of the vorticity: infinitesimal perturbations of wavenumber k grow exponentially on the line

9.7 Donnelly-Glaberson Instability

In the presence of a stationary normal fluid at non-zero temperatures, Kelvin waves are damped by the friction and decay with time. However, if the normal fluid flows in the direction parallel to the vortex line with speed V_n, infinitesimal perturbations in the form of Kelvin waves of wavenumber k grow exponentially on the line with amplitude $A(t) = A(0)e^{\sigma t}$ (see Fig. 9.16) with growth rate,

$$\sigma = \alpha(k V_n - \beta k^2), \tag{9.4}$$

where t is time, $\beta = \kappa \ln(1/(ka_0))$, κ is the quantum of circulation a_0 is the vortex core radius, α is the friction coefficient, and $A(0)$ is the initial amplitude.

The wavenumber k_0 of the fastest growing Kelvin waves is found by setting $d\sigma/dk = 0$. We find $k_0 = V_n/(2\beta)$, hence the growth rate of the fastest growing wave is $\sigma_0 = \sigma(k_0) = \alpha V_n^2/(4\beta)$.

The Donnelly-Glaberson instability was discovered by Donnelly's group [37] by applying a heat flux in the direction of rotation of a vortex lattice, and then explained by Glaberson and collaborators [38]. Numerical simulations of Donnelly's experiment [39] confirmed Glaberson's description.

9.8 Breathers and Hasimoto Solitons

The Kelvin wave dispersion relation (9.3) and the Donnelly-Glaberson growth rate (9.4) are properties of *linear* Kelvin waves (Kelvin waves whose amplitude is smaller than the wavelength). Examples of *nonlinear effects* of Kelvin waves are the slowing down of a perturbed vortex ring shown in the right panel of Fig. 9.15 and the Kelvin wave cascade which will be discussed in Sect. 10.6. It is worth noticing that other nonlinear deformations of a quantum vortex are possible: *Hasimoto solitons* and *breathers*.

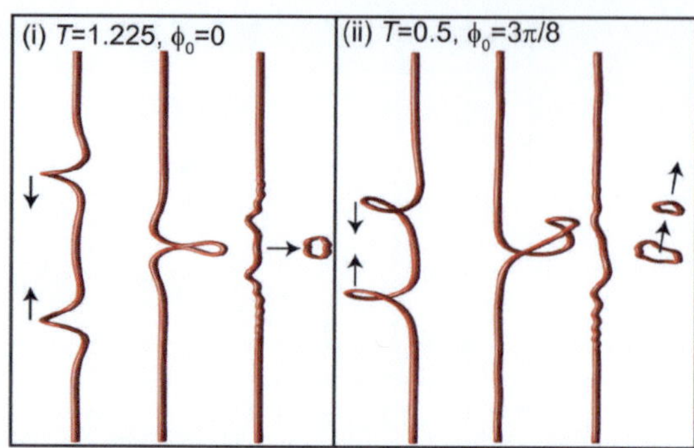

Fig. 9.17 Collision of two Hasimoto solitons propagating in opposite direction along the same vortex filament. Under the LIA (hence under Eq. (9.5) the two solitons pass through each other. Under the Biot-Savart law (implemented by a reconnection algorithm according to the VFM) one (i) or two vortex rings (ii) may be are created in the collision. Reprinted under CC-BY-4.0 license from [43]. Copyright 2020, The Author(s)

A Hasimoto soliton [40] is a localized Kelvin wavepacket on an infinite vortex filament (which we assume being aligned along the z direction) in an incompressible inviscid fluid. For some parameters, this nonlinear wave takes the shape of a twisted loop (the x and y coordinates of the filament being multi-valued functions of z); this shape may recall the strange kink observed on the funnel of a tornado [41]. Hasimoto solitons have been experimentally observed on concentrated vortices [42] in a rotating tank of turbulent water.

It is important to keep in mind that the Hasimoto soliton is only a solution of the Euler equation under the Localized Induction Approximation (LIA), Eq. (8.36), rather than under the full Biot-Savart law. Hasimoto showed that, under the LIA, the shape of the vortex filament can be described by the complex variable $\psi_0(\zeta) = \kappa_0 e^{i \int_0^\zeta \tau_0(\zeta')d\zeta'}$ where ζ is the arc length measured along the filament, κ_0 is the curvature and τ_0 is the torsion (defined in Sect. 8.3). The function ψ_0 (which should not to be confused with the wavefunction) evolves with time according to the one-dimensional *Nonlinear Schrödinger Equation* (NLSE),

$$i\frac{\partial \psi_0}{\partial t} = -\frac{\partial^2 \psi_0}{\partial \zeta^2} - \frac{1}{2}|\psi_0|^2 \psi_0. \tag{9.5}$$

This equation is of the *self-focussing type* (negative nonlinearity) and so is mathematically the same as the attractive, homogeneous 1D GPE we met in Chap. 5 and which supports bright soliton solutions. The Hasimoto soliton is a localized deformation of the vortex filament which propagates at constant speed $c = 2\tau_0$ without changing shape. The soliton curvature is $\kappa_0 = 2\nu_0 \text{sech}(\nu_0(\zeta - ct))$ where ν_0 is a parameter; at infinity, far from the localized perturbance, the vortex filament remains straight ($\kappa_0 \to 0$ at $\zeta \to \infty$).

The natural question is how robust the Hasimoto soliton is, if its initial shape is evolved under the correct Biot-Savart law rather than the LIA. To answer this question, a good test is the collision of two Hasimoto solitons. If the evolution is

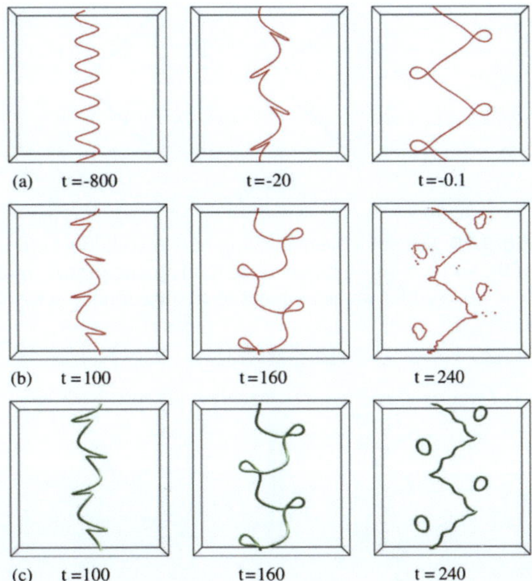

Fig. 9.18 An Akhmediev breather at different times (from left to right) along a quantum vortex computed using the LIA (a, top row), the Biot-Savart law plus the reconnection Ansatz (b, middle row) and the GPE (c, bottom row). Note that in (**b**) and (**c**) the twisted loops do reconnect, emitting vortex rings. Reproduced with permission from Ref. [27]. Copyrights (2013) of the American Physical Society

computed using Eq. (9.5), the two solitons pass through each other (the only effect of the collision being a phase shift). What happens under Biot-Savart evolution was determined by Green and collaborators [43], as shown in Fig. 9.17. Using the VFM (the Biot-Savart law plus the reconnection Ansatz), they found that depending on the parameters two or more vortex rings may be generated by the collision [43]. Green et al. also found that a Hasimoto loop soliton is easily created by the oblique collision of a vortex ring and a vortex line, an event which is probably common in turbulence.

We met *breathers* in Chap. 5—recall that they are localised, periodic solutions supported in nonlinear systems. Salman [27] considered breathers of quantum vortices and showed what happens if a particular breather called the Akhmediev breather, that has a periodic structure along its arclength, evolves according to the LIA, to the correct Biot-Savart law plus reconnection Ansatz), and to the GPE. The three cases are presented in the top, middle and bottom row of Fig. 9.18 respectively. Note that in both the VFM and GPE models the breather emits vortex ring, an event which is relevant to the decay of turbulence at very low temperatures according to Svistunov [28], see Sect. 10.6 on the Kelvin wave cascade. Note that the Akhmediev breather occurs on a continuous background and is, here the displacement of the vortex line, unlike the Satsuma-Yajima breathers we met in Chap. 5 which exist on a zero background.

9.9 Vortex Reconnections

Vortex reconnections, shown schematically in Fig. 9.19, were first conjectured by Feynman [46] and Schwarz [47]. They have been experimentally observed in superfluid helium [30] and in atomic condensates [31] only after Koplik and Levine [48] found reconnecting solutions of the GPE numerically.

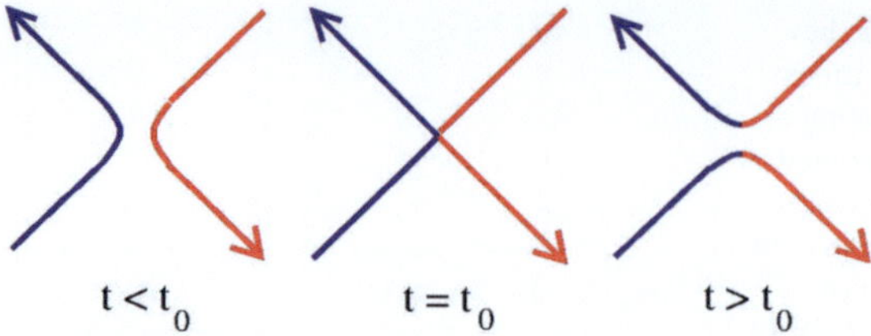

Fig. 9.19 Schematic reconnection of two colliding vortex lines. The arrows indicate the direction of the vorticity (the rotation of the fluid around the axis of the vortex). Left: before the reconnection ($t < t_0$); Middle: at the moment of reconnection, $t = t_0$; Right: after the reconnection ($t < t_0$)

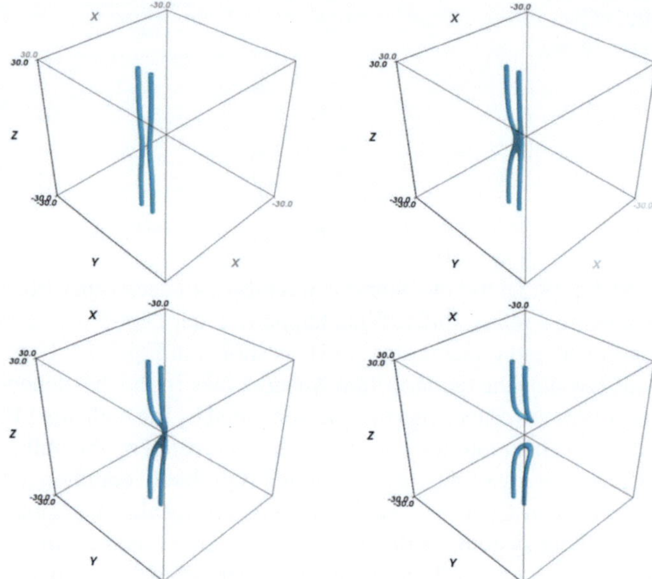

Fig. 9.20 Reconnection of antiparallel vortex lines computed by solving the GPE in a periodic box. Shown is the isosurface of the condensate density $\rho = 0.2$, where $\rho = 1.0$ is the bulk value. Reprinted from [45] with the permission of AIP Publishing

It is interesting to compare vortex reconnections of the GPE with reconnections in classical inviscid fluids (governed by the Euler equation) and in classical viscous fluid (governed by the Navier-Stokes equation). In Euler fluids, reconnections cannot happen, they are prevented by the theorem of conservation of helicity [32]; in Navier-Stokes fluid, reconnections do happen, but involve dissipation of kinetic energy. In contrast, GPE reconnections are smooth solutions that evolve at constant total energy; the transformation of some of the kinetic energy into sound energy introduces irreversibility into the dynamics. Finally, in the VFM, reconnections are performed algorithmically at the smallest arbitrary discretisation distance which is allowed along the vortex filaments.

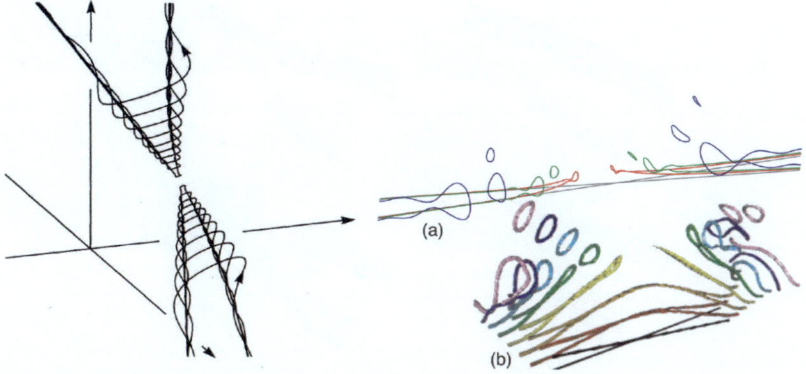

Fig. 9.21 Left: The relaxation of the vortex cusp after a vortex reconnection computed using the VFM [47]. Reprinted with permission from Ref. [47]. Copyright (1985) of the American Physical Society. Right: if the cusp angle is small (here it is 5 degrees) the relaxation triggers a cascade of vortex rings according to both the VFM (**a**) and the GPE (**b**). Reprinted with permission from Ref. [49]. Copyright (2011) of the American Physical Society

Figure 9.20 shows the reconnection of two antiparallel vortices according to the GPE. A vortex-antivortex pair, initially slightly bent,[1] propagates to the right. As the curvature of the vortices at their midpoints quickly increases, they move faster, hit each other, reconnect, and then move away. At the point of reconnection a short (few healing lengths wide) but intense rarefaction density wave (visible in Fig. 9.24) is created and spreads out, turning kinetic energy into sound. The relaxation of the reconnecting cusps has been studied in detail using both the VFM and the GPE, see Fig. 9.21. It is found that if the two vortices collide at a sufficiently small angle, the reconnection triggers a cascade of small vortex rings [49,50].

The minimum distance between two reconnecting vortex lines, $\delta(t)$, obeys a universal scaling law which has been observed in experiments [30] and simulations with both the GPE and the VFM [51],

$$\delta(t) = A^{\pm}\sqrt{\kappa|t - t_0|}, \tag{9.6}$$

where the dimensionless constants A^- and A^+ refer respectively to the pre- ($t < t_0$) and post-reconnection dynamics. The property that $A^+ > A^-$ (i.e. the vortices separate faster than they approach) is related to irreversibility, either due to acoustic emission at $T = 0$ [52], or due to friction effects with the viscous normal fluid at $T \neq 0$ as found in a joint numerical-experimental study [53].

Using dimensional analysis, the authors of Ref. [51] found that, besides the "interaction" scaling ($\delta(t) \sim |t - t_0|^{1/2}$), there is also a second scaling law ($\delta(t) \sim |t - t_0|$) if the vortex motion is "driven" by density gradients (near the edge of a condensate)

---

[1] The two vortices are slightly bent in the middle to make sure that the reconnection takes place in the centre of the computational box.

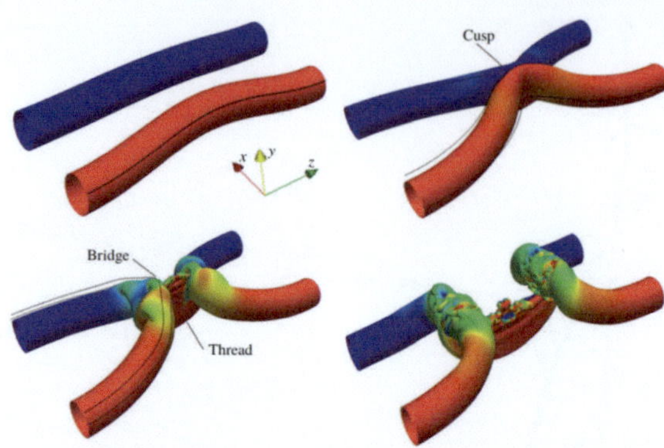

Fig. 9.22 Reconnection of two antiparallel viscous vortex tubes computed by solving the Navier-Stokes equation at Reynolds number Re $= 9000$ [56]: time evolution of vorticity isosurfaces at 40% of maximum initial vorticity with countours of axial vorticity. Under self-induction, the initially curved vortex tubes of opposite polarity approach, forming two sharp cusps, which then move away from the plane of the colliding vortices; simultaneously a bridge and threads form, which undergo further reconnections. Reprinted with permission from [56]. Copyright 2019, Cambridge University Press. All rights reserved

Fig. 9.23 Reconnection of two bundles of quantum vortices of seven vortex lines each, computed by solving the GPE [57]: snapshots of the vortex lines at times **a** $t = 0$ s, **b** $t = 7.13$ s, **c** $t = 23.58$ s, **d** $t = 36.27$ s, **e** $t = 61.49$ s, and **f** $t = 80.35$ s. Notice the bridge which forms between the reconnecting vortex bundles. Reprinted with permission from Ref. [57]. Copyright (2008) of the American Physical Society

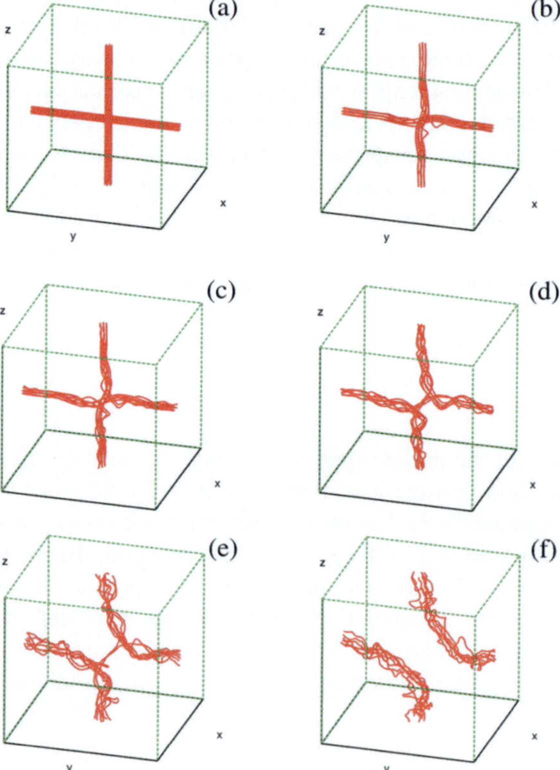

or self-induction (a small vortex ring colliding with a vortex line). In any case, once the distance between the reconnecting vortices is of the order of the vortex core size or less, the nonlinearity of the GPE becomes depleted, the equation becomes linear, and the minimum distance scales as $\delta(t) \sim \sqrt{|t - t_0|}$ as predicted by expanding the reconnecting solution around $\psi = 0$ [55].

How do these result compare to classical fluid dynamics ? Surprisingly, the same interaction scaling $\delta(t) \sim |t - t_0|^{1/2}$ was also found by Yao and Hussain [54] for reconnecting vortices of the Navier-Stokes equation, provided the vortices are slender enough (i.e. the core size is much smaller than the radius of curvature). When studying the reconnection of a vortex ring and a vortex tube, the authors also observed the same crossover from driven ($\delta(t) \sim |t - t_0|$) to interaction ($\delta(t) \sim |t - t_0|^{1/2}$) scaling which has been reported in Ref. [51].

Finally, despite obeying the same scaling, Navier-Stokes reconnections and quantum reconnections differ in an important respect: in ordinary viscous fluids the vorticity is a continuous field, hence there are effectively infinite vortex lines, hence the reconnection between antiparallel vortices involves the formation of smaller vorticity filaments (called bridges and threads) which undergo further reconnections, as shown in Fig. 9.22. However, if we consider not the reconnection of two individual quantum vortices but the reconnection of two bundles of quantum vortices, the formation of something like a bridge is possible, see Fig. 9.23.

In 2D, vortex reconnections become *annihilation* events in which two vortex points of opposite polarity destroy each other. This can occur through the interaction with a third vortex, leaving behind a soliton-like rarefaction pulse of sound called a *Jones-Roberts soliton* [34]. Recently, it has been argued that a fourth vortex is required to turn the rarefaction pulse into sound waves which then spread to infinity [35, 36], making the annihilation a four-vortex process.

9.10 Sound Emission

Even in the absence of thermal effects, moving vortices can lose energy, and they do so by creating sound waves. This occurs when vortex elements accelerate, for example, Fig. 9.24(left) shows the pattern of spiral sound waves emitted outwards by a co-rotating pair of vortices. Sound emission also arises during vortex reconnections, which release a sharp pulse of sound, as seen in Fig. 9.24 (right). In 2D annihilation events leave behind linear sound waves and nonlinear density waves (solitons).

In all of these scenarios, the pattern of the condensate's phase changes. The information about this change can travel outwards from the vortices no faster than the speed of sound. Beyond this "information horizon", the condensate phase has the old pattern. The sound waves act to smooth between the new and old patterns, and prevent discontinuities in the phase at this horizon.

The time evolution of a condensate described by the GPE (that is, a condensate at very small temperatures) conserves the total energy, although the relative proportion of kinetic energy (due to vortices) and sound energy (due to waves) may change. In general, a collection of freely-evolving vortices will decay into sound waves, with the

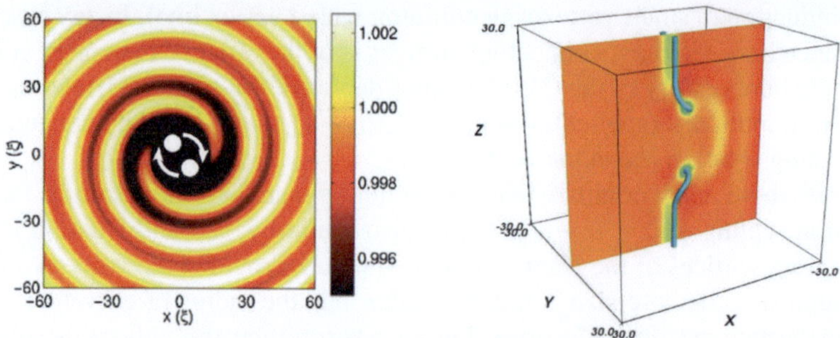

Fig. 9.24 Left: Pattern of sound waves (density variations) on the xy plane generated by a rotating pair of vortices (shown by the white dots). Reprinted with permission from [68]. Copyright 2004, The Author(s). All rights reserved. Note the small amplitude of the sound waves, relative to the background density of one. Right: Rarefaction sound pulse generated by the vortex reconnection of Fig. 9.20, shown as density variations on the central plane. Reprinted from [45] with the permission of AIP Publishing

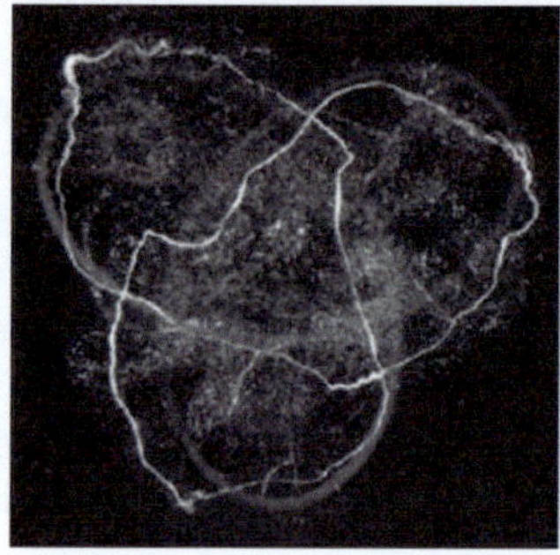

Fig. 9.25 Trefoil vortex knot generated in water by Irvine and collaborators [73]. Visualization via small hydrogen bubbles. Reprinted with permission from [73]. Copyright 2013, Springer Nature Limited. All rights reserved

energy being transferred into the sound field, although this decay is typically slow. The decay can be prohibited, or even reversed, by suitable driving of the system, and under certain conditions, intense sound waves can create vortices [58].

9.11 Vortex Knots

Interest in *vortex knots* dates back to the late XIX century when Lord Kelvin and J.J. Thomson proposed that atoms are stable knotted vortex structures in the aether (a hypothetical perfect fluid that was thought to permeate space). Their aim was to classify knots according to their complexity to explain the periodic table of the elements. The experiment of Michelson-Morley, which proved that the aether does not exist, and the discovery of the electron by J.J. Thomson himself put an end to this *vortex theory of matter*, which still left an important legacy—the mathematical theory of knots—pioneered by Peter Guthrie Tait, Lord Kelvin's collaborator.

Recently, Irwine et al. [73] succeeded in generating vortex knots in water in a controlled way—see Fig. 9.25. This experimental achievement has stimulated renewed interests in vortex knots, particularly in superfluid vortex knots, because (unlike vor-

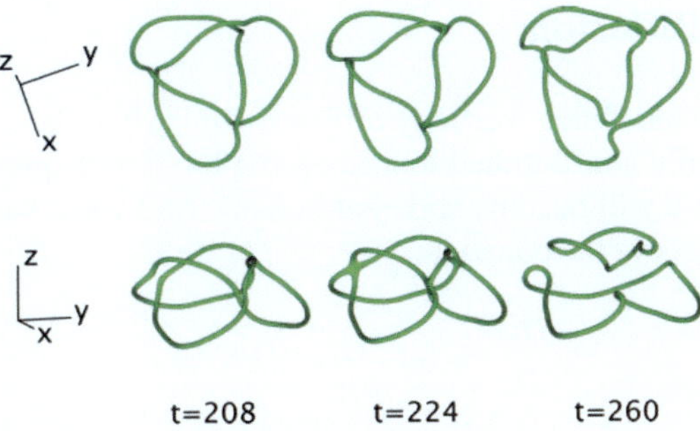

Fig. 9.26 The break-up of a $T_{2,3}$ vortex knot into two vortex rings computed using the GPE [65]. Reprinted figure with permission. Copyright (2012) by the American Physical Society

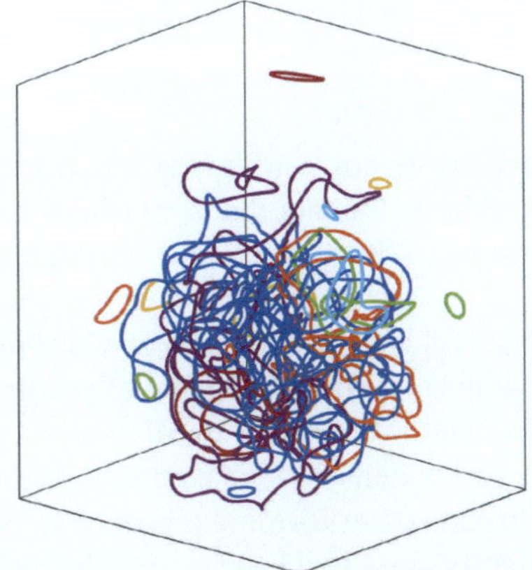

Fig. 9.27 Snapshot of vortex tangle sustained in a statistical steady state by continual forcing [67] computed using the VFM in an infinite domain (the cube is for the sake of visualization only). The tangle thus consists of closed vortex loops. Each loop is characterized by a distinct colour to illustrate the very complex topology which results, with links and some extremely high degree Alexander polynomials. Reprinted under CC-BY-4.0 license from [67]. Copyright 2019, The Author(s)

tex knots in ordinary viscous fluids) they do not decay due to vorticity diffusion. The GPE and the VFM have proven useful computational tools to study the the time evolution of vortex knots; an important question of topological fluid dynamics is in fact a knot's dynamical stability [64]. An example of vortex knot instability is shown in Fig. 9.26 which shows a trefoil vortex knot which splits in two distorted vortex rings. Numerical simulations performed using the GPE suggest that complex knots may decay along special topological pathways, which may be independent of the nature of the physical system [66]. Simulations of turbulent vortex configurations of quantum vortices have confirmed the intuition that turbulent flows contain a complicated topology consisting of vortex knots and links. In particular, it has been found that a vortex tangle consists of a wide spectrum of knots ranging from vortex rings (unknots) to trefoil vortices to vortex knots of extremely high order, as determined by computing their Alexander polynomials [67]. Figure 9.27 shows an example.

9.12   Vortex Diffusion

In an ordinary viscous fluids, vorticity diffuses, similarly to heat. This means that, if vorticity is initially concentrated in a small region, it will spread out spatially in time. This similarity will become apparent in Sect. 10.2 when we shall see that the classical vorticity obeys the equation,

$$\frac{D\boldsymbol{\omega}}{Dt} = (\boldsymbol{\omega} \cdot \nabla)\boldsymbol{v} + \nu\nabla^2\boldsymbol{\omega}, \tag{9.7}$$

where $D/Dt = \partial/\partial t + \boldsymbol{v} \cdot \nabla$ is the material derivative, $\boldsymbol{v}$ is the velocity and ν is the kinematic viscosity (viscosity divided by density). If the flow is two-dimensional, then the vortex stretching term $(\boldsymbol{\omega} \cdot \nabla)\boldsymbol{v}$ vanishes, leaving indeed the heat equation,

$$\frac{D\boldsymbol{\omega}}{Dt} = \nu\nabla^2\boldsymbol{\omega}. \tag{9.8}$$

In quantum fluids, vorticity is not continuous but consists of discrete, isolated vortex lines. Surprisingly, despite the fact that superfluids have zero viscosity, some form of vorticity diffusion can take place. Here we highlight two aspects of this effect.

The first aspect can be called *vortex evaporation*. Consider a random vortex configuration of N_v vortices which at time $t = 0$ is localised in a limited region of area A_0 away from boundaries. The average vortex distance is thus $\ell \approx 1/\sqrt{n_v}$ where $n_v = N_v/A_0$ is the initial vortex density. Occasionally, two vortices of opposite circulation will come close to each other forming a vortex-antivortex pair of separation $d < \ell$. If the pair is sufficiently close to the surface of the region containing vorticity, and if it is oriented outward, it will have a fair probability of escaping the region, moving away ("evaporating") to infinity at speed $\kappa/(2\pi d)$. The loss of these two vortices will increase ℓ, hence the probability of another pair to escape will become larger. In time, the region will become surrounded by a cloud of pairs which evaporate away [59], as shown in Fig. 9.28. Such vortex evaporation takes place also in three-dimensions, as confirmed by experiments with liquid helium [60]. Figure 9.29 shows the evaporation of small vortex loops from a 3D region of vorticity [61].

The second aspect is that, even we neglect the vortices which evaporate away, the region contains vorticity slowly becomes larger with time, spreading out due to vortex interactions. This spreading is quantified by measuring the root mean square deviation of vortices from their initial positions, defined as,

$$d_{rms}(t) = \sqrt{\frac{1}{N(t)} \sum_{i=1}^{i=N(t)} [\Delta x_i^2(t) + \Delta y_i^2(t)]}, \tag{9.9}$$

where $\Delta x_i(t) = x_i(t) - x_i(0)$, $\Delta y_i(t) = y_i(t) - y_i(0)$ and $N(t)$ is the number of vortices at time t left in the region (i.e. evaporating pairs or loops are not included in the count). It is found that the rms deviation is initially ballistic, $d_{rms} \sim t$, followed

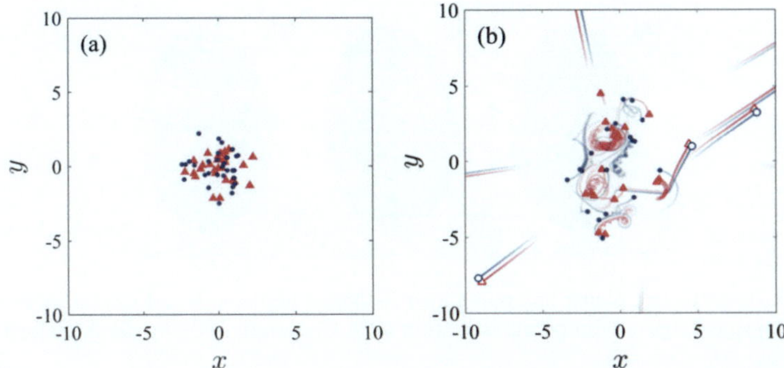

Fig. 9.28 Vortex evaporation in 2D. A random vortex configuration is initially localised at the centre of the infinite xy plane. Positive and negative vortices are denoted by red triangles and blue circles, respectively. Hollow triangles and circles denote the vortices which are parts of a vortex pair. With time, vortex-antivortex pairs which are near the surface of this region escape the region, and move to infinity. The fading comet tails visualize the trajectories of these evaporating pairs. Calculation performed using the PVM [59]. Reproduced with permission from Ref. [59]. Copyright (2018) of the American Physical Society

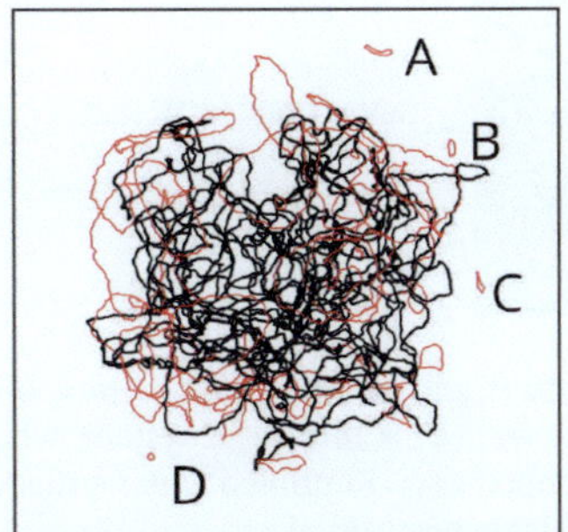

Fig. 9.29 Vortex evaporation in 3D. In this image, calculated using the VFM, four small evaporating vortex loops are visible at A, B, C, and D. From Ref. [61]. (The black lines represent highly knotted vortex loops). Reproduced with permission from Ref. [61]. Copyright (2019) of the American Physical Society

by $d_{rms} \sim t^{1/2}$ at larger t, as each vortex trajectory is altered by continual interactions with other vortices, like particles colliding in a random walk. It is known that the diffusion constant ν of a scaler field $F(\boldsymbol{x}, t)$ which satisfies the diffusion (or heat) equation $\partial F/\partial t = \nu \nabla^2 F$ is related to the rms deviation $d_{rms}(t)$ by,

$$\nu = \frac{d_{rms}^2(t)}{4t}. \tag{9.10}$$

In this way an *effective viscosity* coefficient ν' can be numerically determined for a quantum fluid which is nominally without viscosity. It is convenient to express this effective viscosity in terms of the quantum of circulation, κ. Using the VFM, Rickinson et al. found that $\nu'/\kappa \approx 0.5$ in 3D. In 2D, using both the VPM and the GPE,

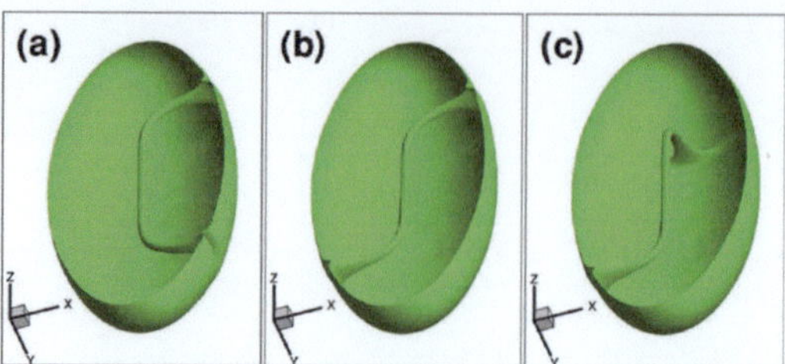

Fig. 9.30 U-vortices (**a**), planer and non-planer S-shaped vortices (**b** and **c**) in a spheroidal condensate. Reprinted figure with permission from [62]. Copyright 2003 by the American Physical Society

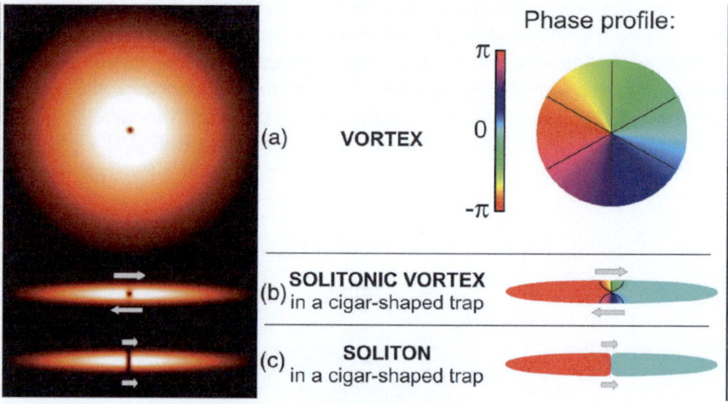

Fig. 9.31 Density (left) and phase (right) profiles of **a** a vortex, **b** a solitonic vortex, and **c** a dark soliton stripe. Note that phase defect (**a**) is in a large system, whereas phase defects (**b**) and (**c**) are confined in a narrow cigar-shaped trap. Reprinted with permission from [70]. Copyright 2014, American Physical Society. All rights reserved

they found $0.3 < \nu'/\kappa < 0.5$ (the smaller values were found in confined geometries such as disks and squares where image vortices may play a role).

9.13 Solitonic Vortices and Other Vortex States

Besides lattices, Kelvin waves, vortex rings and vortex knots, other complex vortex states have been studied. Examples are *U-shaped* and *S-shaped vortices* [62], which have also been identified experimentally [63]. U-shaped vortices form spontaneously near the edges of a shaken atomic condensate as in Fig. 9.30.

Another example are the *solitonic vortices* [69,70]. In Chap. 5 we introduced the 1D analog of a vortex - the dark soliton. The dark soliton is a localised region of low density and with a linear phase step, and exists in quasi-one-dimensional repulsive condensates which are too narrow to support vortices. Solitonic vortices are phase defects which sit at the boundary of 1D and higher-dimensions, and whose nature bridges both dark solitons and vortices. They have been created and observed experimentally as the decay product of phase defects of the order parameter ψ after a rapid thermal quench across the BEC transition in a highly-elongated harmonic trap [69,70].

The crossover between vortices, solitonic vortices and dark solitons is illustrated in Fig. 9.31 for condensates with different aspect ratios of harmonic confinement:

- *Vortex regime:* In a wide system (panel (a)) the phase around an isolated vortex line increases uniformly from $-\pi$ to π; if positioned off-centre, the vortex line will precess about the origin, following the path of constant potential as discussed in Sect. 9.1.
- *Solitonic vortex:* For a geometry which is squeezed but not strictly 1D (panel (b)), a vortex line can exist but it becomes heavily modified by the constrained boundaries; the phase increases non-uniformly and is essentially constant in front and behind the phase defect but changes rapidly near the boundaries. This is a solitonic vortex. The motion of the vortex is constrained to follow highly elongated, almost linear, trajectories through the condensate.
- *Dark soliton:* For a geometry which is squeezed further into the 1D regime, vortices cannot be supported and instead the defect is a dark soliton (panel (c)); this is a phase defect and divides the condensate into has two different values in front and behind. The motion of the dark soliton is limited to linear oscillations along the long axis of the condensate.

As the confinement is squeezed, the frequency of the oscillatory motion smoothly increases from a low-frequency precession characteristic of vortices towards the characteristic oscillation frequency of a dark soliton oscillating in a harmonic potential [72]. Brand and Reinhardt [71] compared dark soliton stripes and solitonic vortex solutions in a waveguide geometry and showed that when the waveguide width exceeds approximately 6ξ the solitonic vortex becomes energetically favoured over the dark soliton stripe.

Problems

9.1 *Vortex-antivortex pair with friction:*

Consider a 2D vortex-antivortex pair (x_1, y_1) and (x_2, y_2) where $x_1 = x_2 = X(t)$, $y_1 = Y(t)$ and $y_2 = -Y(t)$. Assume that the first vortex has positive circulation k, and the second vortex has negative circulation $-k$. The initial condition at $t = 0$ is $X = 0$. Assume that the normal fluis is at rest but provides friction parameterised by the usual friction coefficients α and α'. Shows that because of the friction the separation $2X(t)$ between the two vortices of the pair decreases with time until the two vortices annihilate at time $t = \tau$. What is the lifetime τ of the vortex pair and the distance they have travelled?

9.2 *Donnelly-Glaberson instability on a vortex line:*

Consider a straight vortex line aligned in the z-direction; its points have position $s = (0, 0, z)$. line. Assume that a uniform normal fluid flows with velocity $\boldsymbol{v}_n = (0, 0, U)$ parallel to the vortex line.

Assume that, because of thermal or mechanical noise, the vortex line has infinitesimal Kelvin wave perturbations of the form $s = (A \cos\phi, A \sin\phi, z)$ where $A \ll 1$ is the amplitude, $\phi = kz - \omega t$, k is the wavenumber, and ω is the angular frequency. Using the Schwarz equation

$$\frac{d}{dt}\boldsymbol{s} = \boldsymbol{v}_i + \alpha \boldsymbol{s}' \times (\boldsymbol{v}_n - \boldsymbol{v}_i) - \alpha' \boldsymbol{s}' \times \left(\boldsymbol{s}' \times (\boldsymbol{v}_n - \boldsymbol{v}_i)\right), \qquad (9.11)$$

where a prime denotes derivative with respect to the arclength ζ, $\boldsymbol{v}_i = \beta \boldsymbol{s}' \times \boldsymbol{s}''$ is the self-induced velocity, $\beta = \kappa/(4\pi) \ln 1/(ka_0)$, a_0 is the vortex core radius, and α, α' are temperature-dependent friction coefficients. Since the amplitude A is very small, we have $\zeta \approx z$.

1. Show that the time dependence of the amplitude of the Kelvin waves of wavenumber k is $A(t) = A(0)e^{\sigma t}$, where the growth rate is,

$$\sigma = \alpha k(U - \beta k).$$

This means that, if $U > \beta k$, the perturbation of wavenumber k is unstable and grows exponentially with time.
2. What is the most rapidly growing wavenumber ?

9.3 *Donnelly-Glaberson instability on a lattice:*

Consider a vortex lattice rotating at angular velocity $\boldsymbol{\Omega} = \Omega\widehat{\boldsymbol{z}}$ about the z-axis in the presence of an axial normal flow $\boldsymbol{v}_n = U\widehat{\boldsymbol{z}}$. In the rotating frame, the HVBK Eq. (8.46) for the superfluid velocity $\boldsymbol{v}$ (we drop the subscript s for simplicity) is,

$$\frac{\partial \boldsymbol{v}}{\partial t} + (\boldsymbol{v} \cdot \nabla)\boldsymbol{v} = -\nabla\phi - 2\boldsymbol{\Omega} \times \boldsymbol{v} - \alpha\widehat{\boldsymbol{\lambda}} \times (\boldsymbol{\lambda} \times \boldsymbol{q})$$

$$-\alpha\nu_s\boldsymbol{\lambda} \times (\boldsymbol{\lambda} \cdot \nabla)\widehat{\boldsymbol{\lambda}} + \nu_s(\boldsymbol{\lambda} \cdot \nabla)\boldsymbol{\lambda},$$

where $\boldsymbol{q} = \boldsymbol{v}_n - \boldsymbol{v}$, ν_s is the vortex tension parameter, α is the friction parameter; we assume $\nabla \cdot \boldsymbol{v} = 0$ and we have neglected the smaller friction parameter α'. Since we are in the rotating frame, the term with ϕ includes centrifugal effects besides temperature and pressure gradients, the Coriolis acceleration $-2\boldsymbol{\Omega} \times \boldsymbol{v}$ was added, and the total vorticity is $\boldsymbol{\lambda} = 2\boldsymbol{\Omega} + \boldsymbol{\omega}$ where $\boldsymbol{\omega} = \nabla \times \boldsymbol{v}$. Here $\widehat{\boldsymbol{\lambda}} = \boldsymbol{\lambda}/\lambda$ is the unit vector along the direction of $\boldsymbol{\lambda}$ and $\lambda = |\boldsymbol{\lambda}|$ is the magnitude of $\boldsymbol{\lambda}$.

Assume that ϕ, the components of $\boldsymbol{v} = (v_x, v_y, v_z)$ and $\boldsymbol{\omega} = (\omega_z, \omega_y, \omega_z)$ are all small perturbations.

1. Show that

$$\widehat{\lambda} \approx \frac{1}{2\Omega}\,(\omega_x,\,\omega_y,\,2\Omega).$$

2. Show that the linearised equations for the perturbations are

$$
\begin{pmatrix}\partial_t v_x\\ \partial_t v_y\\ \partial_t v_z\end{pmatrix}
=
\begin{pmatrix}\partial_x \phi\\ \partial_y \phi\\ \partial_z \phi\end{pmatrix}
-
\begin{pmatrix}-2\Omega v_y\\ 2\Omega v_x\\ 0\end{pmatrix}
-\alpha
\begin{pmatrix}2\Omega v_x + U(\partial_y v_z - \partial_z v_y)\\ 2\Omega v_y + U(\partial_z v_x - \partial_x v_z)\\ 0\end{pmatrix}
+
$$

$$
-\alpha\nu_s
\begin{pmatrix}-\partial_{zz}^2 v_x + \partial_{xz}^2 v_z\\ \partial_{yz}^2 v_z - \partial_{zz}^2 v_y\\ 0\end{pmatrix}
+\nu_s
\begin{pmatrix}\partial_{yz}^2 v_z - \partial_{zz}^2 v_y\\ \partial_{zz}^2 v_x - \partial_{xz}^2 v_z\\ 0\end{pmatrix},
$$

$$\partial_x v_x + \partial_y v_y + \partial_z v_z = 0.$$

3. Assuming that v_x, v_y, v_z, and ϕ depend only on z and t as $e^{ik_z z + i\omega t}$ where k_z is the wavenumber and ω is the angular velocity, show that

$$
\begin{aligned}
i\omega v_x &= 2\Omega v_y - \alpha(2\Omega v_x - ik_z U v_y) - \alpha\nu_s k_z^2 v_x + \nu_s k_z^2 v_y,\\
i\omega v_y &= -2\Omega v_x - \alpha(2\Omega v_y + ik_z U v_x) - \alpha\nu_s k_z^2 v_y - \nu_s k_z^2 v_x,\\
i\omega v_z &= -ik_z \phi,\\
ik_z v_z &= 0.
\end{aligned}
$$

4. Reduce the equations to a liner system for v_x and v_y which has non-trivial solution only if,

$$(i\omega + \alpha\eta_0)^2 + (\eta_0 + i\alpha k_z U)^2 = 0,$$

where $\eta_0 = 2\Omega + \nu_s k_z^2$.

5. Since $\alpha < 1$, neglect terms proportional to α^2 and conclude that

$$\omega = i\alpha\eta_0 \pm \sqrt{1 + 2i\alpha k_z U/\eta_0}.$$

6. Assuming $\alpha k_z U/\eta_0 \ll 1$ show that $\omega = \omega_R + i\omega_I$ where $\omega_R = \pm\eta_0$ and $\omega_I = \alpha(\eta_0 \pm k_z U)$. Perturbations grow exponentially (making the lattice unstable) if $\omega_I < 0$: show that this happens when we choose the minus sign and $U > (2\Omega + \nu_s k_z^2)/k_z$.

9.4 *Hamilton equations for a vortex ring*:

Consider a vortex ring of radius R, vortex core radius a_0 and circulation κ in a fluid of density ρ. Assuming a hollow vortex core at constant pressure, the energy, velocity and momentum of ring are respectively [74]

$$E = \frac{\rho \kappa^2 R}{2} \left(\ln (8R/a_0) - \frac{3}{2} \right), \tag{9.12}$$

$$V = \frac{\kappa}{4\pi R} \left(\ln (8R/a_0) - \frac{1}{2} \right), \tag{9.13}$$

$$P = \rho \kappa \pi R^2. \tag{9.14}$$

Verify that Hamilton equation $V = \partial E / \partial P$ is satisfied.

References

1. D.V. Freilich, D.M. Bianchi, A.M. Kaufman, T.K. Langin, D.S. Hall, Science **329**, 1182 (2010)
2. B. Jackson, J.F. McCann, C.S. Adams, Phys. Rev. A **61**, 013604 (1999)
3. T.P. Simula, P.B. Blakie, Phys. Rev. Lett. **96**, 020404 (2006)
4. H. Borner, T. Schmeling, D. Schmidt, Phys. Fluids **26**, 1410 (1983)
5. G. Rayfield, G.F. Reif, Phys. Rev. **136** A1194 (1964)
6. H. von Helmholtz, J. Reine Angew. Math. **55**, 25 (1858)
7. D.H. Wacks, A.W. Baggaley, C.F. Barenghi, Phys. Fluids **26**, 027102 (2014)
8. L. Galantucci, G. Krstulovic, C.F. Barenghi, Phys. Rev. Fluids **8**, 014702 (2023)
9. L. Galantucci, M. Sciacca, N.G. Parker, A.W. Baggaley, C.F. Barenghi, J. Fluid Mech. **912**, A9 (2021)
10. B.P. Anderson, P.C. Haljan, C.A. Regal, D.L. Feder, L.A. Collins, C.W. Clark, E.A. Cornell, Phys. Rev. Lett. **86**, 2926 (2001)
11. A.E. Leanhardt, A. Görlitz, A.P. Chikkatur, D. Kielpinski, Y. Shin, D.E. Pritchard, W. Ketterle, Phys. Rev. Lett. **89**, 190403 (2001)
12. K.W. Madison, F. Chevy, W. Wohlleben, J. Dalibard, Phys. Rev. Lett. **84**, 806 (2000)
13. E.C. Samson, K.E. Wilson, Z.L. Newman, B.P. Anderson, Phys. Rev. A **93**, 023603 (2016)
14. T. Frisch, Y. Pomeau, S. Rica, Phys. Rev. Lett. **69**, 1644 (1992)
15. G.W. Stagg, A.J. Allen, C.F. Barenghi, N.G. Parker, Bose-Einstein condensates. J. Phys. Conf. Ser. **594** 012044 (2015)
16. T.W. Neely, E.C. Samson, A.S. Bradley, M.J. Davis, B.P. Anderson, Phys. Rev. Lett. **104**, 160401 (2010)
17. G.W. Stagg, N.G. Parker, C.F. Barenghi, J. Phys. B **47**, 095304 (2014)
18. N.A. Keepfer, G.W. Stagg, L. Galantucci, C.F. Barenghi, N.G. Parker, Phys. Rev. B **102**, 144520 (2020)
19. T. Frisch, S. Nazarenko, S. Rica, Phys. Rev. Fluids **9**, 024701 (2024)
20. W.J. Kwon, G. Moon, J. Choi, S.W. Seo, Y. Shin, Phys. Rev. A **90**, 063627 (2014)
21. W.J. Kwon, S.W. Seo, Y. Shin, Phys. Rev. A **92**, 033613 (2015)
22. W. Thomson (Lord Kelvin), Phil. Mag. **10**, 155 (1880)

23. C. Pocklington, Phil. Trans. Roy. Soc. London A **186**, 603 (1895)
24. R.A. Ashton, W.I. Glaberson, Phys. Rev. Lett. **42**, 1062 (1979)
25. C.F. Barenghi, R.J. Donnelly, W.F. Vinen, Phys. Fluids **28**, 498 (1985)
26. T. Winiecki, C.S. Adams, Europhys. Lett. **52**, 257 (2000)
27. H. Salman, Phys. Rev. Lett. **16**, 165301 (2013)
28. B. Svistunov, Phys. Rev. B **52**, 3647 (1995)
29. C.F. Barenghi, R. Hänninen, M. Tsubota, Phys. Rev. E **74**, 046303 (2006)
30. G.P. Bewley, M.S. Paoletti, K.R. Sreenivasan, D.P. Lathrop, Proc. Nat. Acad. Sci. USA **105**, 13707 (2008)
31. S. Serafini, M. Barbiero, M. Debortoli, S. Donadello, F. Larcher, F. Dalfovo, G. Lamporesi, G. Ferrari, Phys. Rev. Lett. **115**, 170402 (2015)
32. H.K. Moffatt, J. Fluid Mech. **35**, 117 (1969)
33. G.W. Stagg, A.J. Allen, N.G. Parker, C.F. Barenghi, Phys. Rev. A **91**, 013612 (2015)
34. N. Meyer, H. Proud, M. Perea-Ortiz, C. O'Neale, M. Baumert, M. Holynski, J. Kronjäger, G. Barontini, K. Bongs, Phys. Rev. Lett. **119**, 150403 (2017)
35. A.J. Groszek, T.P. Simula, D.M. Paganin, K. Helmerson, Phys. Rev. A **93**, 043614 (2016)
36. A. Cidrim, F.E.A. dos Santos, L. Galantucci, V.S. Bagnato, C.F. Barenghi, Phys. Rev. A **93**, 033651 (2016)
37. D.K. Cheng, M.W. Cromar, R.J. Donnelly, Phys. Rev. Lett. **31**, 433 (1973)
38. W.I. Glaberson, W.J. Johnson, R.M. Ostermeier, Phys. Rev. Lett. **33**, 1197 (1974)
39. M. Tsubota, T. Araki, C.F. Barenghi, Phys. Rev. Lett. **90**, 205301 (2003)
40. H. Hasimoto, J. Fluid Mech. **51**, 477 (1972)
41. H. Aref, E.P. Flinchem, J. Fluid Mech. **148**, 477 (1984)
42. E.J. Hopfinger, F.K. Browand, Nature **295**, 393 (1982)
43. P.J. Green, M.J. Grant, J.W. Nevin, P.M. Walmsley, A.I. Golov, J. Low Temp. Phys. **201**, 11 (2020)
44. M.J. Ablowitz, D.J. Kaup, A.C. Newell, H. Segur, Stud. Appl. Maths. **53**, 249 (197)
45. S. Zuccher, M. Caliari, A.W. Baggaley, C.F. Barenghi, Phys. Fluids **24**, 125108 (2012)
46. R.P. Feynman, Application of quantum mechanics to liquid helium, in *Progress in Low Temperature Physics*, vol. 1, ed. by C.J. Gorter (North Holland, Amsterdam, 1955)
47. K.W. Schwarz, line-line and line-boundary interactions. Phys. Rev. B **31**, 6782 (1985)
48. J. Koplik, H. Levine, Phys. Rev. Lett. **71**, 1375 (1993)
49. M. Kursa, K. Bajer, T. Lipniacki, Phys. Rev. B **83**, 014515 (2011)
50. R.M. Kerr, Phys. Rev. Lett. **106**, 224501 (2011)
51. L. Galantucci, A.W. Baggaley, N.G. Parker, C.F. Barenghi, Proc. Nat. Acad. Sci. USA **116**, 12204 (2019)
52. A. Villois, D. Proment, G. Krstulovic, Phys. Rev. Lett. **125**, 164501 (2020)
53. P.Z. Stasiak, Y. Xing, Y. Alihosseini, C.F. Barenghi, A. Baggaley, W. Guo, L. Galantucci, and G. Krstulovic, Proc. Nat. Acad. Sci. USA **122**, No. 21 e2426064122 (2025)
54. J. Yao, F. Hussain, J. Fluid Mech. **900**, R4 (2020)
55. S. Nazarenko, R. West, J. Low Temp. Phys. **132**, 1 (2003)
56. J. Yao, F. Hussain, J. Fluid Mech. **883**, A51 (2020)
57. S.Z. Alamri, A.J. Youd, C.F. Barenghi, Phys. Rev. Lett. **101**, 215302 (2008)
58. N.G. Berloff, C.F. Barenghi, Phys. Rev. Lett. **93**, 090401 (2004)
59. E. Rickinson, N.G. Parker, A.W. Baggaley, C.F. Barenghi, Phys. Rev. A **98**, 023608 (2018)
60. C.F. Barenghi, D.C. Samuels, Phys. Rev. Lett. **89**, 155302 (2002)
61. E. Rickinson, N.G. Parker, A.W. Baggaley, C.F. Barenghi, Phys. Rev. B **99**, 224501 (2019)
62. A. Aftalion, I. Danaila, Phys. Rev. A **68**, 023603 (2003)
63. P. Rosenbusch, V. Bretin, J. Dalibard, Phys. Rev. Lett. **89**, 200403 (2002)
64. R.L. Ricca, D.C. Samuels, C.F. Barenghi, J. Fluid Mech. **391**, 29 (1999)
65. D. Proment, M. Onorato, C.F. Barenghi, Phys. Rev. E **85**, 036306 (2012)
66. D. Kleckner, L.H. Kauffman, and W.T.M. Irvine, Nature Phys. **12**, 650 (20160
67. R.G. Cooper, M. Mesgarnezhad, A.W. Baggaley, C.F. Barenghi, Sci. Reports **9**, 10545 (2019)

68. N.G. Parker, *Numerical Studies of Vortices and Dark Solitons in Atomic Bose-Einstein Condensates*, Ph.D. thesis, University of Durham (2004)
69. G. Lamporesi, S. Donadello, S. Serafini, F. Dalfovo, G. Ferrari, Nat. Phys. **9**, 656 (2013)
70. S. Donadello, S. Serafini, M. Tylutki, L.P. Pitaevskii, F. Dalfovo, G. Lamporesi, G. Ferrari, Phys. Rev. Lett. **113**, 065302 (2014)
71. J. Brand, W.P. Reinhardt, Phys. Rev. A **65**, 043612 (2002)
72. M.C. Tsatsos, M.J. Edmonds, N.G. Parker, Phys. Rev. A **94**, 023627 (2016)
73. D. Kleckner, W.T.M. Irvine, Nat. Phys. **9**, 253 (2013)
74. P.H. Roberts, R.J. Donnelly, Phys. Lett. **31A**, 137 (1970)

Quantum Turbulence

10

Abstract

This chapter introduces quantum turbulence - the turbulence of a quantum fluid. It is in this context that the dynamical effects described in the previous chapter (Kelvin waves, vortex reconnections, sound emission, etc.) come together. We start by describing turbulence in superfluid helium generated by heat transfer. This is the form of quantum turbulence which was discovered first, and is clearly quite different from ordinary turbulence. Then, after reviewing the main properties of three-dimensional (3D) turbulence in ordinary classical viscous fluids governed by the Navier-Stokes equation, we highlight the similarities and the differences between classical turbulence and quantum turbulence, focusing on the consequences of the quantisation of the circulation. In particular, we identify three different types of quantum turbulence: the Kolmogorov type, the Vinen type and turbulence in atomic Bose-Einstein condensates. Finally, we consider two-dimensional (2D) turbulence, which is very different from its 3D counterpart. In ordinary fluids this difference manifests itself as the tendency of large-scale structures to emerge from the small-scale disorder, a well-known example being Jupiter's Great Red Spot. In the context of quantised vorticity, this phenomenon is understood in terms of negative absolute temperatures, as proposed by Onsager and recently verified in experiments with atomic condensates.

10.1 Counterflow Turbulence

Quantum turbulence was discovered in the thermal counterflow of liquid helium II by W.F. (Joe) Vinen in a pioneering sequence of experiments [1]. The counterflow set-up, already introduced in Sect. 6.6 to explain heat transfer in helium, consists simply of a channel with no moving parts (unlike wind tunnels used to study ordinary turbulence). One end of the channel is closed, and the other end is open to the helium bath. At the closed end, an electrical heater dissipates a known heat flux, W, which is carried by the normal fluid towards the bath. In order to conserve mass (i.e. to

C. F. Barenghi et al., *Quantum Fluids, Solitons, and Vortices*, Lecture Notes in Physics 1050, https://doi.org/10.1007/978-3-032-20171-3_10

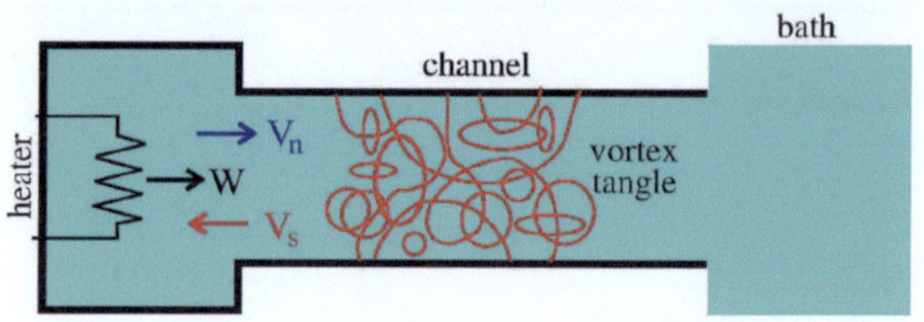

Fig. 10.1 Schematic vortex tangle in counterflow channel

enforce the condition $\rho_n V_n + \rho_s V_s = 0$, where V_n and V_s are the normal fluid and superfluid velocities averaged across the cross section of the channel), the superfluid component flows in the opposite direction, from the bath to the heater (Fig. 10.1).

In this way a relative velocity of the the two fluid components (a counterflow) $V_{ns} = V_n - V_s$ is generated which is proportional to the applied heat flux W. If this heat flux exceeds a critical value, $W > W_c$, this ideal form of heat transfer breaks down, and the channel fills with a disordered tangle of vortex lines which Vinen detected by monitoring the extra attenuation of a second sound signal sent across the channel (see Sect. 6.5). The vortex lines can also be detected by measuring the extra temperature gradient which they induce along the channel.

Vinen showed that, for $W > W_c$, the vortex line density L (defined as the length of vortex lines per unit volume) is proportional to the square of the heat flux. Since W is proportional to the average counterflow velocity V_{ns}, this means that, for $V_{ns} > V_c$ (where V_c is a critical value) the vortex line density is,

$$L = \gamma^2 V_{ns}^2, \tag{10.1}$$

where γ is a temperature dependent constant. From the measured vortex line density L, which is usually interpreted as a measure of the intensity of the turbulence, one estimates that the average distance between vortex lines is $\ell \approx L^{-1/2}$.

By modelling the growth and decay of the vortex tangle, Vinen proposed an evolution equation for the vortex line density of the form,

$$\frac{dL}{dt} = \frac{1}{2}\frac{\rho_n}{\rho} V_{ns} L^{3/2} - \frac{\chi_2}{2\pi}\kappa L^2, \tag{10.2}$$

where χ_1 and χ_2 are phenomenological parameters of order unity, B is the mutual friction coefficient defined in Eq. (8.47) and κ is the quantum of circulation.

Numerical simulations by Klaus Schwarz [2], who developed the Vortex Filament Method for this purpose—see Sect. 8.3—confirmed Vinen's finding. Lacking direct flow visualization, Schwarz's early work provided physical insight in this new form of turbulence of quantum vortex lines. In particular, Schwarz highlighted the key role played by vortex reconnections. Figure 10.2 shows a snapshot of the vortex tangle computed using the VFM generated by an uniform counterflow velocity V_{ns}. The figure clearly suggests that the superfluid flow around the vortex lines is very complicated in both space and time, as the vortex lines move, develop Kelvin waves, collide and reconnect. It is important to appreciate that an image of the instantaneous configuration of vortex lines such as Fig. 10.2 lacks information about the orientation

Fig. 10.2 Snapshot of vortex tangle in a statistically steady-state regime [3] computed using the VFM in a periodic cubic box of size D, driven by a uniform counterflow velocity V_{ns}. The parameters are: temperature $T = 1.9\,\mathrm{K}$, box size $D = 0.1\,\mathrm{cm}$, $V_{ns} = 1\,\mathrm{cm/s}$, $L \approx 2 \times 10^4\,\mathrm{cm}^{-2}$. Reproduced with permission from Ref. [3]. Copyright (2015) of the American Physical Society

of the vortex lines. Later we shall see how this orientation is important to understand the properties of the turbulent velocity field.

Further investigations of counterflow turbulence were performed by the experimental groups of van Beelen [4], Donnelly [5], Tough [6], and, more recently, by the groups of Skrbek [7] and Guo [8]. By suitable arrangements of superleaks along the channel, a *pure superflow* can also be produced in which $V_n = 0$ [9].

In summary, the experiments of Vinen opened the way to study quantum turbulence in superfluid helium under three different conditions:

- counterflow driven by a heater, in which normal fluid and superfluid components flow in opposite directions with $V_s = -\rho_n V_n / \rho_s$;
- co-flow driven mechanically e.g. by bellows, in which $V_n = V_s$;
- pure superflow, in which only the superfluid component moves along the channel.

It is important to notice that at the early stages of investigation there was no contact between the study of turbulence in superfluid helium and the study of turbulence in ordinary fluids, such as air or water. Contact was achieved only in the late 1990's by the seminal experiments of Donnelly [10] and Tabeling [12]. Before we describe these important experiments, in the next section we review the basic notions of classical turbulence theory.

10.2 Review of 3D Classical Turbulence

The motion of an ordinary viscous fluid, either laminar or turbulent, is described by two equations: the continuity equation (which expresses conservation of mass) and the Navier-Stokes equation (which expresses Newton's second law). The continuity equation is,

$$\frac{\partial \rho}{\partial t} + \nabla \cdot (\rho \boldsymbol{u}) = 0, \tag{10.3}$$

where $\boldsymbol{u}(\boldsymbol{x}, t)$ is the fluid's velocity at position $\boldsymbol{x}$ and time t, and ρ is the density. Assuming for simplicity that the density is constant, the continuity equation reduces to the incompressibility condition $\nabla \cdot \boldsymbol{u} = 0$, and the Navier-Stokes equation becomes,

$$\frac{\partial \boldsymbol{u}}{\partial t} + (\boldsymbol{u} \cdot \nabla)\boldsymbol{u} = -\frac{1}{\rho}\nabla p + \nu\nabla^2\boldsymbol{u}, \tag{10.4}$$

where $p(\boldsymbol{x}, t)$ is the pressure, $\nu = \mu/\rho$ is the kinematic viscosity, and μ is the viscosity.

Sometimes it is convenient to introduce the *material derivative* $D/Dt = \partial/\partial t + \boldsymbol{u} \cdot \nabla$, which describes the rate of change of a property of a fluid parcel as the fluid parcel moves (rather than the rate of change of a property of the fluid at a fixed location), and write Eq. (10.3) and (10.4) as,

$$\frac{D\rho}{Dt} = -\rho\nabla \cdot \boldsymbol{u}, \qquad \frac{D\boldsymbol{u}}{Dt} = -\frac{1}{\rho}\nabla p + \nu\nabla^2\boldsymbol{u}. \tag{10.5}$$

If D_0 and U_0 are the characteristic length scale and velocity scale of the flow, it is convenient to introduce the following dimensionless variables (indicated by primes): $\boldsymbol{x}' = \boldsymbol{x}/D_0$, $t' = tU_0/D_0$, $\boldsymbol{u}' = \boldsymbol{u}/U_0$ and $p' = p/(\rho U_0^2)$, and re-write the Navier-Stokes equation in dimensionless form,

$$\frac{\partial \boldsymbol{u}'}{\partial t'} + (\boldsymbol{u}' \cdot \nabla')\boldsymbol{u}' = -\nabla' p' + \frac{1}{\mathrm{Re}}\nabla'^2\boldsymbol{u}', \tag{10.6}$$

where ∇' and ∇'^2 contain derivatives with respect to $\boldsymbol{x}'$. The dimensionless Navier-Stokes equation (10.6) depends on a single (dimensionless) parameter,

$$\mathrm{Re} = \frac{U_0 D_0}{\nu}, \tag{10.7}$$

called the *Reynolds number*. Flows with different values of U_0, D_0 and ν but the same Reynolds number are therefore identical (a result which has many practical applications, for example in wind tunnels). Physically, the Reynolds number estimates the ratio of inertial and viscous forces in Eq. (10.4),

$$\frac{\mathrm{inertial}}{\mathrm{viscous}} \approx \frac{|(\boldsymbol{u} \cdot \nabla)\boldsymbol{u}|}{|\nu\nabla^2\boldsymbol{u}|} \approx \frac{U_0^2/D_0}{\nu U_0/D_0^2} = \frac{U_0 D_0}{\nu} = \mathrm{Re}. \tag{10.8}$$

This expression shows that turbulence requires large values of the Reynolds number, so that inertial forces overcome viscous forces which tend to bring the fluid to rest. Typically, flow in a pipe becomes turbulent at Reynolds number larger than ≈ 2000.

Turbulence takes many forms. The simplest is homogeneous isotropic turbulence (HIT) held in a statistical steady-state by continual forcing. Statistical steady-state

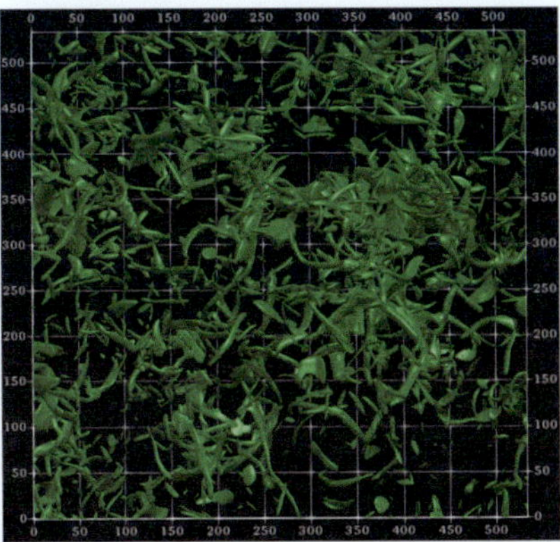

Fig. 10.3 Numerical simulation of the Navier-Stokes equation [14]. Isosurface of $|\omega|$ are plotted corresponding to the volume that is occupied by the 1% most intense vorticity. The domain that is plotted corresponds to $512\eta_K \times 512\eta_K \times 512\eta_K$. Notice the worms. Reproduced with permission from Ref. [14]. Copyright (2022) of the American Physical Society

means that although the properties of the flow fluctuate irregularly in space and time, their average values are constant. The forcing is necessary to compensate for viscous losses of energy—without forcing turbulence would decay—and is usually modelled by adding a suitable term $\boldsymbol{f}(\boldsymbol{x}, t)$ at the right hand side of Eq. (10.4). Without forcing, the kinetic energy of the fluid (per unit mass), defined as,

$$E = \frac{1}{\mathcal{V}} \int_{\mathcal{V}} \frac{|\boldsymbol{u}|^2}{2} d^3 \boldsymbol{r}, \qquad (10.9)$$

where $|\boldsymbol{u}|^2 = \boldsymbol{u} \cdot \boldsymbol{u}$ and $\mathcal{V}$ is the volume of the fluid, would decrease with time according to

$$\frac{dE}{dt} = -\frac{\nu}{\mathcal{V}} \int_{\mathcal{V}} |\boldsymbol{\omega}|^2 d^3 \boldsymbol{r}, \qquad (10.10)$$

where $\boldsymbol{\omega} = \nabla \times \boldsymbol{u}$ is the vorticity (the local rotation).

HIT is typically achieved in a wind tunnel away from boundaries. Driven by a propeller, the fluid enters the test section of the wind tunnel through a mesh, and exits at the opposite end, where it is re-circulated. Our understanding of HIT is based on the pioneering work of the meteorologist Lewis Richardson in the 1920s and the mathematician Andrey Kolmogorov in the 1940s. It is clear from experiments [13] that turbulence comprises eddies or vortices which have a wide range of sizes and strengths. Richardson realized that the largest eddies, created by instabilities of the mean flow, are themselves subject to instabilities, breaking up into smaller eddies. The smaller eddies are subject to further instabilities, and so on, thus creating a continual *cascade of energy* from the largest eddies of size D_0 to the smallest eddies of size η_K, a quantity called the *Kolmogorov dissipation length*. In the range of length scales from D_0 to η_K, called the *inertial range,* viscosity plays no role. At the scale η_K, eddies are small enough that viscous forces become as large as inertial

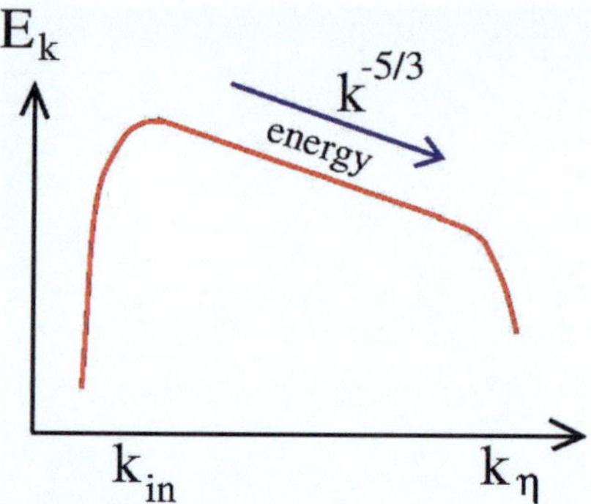

Fig. 10.4 Schematic Kolmogorov energy spectrum of 3D classical turbulence. Energy is injected at wavenumber $k_{in} = 2\pi/D_0$ and extracted at $k_\eta = 2\pi/\eta_K$ where η_K is the Kolmogorov dissipation length.   A direct cascade transfer energy to smaller length scales (larger k). The figure must be compared to the analog spectrum of 2D classical turbulence in Fig. 10.22

forces in the balance expressed by Eq. (10.8), and can dissipate the kinetic energy of eddies into heat.

Numerical simulations of the Navier-Stokes equation confirm that turbulent flows contain vortex tubes down to the Kolmogorov scale η_K, as shown in Fig. 10.3. These tubes, also called *worms*, are filamentary regions where the vorticity is concentrated. It is thought [23] that *vortex stretching* is the physical mechanism which is responsible for shifting the energy to smaller and smaller length scales; this is done by thinning vortex tubes in the direction perpendicular to their axes, reducing the radial length scales and intensifying the rotation (vorticity). To identify the mathematical origin of this mechanism we take the curl of the Navier-Stokes equation (10.4) (neglecting forcing for simplicity), finding

$$\frac{D\boldsymbol{\omega}}{Dt} = (\boldsymbol{\omega} \cdot \nabla)\,\boldsymbol{v} + \nu\nabla^2\boldsymbol{\omega}. \tag{10.11}$$

The term $(\boldsymbol{\omega} \cdot \nabla)\boldsymbol{v}$ at the right hand side of Eq. (10.11) is the source of vortex stretching.

An important property of HIT is the distribution of kinetic energy over the length scales. In this context it is convenient to Fourier-transform the velocity field and consider the distribution of energy over the wavenumbers k (strictly speaking, k is the magnitude of the 3D wavevector $\boldsymbol{k}$). We represent the kinetic energy (per unit mass) of the fluid in volume $\mathcal{V}$ as a distribution $\hat{E}(k)$ over the wavenumbers:

$$E = \frac{1}{\mathcal{V}}\int_{\mathcal{V}} \frac{1}{2}|\boldsymbol{u}|^2 d^3\boldsymbol{r} = \int \hat{E}(k)dk, \tag{10.12}$$

where $\hat{E}(k)$ is called the *energy spectrum*. At the right hand side, the integration in k extends not from zero to infinity, but, in practice, from $k = 2\pi/D_0$ (corresponding to the largest length scale available, D_0, the channel size) to $k = 2\pi/\eta_K$ (corresponding to the smallest eddy).

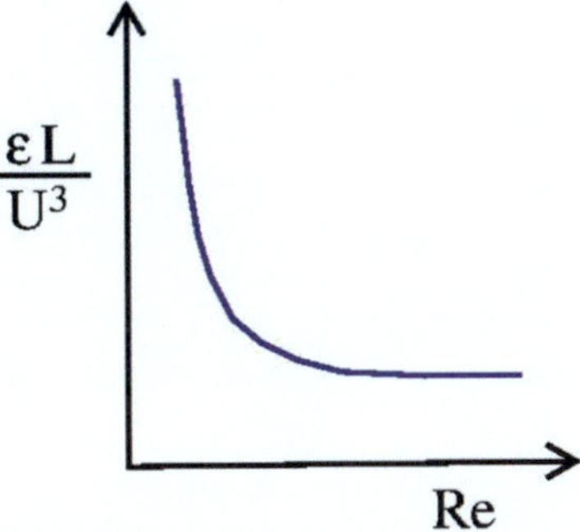

Fig. 10.5 Schematic dissipation anomaly of classical turbulence. Normalised energy dissipation rate $\epsilon L/U^3$ versus Reynolds number Re where L and U are length and velocity scales characteristic of large-scale motion. According to experiments and numerical simulations, the normalised rate of kinetic energy dissipation tends to a nonzero value for $1/\nu \to 0$ (i.e. Re $\to \infty$) [16]

We define the energy dissipation rate (per unit mass) as

$$\epsilon = -\frac{dE}{dt}. \tag{10.13}$$

Then, assuming that in the wavenumber range of the energy cascade $2\pi/D_0 \ll k \ll 2\pi/\eta_K$ the energy spectrum $\hat{E}(k)$ depends only on k and ϵ (but not on ν), a simple dimensional argument shows that the energy spectrum is

$$\hat{E}(k) = C\epsilon^{2/3}k^{-5/3}, \tag{10.14}$$

where C is a dimensionless constant of order unity, see Fig. 10.4. This scaling is called the *Kolmogorov 5/3 law* and is the most robust result of turbulence theory. It appears to be approximately valid even in situations when the turbulence is nor quite steady or homogeneous or incompressible [15].

An important property of turbulence is the fact that, in the limit of vanishing viscosity $1/\nu \to \infty$ (i.e. infinite Reynolds number), the kinetic energy dissipation rate does not tend to zero (as for laminar flows) but tends to a constant [16]. This property, shown in Fig. 10.5, is called the *dissipation anomaly* or the *zeroth law of turbulence,* and means that in this limit the velocity field creates smaller and smaller structures, becoming non-smooth, hence non-differentiable.

10.3 Quantum Turbulence of the Kolmogorov Type

The first experiment which showed a connection between turbulence of a quantum fluid and ordinary (classical) turbulence was performed by Donnelly and collaborators and consisted of towing a grid along a $1 \times 1 \times 56$ cm^3 channel containing a stationary sample of liquid helium [10,11] - see Fig. 10.6. Monitoring the attenuation of second sound allowed direct measurement of the turbulent superfluid vorticity

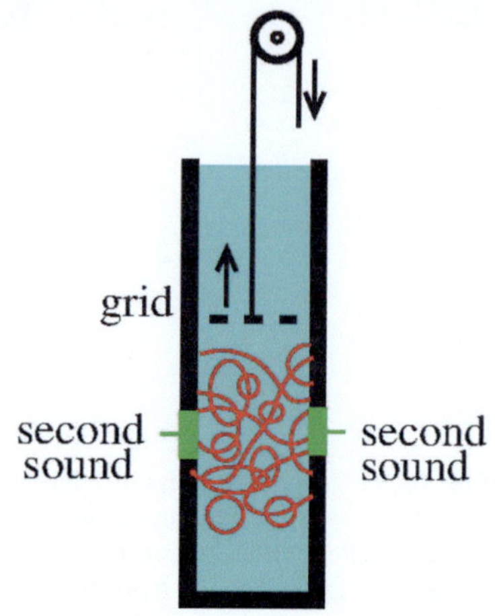

Fig. 10.6 Schematic towed grid experiment [10,11]. Turbulence is produced by suddenly moving a grid and is detected by monitoring the attenuation of second sound. The red lines represent the vortex tangle

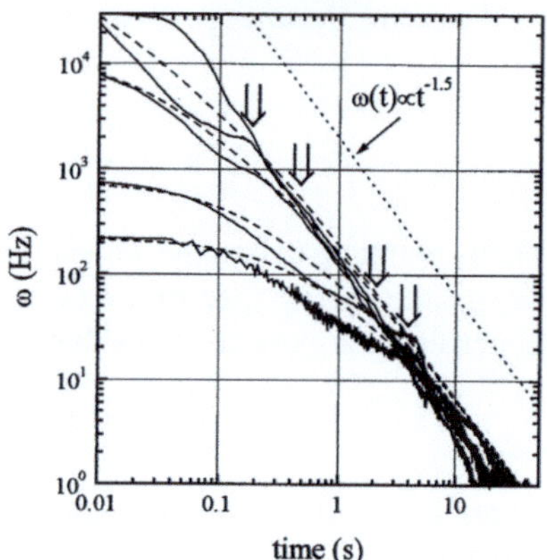

Fig. 10.7 Temporal decay of superfluid vorticity, ω vs t, generated by a towed grid [11]. The data correspond to grid velocities 5, 10, 50, 100 and 200 cm/s at $t = 15$ K. Reproduced with permission from Ref. [11]. Copyright (1999) of the American Physical Society

(defined as $\omega = \kappa L$ where L is the measured vortex line density) and its decay in time. The kinematic viscosity ν of helium II, based on the normal fluid's viscosity and the total density, is almost three orders of magnitude smaller than air, generating Reynolds numbers of order 10^5 based on the mesh size of the grid.

The result of the experiment was striking. It was found that the decay of the superfluid vorticity obeys the scaling $\omega \sim t^{-3/2}$ over almost four orders of magnitude of vorticity, see Fig. 10.7. Donnelly and collaborators argued that this behaviour marks the decay of the classical Kolmogorov spectrum. Initially, the size of the energy-containing eddies increases from the mesh size to the channel size D_0. By integrating Eq. (10.12) from $k = 2\pi/D_0$ to $k = 2\pi/\eta_K$ and then using Eqs. (10.13) and (10.10), one can show that [11], at later times, the vorticity decays as

$$\omega(t) \approx \frac{(3C)^{3/2} D_0}{2\pi\nu^{1/2}} t^{-3/2}, \qquad (10.15)$$

where ν is some efficient kinematic viscosity.

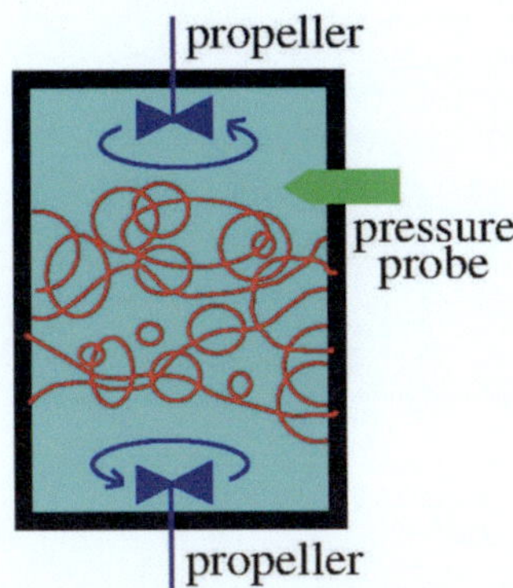

Fig. 10.8 Schematic apparatus of Maurer and Tabeling [12]. The two counter-rotating propellers induce a turbulent von Karman flow. The vortex tangle is represented by the red lines

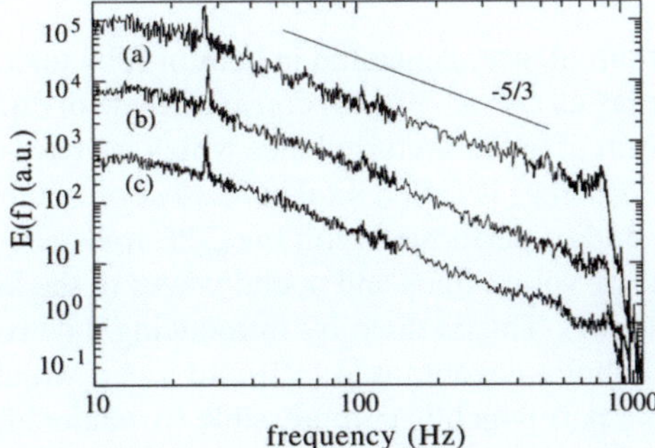

Fig. 10.9 Experiment of Maurer and Tabeling [12]. Energy spectra measured at $T = 2.30$ K (**a**); $T = 2.08$ K (**b**); and $T = 1.4$ K (**c**). The spectra have been shifted vertically for clarity. The solid line marks the Kolmogorov scaling. Note the Kolmogorov scaling in both (**a**) helium I and (**b**, **c**) helium II. Reprinted with permission from [12]. Copyright 1998, EDP Sciences. All rights reserved

The second experiment was performed by Maurer and Tabeling [12] who stirred liquid helium in a cylindrical container (diameter 8 cm, height 20 cm) using two counter-rotating propellers as shown in Fig. 10.8. The turbulent velocity was measured by a local pressure probe. They found that above the lambda-point, at $T = 2.3$ K, the energy spectrum obeys the classical Kolmogorov scaling, see Fig. 10.9a. This is expected because helium I is a classical fluid (albeit very cold and with very small viscosity). Surprisingly, they also found that the Kolmogorov scaling holds true at lower temperatures below the lambda point, see Fig. 10.9b, c, even at $T = 1.4$ K, a temperature at which the normal fluid fraction is only $\rho_n/\rho \approx 7\%$.

Better, larger scale experiments [20–22]—see Fig. 10.10—and numerical simulations performed using both the GPE [17] and the VFM [19] confirmed the presence of Kolmogorov scaling (signature of the energy cascade) in quantum turbulence. Further evidence for the Kolmogorov scenario came from experiments in which the

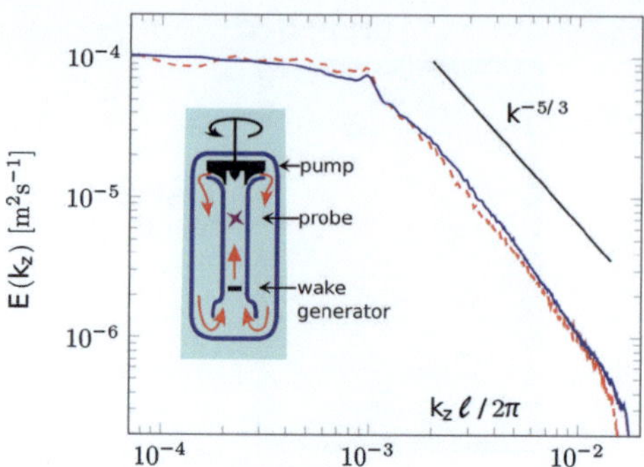

Fig. 10.10 Kolmogorov energy spectrum measured in the TOUPIE wind tunnel (inset), a superfluid wind tunnel at CNRS Grenoble [22]. The solid blue line and the dashed red line are the measurements in helium II (at $T = 1.56$ K) and in helium I respectively. The solid black line is a guide to the eye to represent the $k^{-5/3}$ Kolmogorov scaling

vortex line density of turbulence, generated in helium II by injecting charged vortex rings, was found to decay as $L \sim t^{-3/2}$ [18] consistently with Eq. (10.15) as $\omega \approx \kappa L$.

In summary, the form of quantum turbulence which we have described (typically created by large scale stirring) is called *quasi-classical* or *of the Kolmogorov type*.

Notice that in the studies performed with the GPE it is necessary to identify the separate contributions of vortex lines and sound waves to the kinetic energy of the condensate, see Eq. (3.14). This is done by introducing a density-weighted velocity field which is Helmholtz-decomposed [17] into a part which is incompressible (divergence free) and a part which is compressible (irrotational). It is the spectrum of the incompressible kinetic energy which must be compared to the Kolmogorov scaling.

The natural question is then: what is the origin of the energy cascade in the Kolmogorov type of quantum turbulence ? In classical turbulence it is thought that the answer is vortex stretching, as we have seen. However vortex stretching is not possible for a quantum vortex: its core size and circulation are constrained by quantum mechanics.

The puzzle of the origin of Kolmogorov-type quantum turbulence is solved if the turbulent flow contains bundles of vortex lines, at least statistically. In fact, if the vortex lines within a bundle become closer to each other, the bundle undergoes an intensification of the rotation (vorticity) similar to classical vortex stretching. The orientation of the vortex lines is therefore crucial: if they are aligned in the same direction the average superfluid velocity which is induced is large, and the bundle has a large kinetic energy; vice versa, if the vortex lines are oriented randomly with respect to each others, the average velocity is very small, corresponding to low energy, see Fig. 10.11.

Numerical simulations reveal that, in the Kolmogorov regime, the vortex tangle is partially polarised, and contains bundle of vortex lines amid random vortex lines. To visualize this polarization it is necessary to take into account the orientation of the vortex lines. This is done by computing the macroscopic (smoothed, coarse-grained)

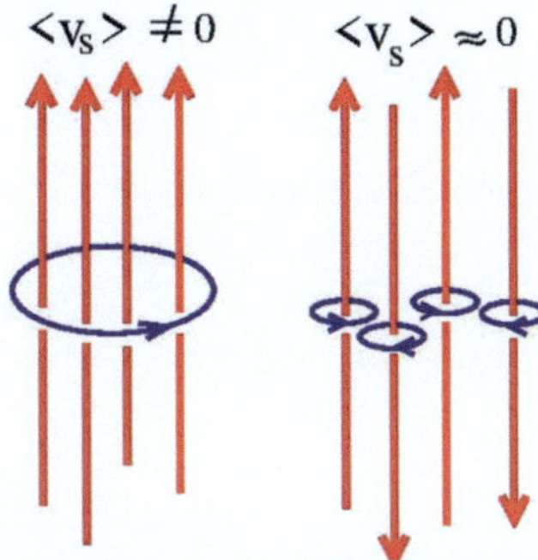

Fig. 10.11 Schematic vortex bundle. The vortex lines are in red, the arrows represent the vorticity direction. The flow around the lines is in blue. Left: large average velocity and energy corresponding to bundle of vortex lines aligned in the same direction. The velocity fields of the individual vortex lines tend to add up, creating an energetic velocity field which increases with the distance from the axis of the bundle. Right: if the vortex lines are at random orientation to each others, the velocity fields of the individual vortex line tend to cancel each other out, creating a weak velocity field with very littleenergy

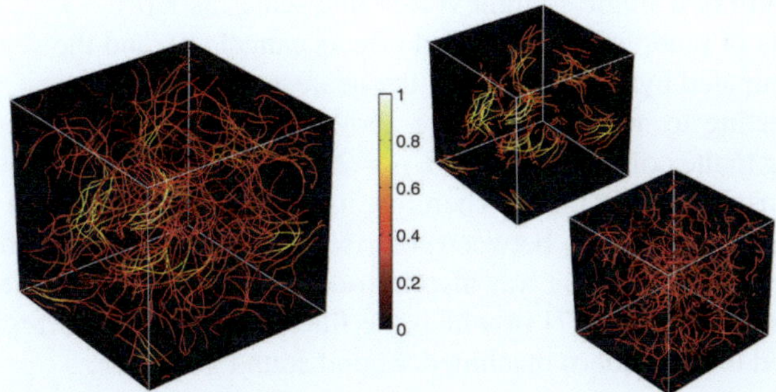

Fig. 10.12 Left: Snapshot of the vortex tangle in the statistical steady state computed in a periodic box of size $D_0 = 0.1$ cm [24]. The vortex lines are coloured according to the local smoothed vorticity ω_s (yellow corresponds to large values, red to low values). Right: the same snapshot is split into locally polarised lines (top) whose smoothed vorticity ω_s has magnitude which exceeds $1.4\omega_{rms}$, and unpolarised (random) lines (bottom) whose smoothed vorticity has magnitude less than this threshold. Reproduced with permission from Ref. [24]. Copyright (2012) of the American Physical Society

continuous vorticity field from the individual vortex lines, essentially averaging over distances larger than the average inter-vortex distance using local spline fits [24] or Gaussian kernels [25]. An example of this exercise is shown in Fig. 10.12. Here the vortex lines are locally coloured according to the smoothed vorticity field ω_s: yellow corresponds to large ω_s, red to small ω_s.

Figure 10.13 shows the energy spectra corresponding to vortex lines whose smoothed vorticity which is, respectively, (a) below the threshold value $1.4\omega_{rms}$, and (b) above this threshold. The dashed lines are guides to the eye to identify the k^{-1} scaling of the unpolarised lines and the $k^{-5/3}$ Kolmogorov scaling of the

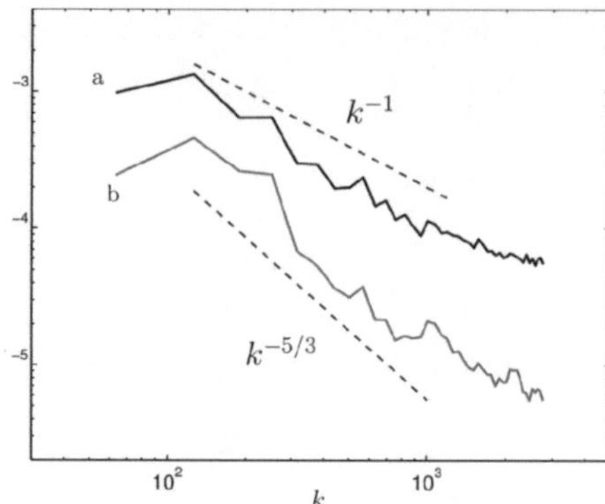

Fig. 10.13 Energy spectra [24] corresponding to vortex lines with magnitude of smoothed vorticity which is respectively (**a**) below and (**b**) above the threshold value $1.4\omega_{rms}$, i.e. to unpolarised lines (**a**) and polarised lines (**b**). Reproduced with permission from Ref. [24]. Copyright (2012) of the American Physical Society

polarised lines. The origin of the k^{-1} scaling of the unpolarised lines is simple: the energy spectrum of a single straight vortex line scales as $\hat{E}(k) \sim k^{-1}$. If the vortex tangle consists of random lines, far field effects cancel out, and the velocity field at a point is dominated by the presence of the nearest vortex line.

It is interesting to notice that in the above calculation the vortex line density corresponding to the polarised lines is $L_{\parallel} \approx 2 \times 10^3$ cm^{-2}, much less than the total vortex line density $L = 10^4$ cm^{-2}. Therefore most of the vortex lines are unpolarised. The distinction $L = L_{\parallel} + L_{\times}$ between polarised lines (of length $L_{\parallel}$) and unpolarised (random) lines (of length $L_{\times}$) was also introduced by Roche and Barenghi [26] in discussing an experiment [27] in which the fluctuations of the vortex line density were measured using a micro machined second sound generator.

10.4   Quantum Turbulence of the Vinen Type

A different type of quantum turbulence, first envisaged by Volovik [28], was identified by Walmsley and Golov [29] at low injection rate in their experiments with charged vortex rings. When unforced, this turbulence is characterized by a slower decay ($L \sim t^{-1}$) then the Kolmogorov type ($L \sim t^{-3/2}$). This turbulence is called *ultraquantum* or *quantum turbulence of the Vinen type* since the $L \sim t^{-1}$ behaviour is consistent to the decaying solution of Vinen's Eq. (10.2) (see Exercise 8.3).

Numerical simulations [30] show that the energy spectrum of Vinen turbulence is quite different from that of Kolmogorov turbulence, lacking both the typical energy concentration at large length scales and the $k^{-5/3}$ tail (which implies the presence of an energy cascade). Instead, most of the energy is broadly concentrated at wavenumbers of the order $k_{\ell} = 2\pi/\ell$, where $\ell \approx L^{-1/2}$ is the average inter vortex spacing. If enough k-range is available, at larger wavenumbers the spectrum becomes proportional to k^{-1}. The energy spectrum of Vinen turbulence thus looks like the spectrum

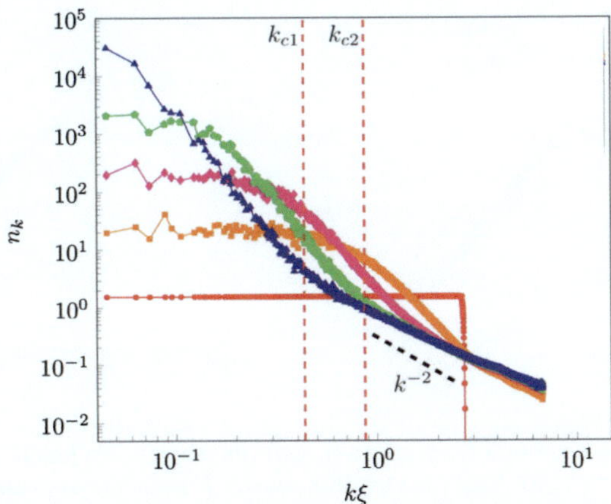

Fig. 10.14 Thermal quench of a Bose gas. Occupation numbers n_k as a function of the dimensionless wavevector $k\xi$ at different times t/τ as the Bose gas decays to equilibrium [32]: $t/\tau = 0$ (red circles); $t/\tau = 25$ (orange squares); $t/\tau = 75$ (magenta diamonds); $t/\tau = 200$ (green pentagons) and $t/\tau = 1000$ (blue triangles). Reprinted under CC-BY-3.0 license from [32]. Copyright 2016, The Author(s)

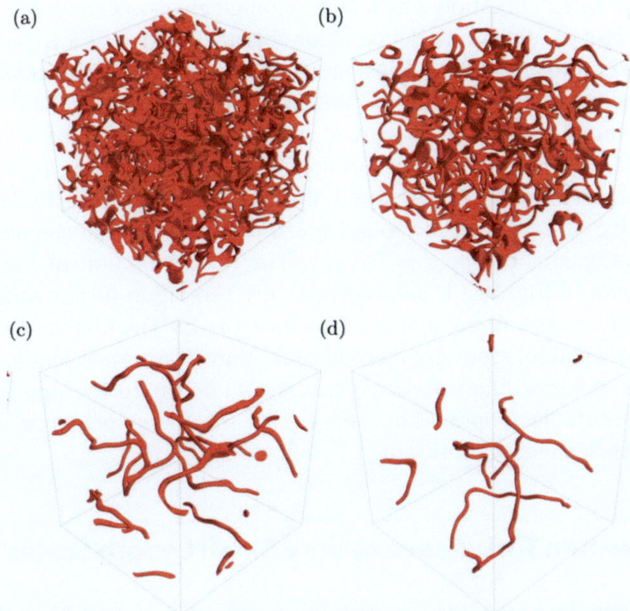

Fig. 10.15 Thermal quench of a Bose gas. Evolution of the turbulent vortex tangle as the Bose gas decays to equilibrium [32]. Shown are isosurfaces of the condensate at times $t/\tau = 0$ (**a**), 250 (**b**), 1250 (**c**) and 2500 (**d**). Reprinted under CC-BY-3.0 license from [32]. Copyright 2016, The Author(s)

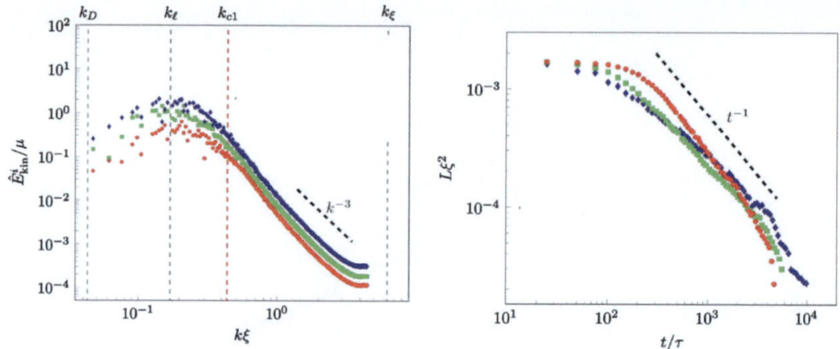

Fig. 10.16 Thermal quench of a Bose gas [32]. Left: incompressible kinetic energy spectrum of the system at time $t/\tau = 300$. Right: Vortex line density L over time for condensate fractions of 22% (red markers), 48% (green markers) and 77% (blue markers). Reprinted under CC-BY-3.0 license from [32]. Copyright 2016, The Author(s)

of a gas of random vortex rings, which is concentrated at $k_R = 2\pi/R$ where R is the average radius, and decays as k^{-1} for $k \gg k_R$. The key ingredient to create steady Vinen turbulence is the lack of forcing at large enough length scales, or, for decaying turbulence, an initial condition which lacks enough energy at small k.

The paradigm of Vinen turbulence is the thermal quench of a Bose gas, that is the formation of a coherent Bose-Einstein condensate from a highly non equilibrium initial condition. This is studied by solving the GPE for a classical field $\psi(\boldsymbol{x}, t)$ with initial condition $\psi(\boldsymbol{x}, 0) = \sum_k a_k \exp(i\boldsymbol{k} \cdot \boldsymbol{x})$ where the coefficients a_k are uniform up to a cutoff wavenumber k_c (the occupation of mode $\boldsymbol{k}$ being $n_k = |a_k|^2$) and the phases are random [31,32]. As usual, the GPE is made dimensionless using the healing length $\xi = \hbar/\sqrt{mg\rho}$, the sound speed $c = \sqrt{\rho g/m}$, the chemical potential μ and the characteristic time $\tau = \hbar/(g\rho)$. The time evolution of the occupation numbers, shown in Fig. 10.14, demonstrates the formation of the condensate: the modes at low k acquire macroscopic occupation ($n_k \gg 1$) while the modes at high k, with low occupation, represent thermal excitations. Correspondently, the initially intense tangle of vortex filaments decays, as shown in Fig. 10.15. Figure 10.16 show the spectrum of the incompressible energy (note the energy peak near k_ℓ), and the decay of the vortex length (notice the t^{-1} behaviour).

10.5   Quantum Turbulence at Very Small Length Scales

The friction with the normal fluid damps the energy of the vortex lines. Particularly affected are the vortex structures with the shortest length scales (short Kelvin waves and small vortex rings) which quickly pass their energy to the normal fluid, hence to phonons (heat). But as the temperature is reduced towards absolute zero, the smoothing effect of the friction disappears and the vortex lines become very wiggly, as found in numerical simulations by Tsubota and collaborators [33], see Fig. 10.17. The kinks and cusps left over by reconnection events last longer hence become

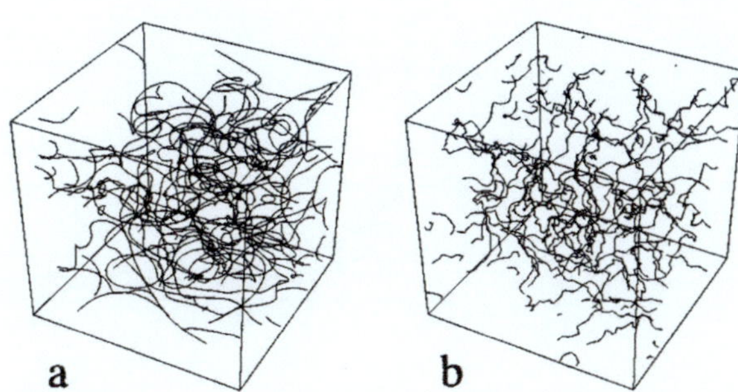

Fig. 10.17 Examples of vortex tangle at **a** $T > 0$ and **b** $T \approx 0$ [33]. Note that in (a) the vortex lines are smoother because of the friction. Reproduced with permission from Ref. [33]. Copyright (2000) of the American Physical Society

more visible. These findings suggest that quantum turbulence has new interesting properties in the low temperature limit.

Since the superfluid component has zero viscosity, the classical definition of Reynolds number given in Eq. (10.7) does not apply to helium II—with $\nu = 0$ the superfluid Reynolds number would be infinity. But we have seen that in fluid dynamics the Reynolds number can also be interpreted as the ratio of inertial and viscous forces, see Eq. (10.8). In the same spirit, by taking the ratio of inertial and viscous forces in Eq. (8.41), we can define the following *superfluid Reynolds number* [41]

$$\mathrm{Re}_s = \frac{1 - \alpha'}{\alpha}. \tag{10.16}$$

Notice that Re_s does not depend on U or D but only on the temperature T, as verified experimentally [41]. Since α and α' tend to zero for $T \to 0$, we conclude that $\mathrm{Re}_s \to \infty$ for $T \to 0$. Physically, the lower the temperature, the more freely the Kelvin waves propagate, and the larger the curvature of the vortex lines become. Recent numerical simulations [42] which carefully resolved even the shortest Kelvin waves which were excited showed that, as Re_s increases, the normalised rate of kinetic energy dissipation flattens and becomes constant, as in classical turbulence: compare Fig. 10.5 and 10.18. As in classical turbulence, as the turbulence becomes more intense (larger Reynolds number), the generation of small-scale vortex structures prevents the dissipation from vanishing, unlike what happens in laminar flows. This is another remarkable similarity between classical turbulence and quantum turbulence.

10.6 The Kelvin Wave Cascade

If the temperature is further reduced, another route to dissipate kinetic energy becomes possible, as first suggested by Svistunov [35]. He showed that the non-linear interaction of finite-amplitude Kelvin waves creates shorter and shorter waves which rotate more and more rapidly according to Eq. (9.3). This *Kelvin wave cascade* can transfer energy to wavenumbers much larger than k_ℓ (an example is shown in Fig. 10.19), until k is large enough that the waves efficiently radiate sound (phonons). The cascade can be initiated by a reconnection—an example is shown in Fig. 10.19 or more in general simply by vortex interactions.

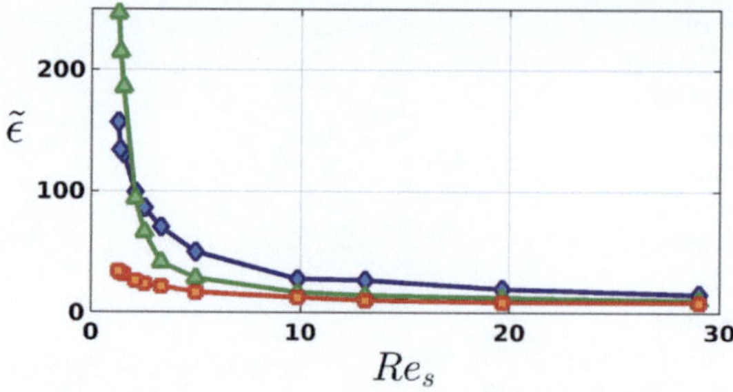

Fig. 10.18 Dissipation anomaly in a quantum fluid. Normalised energy dissipation rate versus superfluid Reynolds number; note that it tends to a constant for large Reynolds number, as in classical turbulence. Reprinted under CC-BY-4.0 license from [42]. Copyright 2023, The Author(s)

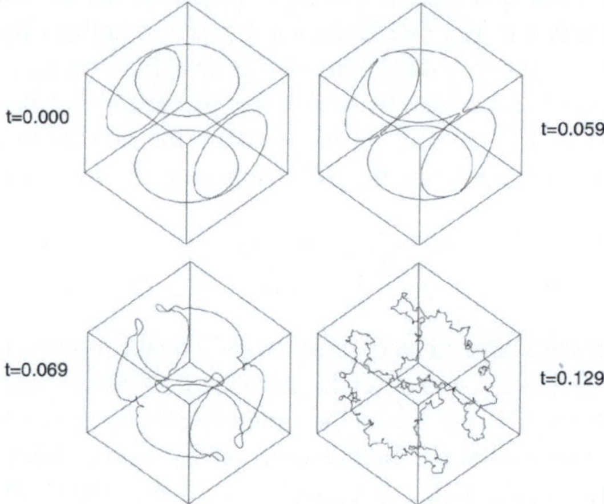

Fig. 10.19 Collision of four vortex rings at $T = 0$ computed using the VFM [34]. Notice how, in the absence of friction, the four reconnection cusps relax into Kelvin waves which interact and become shorter and shorter (Kelvin wave cascade). Reproduced from Ref. [34] with permission. Copyright (2001) of the American Physical Society

The typical crossover from friction dissipation to sound dissipation occurs at $T \approx 0.5$ K [36]. Therefore, at sufficiently low temperatures, quantum turbulence may consists of two energy cascades: a Kolmogorov cascade of bundled vortices (analog to classical eddies) in the range $k_D \ll k \ll k_\ell$, and a Kelvin wave cascade on individual vortex lines for $k \gg k_\ell$ (with no classical analog). Large GPE simulations of quantum turbulence suggest that the two cascades are separated by a bottleneck region of k-space [37] and confirmed [38] the predicted scaling behavior of the Kelvin cascade [39]. Further experimental evidence for the Kelvin wave cascade is provided by experiments in ^{3}He [40].

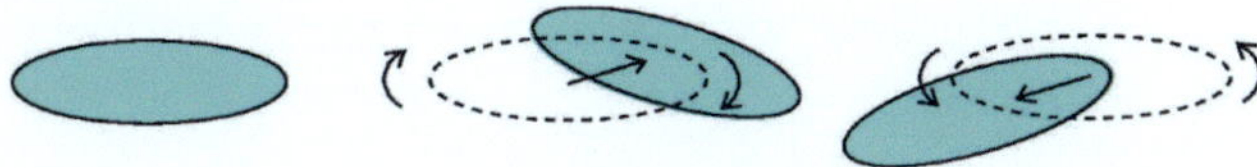

Fig. 10.20 Schematic shaking of a condensate confined in a harmonic trap to induce turbulence in the experiments of Ref. [44–46]

10.7   Quantum Turbulence in Atomic Condensates

When considering turbulence in atomic condensates [43], it is important to appreciate the small size of these systems compared to the size of experiments in superfluid helium or in ordinary fluids. The small size means that the range of length scales which is available to verify any theoretical prediction of scaling laws is very limited.

To quantify this comment, recall that in helium II the thickness of the vortex core is $a_0 \approx 10^{-10}$ m (of the order of the healing length); the average distance between the vortex lines in typical experiments varies between $\ell \approx 10^{-4}$ and 10^{-6} m, and the size of the helium sample varies from $D_0 \approx 10^{-3}$ m for capillary tubes to $D_0 = 0.8$ m for the SHREK experiment [52]. Thus the length scale ratios are as large as $D_0/a_0 \approx 10^{10}$ and $\ell/a_0 \approx 10^6$. Now consider ordinary turbulence. One of the earliest measurement which confirmed the Kolmogorov 5/3 law in ordinary fluids was performed in the Seymour Narrows [53], a tidal channel near Vancouver. Since the size of the smallest eddies is of the order of the Kolmogorov length, even for this classical experiment we have a large ratio: $D_0/\eta_K \approx 10^6$. In comparison, in a typical turbulence experiment with atomic gases, the length ratios are only $D_0/\ell \approx 1$ to 10 and $D_0/\xi \approx 10$ to 100. Unfortunately atomic condensates are small because the last stage of evaporative cooling greatly reduces the size of the cloud of trapped atoms.

Turbulence in atomic Bose-Einstein condensates has been experimentally generated by shaking a harmonic trap [44–46], see Fig. 10.20, or by oscillating a boxtrap [47–49]. Provided the temperature is sufficientlyfiiciently small compared to the critical temperature, the dynamics of these systems can be accurately simulated by the GPE.

The simulations [50,51] and the absorption images of the expanded condensates reveal the presence of large density waves; the shaking induce solitonic structures which quickly become vortices. Insight into the vortex tangle is provided only by simulations, because there are no suitable experimental techniques to directly visualize and measure turbulent vortices in 3D (unlike the situation of 2D atomic condensates, as we shall see in the next section). It appears that the tangle consists mainly of very small vortex loops which are non-uniformly distributed in the midst of large waves, as shown in Fig. 10.21. The figure also shows that some vortices are closed loops, and other vortices are U-vortices that terminate at the fragmented boundaries of the. The combination of small vortex loops and large density waves is unlike what is observed in turbulent helium. Moreover, the spectrum of the incompressible energy density does not display any particular scaling in the small range of k-space which is available, and the decay of the vortex length is roughly consistent with the t^{-1} behaviour of Vinen turbulence.

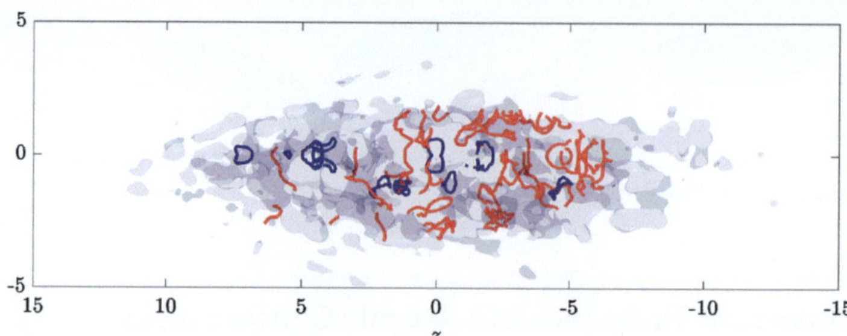

Fig. 10.21 Turbulence in a shaken atomic condensate computed using the GPE [50] to model experiments by Bagnato and collaborators. Red and blue tubes denote U-vortices and closed vortex loops. The grey surface is an isosurface of the density; where it is dark grey the line of sight has intersected the isosurface more than once (this happens because the condensate's edge are rather fragmented). Reprinted under CC-BY-4.0 license from [50]. Copyright 2023, the Author(s)

Experimentally, most of the attention has been on the momentum distribution instead (obtained from time of flight expansion of the condensate). The range of k-space available in these systems is quite small, as we have remarked, but the computed momentum appears to scale as $n_k \sim k^{-2.6}$ at large k [51], in agreement with experiments. A slightly different scaling, $n_k \sim k^{-3.5}$, was found in the experiments with oscillating condensates [47,49], which compares fairly well with Zakharov's classical theory of *wave turbulence* [54].

10.8   Review of 2D Classical Turbulence

Provided we are interested only in sufficiently large motions, many flows in nature can be considered two-dimensional (2D). An example is the atmosphere: its height is approximately 10 km, but weather patterns (e.g. cyclons, anticyclons, jets, etc.) extends for hundreds or even thousands of kilometres. The 2D aspect of atmospheric motion is strengthened by the Coriolis effect induced by the Earth's rotation, which, by balancing horizontal pressure gradients, forces the wind to blow parallel to isobars, and by the suppression of large vertical motions by density stratifications. The resulting turbulence is approximately 2D and its fundamental properties are quite different from the 3D Kolmogorov scenario that we have seen in Sect. 10.2. Recall the vorticity equation (10.11) and notice that in 2D, since ω is perpendicular to the plane of the velocity, the *vortex stretching* term at the right hand size of the equation is always zero.

Because of the lack of vortex stretching, the dynamics of the 2D Euler equation is quite different from 3D. In 3D, the inviscid invariants are energy ($(1/2) \int v^2 \, d^3 r$) and helicity ($\int \omega \cdot d^3 r$), whereas in 2D they are energy and enstrophy ($\int \omega^2 \, d^3 r$). The implications for turbulence are significant. In the presence of dissipation, a classical 2D turbulent flow, forced at the length scale η_{in} to maintain a statistical steady state, undergoes a *double cascade*: a *direct cascade* of enstrophy from η_{in} to

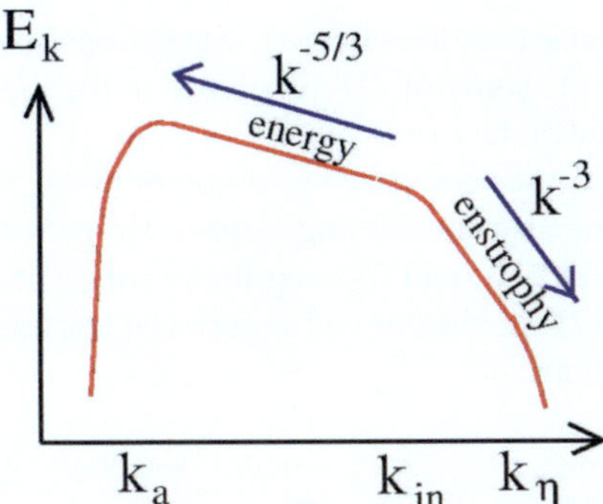

Fig. 10.22 Schematic energy spectrum of 2D classical turbulence. Energy is injected at wavenumber k_{in} and extracted at $k_{\eta_K} = 2\pi/\eta_K$ (where η_K is the Kolmogorov dissipation length) and k_a (corresponding to the length scale of large scale boundary friction). A direct cascade transfers enstrophy to smaller length scales (larger k) and an inverse cascade moves energy to larger length scales (smaller k). Compare to the analog spectrum for 3D turbulence in Fig. 10.4

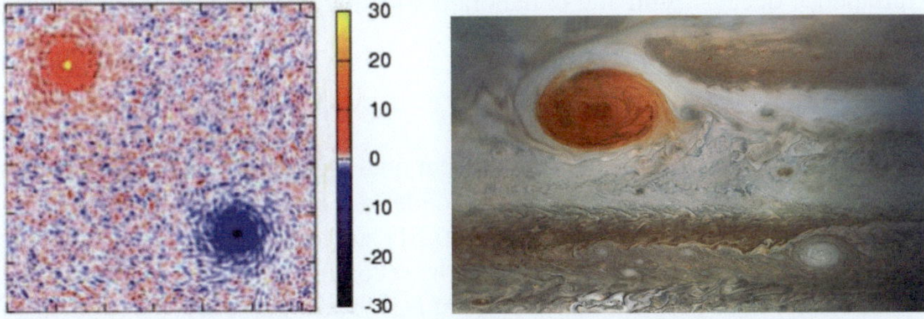

Fig. 10.23 Left: Vorticity field after the emergence of an Onsager condensate from the solution of the classical 2D Navier-Stokes equation [56]. Two large-scale opposite signed vortex structures appear rather than one because the calculation imposed the condition of zero mean vorticity. Reproduced with permission from Ref. [56]. Copyright (2007) of the American Physical Society. Right: Jupiter's Great Red Spot, a gigantic anticyclone in Jupiter's atmosphere, probably first observed in 1665 by Giovanni Cassini. Image Credit: NASA/JPL-Caltech/SwRI/MSSS/Gerald Eichstädt/Sean Doran

the Kolmogorov scale η_K, and an *inverse cascade* of energy from η_{in} to a large scale η_a where friction with the boundaries is the energy sink.

The corresponding energy spectra of the direct and inverse cascade are respectively $E_k \sim k^{-3}$ and $k^{-5/3}$; both have been observed in experiments with soap films [55]. The large jet streams and Great Red Spot of Jupiter's atmosphere, see Fig. 10.23, whose large-scale dynamics is 2D due to the rapid rotation of the planet, are powered by the inverse energy cascade. In the absence of large length scale dissipation, energy can accumulate into the mode with the lowest wavenumber, $k = 2\pi/D_0$, leading to a stable large-scale coherent structure containing most of the energy which emerges from the chaotic motion of the surrounding flow; this structure is called an *Onsager condensate* in analogy with Bose-Einstein condensation. An example is shown in Fig. 10.23 (left).

Due to the ability to engineer the effective dimensionality, atomic condensates at low temperatures allow the study of 2D turbulence in the limit of vanishing viscous dissipation. Their great advantage over superfluid helium films is vortex visualization. These are remarkable features compared to ordinary flows, that are never really 2D and dissipation-free. Only by considering large-scale patterns the atmosphere can be approximated by a 2D flow, and rotating fluids suffer Ekman friction; similarly thin layer electrolytes [57] are viscous and experience Hartmann friction, while soap films suffer friction with air.

10.9 Quantum Turbulence in 2D

A simple way to create turbulence in a 2D atomic condensate is to stir it with a laser beam to create a large number of vortices as described in Sect. 9.5, then remove the stirrer, leaving behind a disordered vortex configuration consisting of N_v vortices, as shown in Fig. 10.24 and in the previous chapter. The natural question is how this 2D turbulence decays. Shin and collaborators [58] argued that in their experiment N_v obeys the equation,

$$\frac{dN_v}{dt} = -c_1 N_v - c_2 N_v^2, \tag{10.17}$$

where the rate coefficients c_1 and c_2 represent one-vortex and two-vortex processes respectively: the drift of vortices out of the condensate and the annihilation of vortex-antivortex pairs. Numerical experiments [60,61] suggested that the second process involves four vortices, not just two (hence the second term at the right hand side of Eq. (10.17) should be proportional to N_v^4, not N_v^2). The argument is the following: in the absence of dissipation, a vortex-antivortex pair would be stable. A third vortex is needed to bring the two vortices of the pair together, destroying the circulation and creating a stable nonlinear wave called *crescent-shaped wave* in Ref. [58] and *vortex-onium* in Ref. [61], and identified [63] as a *Jones-Roberts soliton*. Finally, the fourth vortex collides with the soliton, eventually radiating sound away. Such four-vortex process is frequent during the decay of 2D quantum turbulence, as demonstrated in Fig. 10.25. A more controlled setting of the generation of crescent-shaped waves and their disintegration in shown in Fig. 10.26.

To better identify the decay mechanism, numerical simulations [62] were performed in a periodic domain in order to remove the loss of vortices at the edge of the condensate; the result showed that in the absence of dissipation $N_v \sim t^{-1/3}$ for large t, which is consistent with the four-vortex process ($dN_v/dt = -c_2 N_v^4$). By solving the modified GPE (3.65) with a small parameter γ to model phenomenologically thermal effects, the authors showed that when the dissipation becomes the dominant mechanism the scaling $N_v \sim t^{-1}$ of two-vortex process ($dN_v/dt = -c_2 N_v^2$) emerges.

The Jones-Roberts soliton mentioned above is a nonlinear solution of the GPE equation [64,65] which has been observed in atomic condensate [66] by Bongs and collaborators. What makes the Jones-Roberts soliton remarkable is its confined

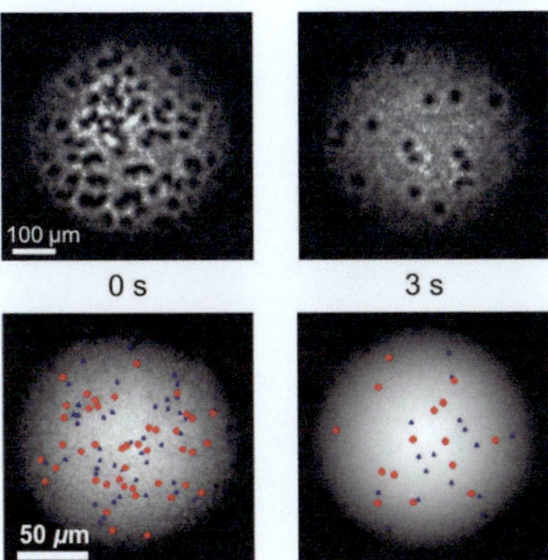

Fig. 10.24 Two-dimensional quantum turbulence. Top: experimental absorption images of a condensate in a state of 2D turbulence [58]. Reproduced with permission from Ref. [58]. Copyright (2014) of the American Physical Society. Bottom: the corresponding images of the unexpanded condensate density from GPE simulations of the experiment [59]. Vortices with positive (negative) circulation are highlighted by red circles (blue triangles). The vortices appear much smaller since the condensate has not been expanded. Reproduced with permission from Ref. [59]. Copyright (2015) of the American Physical Society

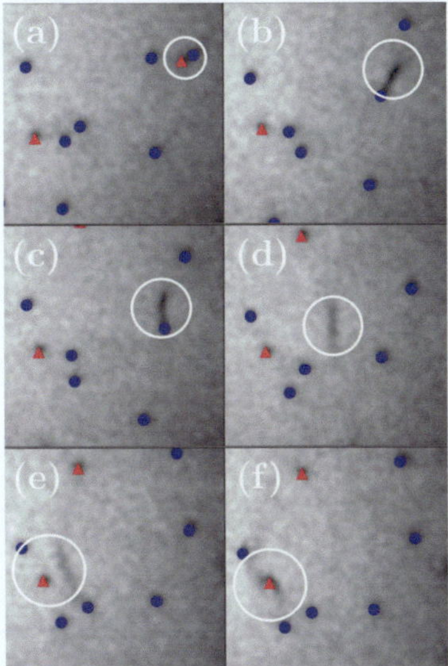

Fig. 10.25 Decay of 2D turbulence, four-vortex process sequence computed by the GPE [60]. Vortices and antivortices are marked by red and blue symbols. **a** the white circle highlights a vortex-antivortex pair; **b** the pair annihilates, creating a strong nonlinear wave (its dark colour represents reduced density); **c** the wave deflects a negative vortex and moves away (**d**), until **e, f** it hits a positive vortex and disintegrates in smaller sound waves. Reproduced with permission from Ref. [60]. Copyright (2016) of the American Physical Society

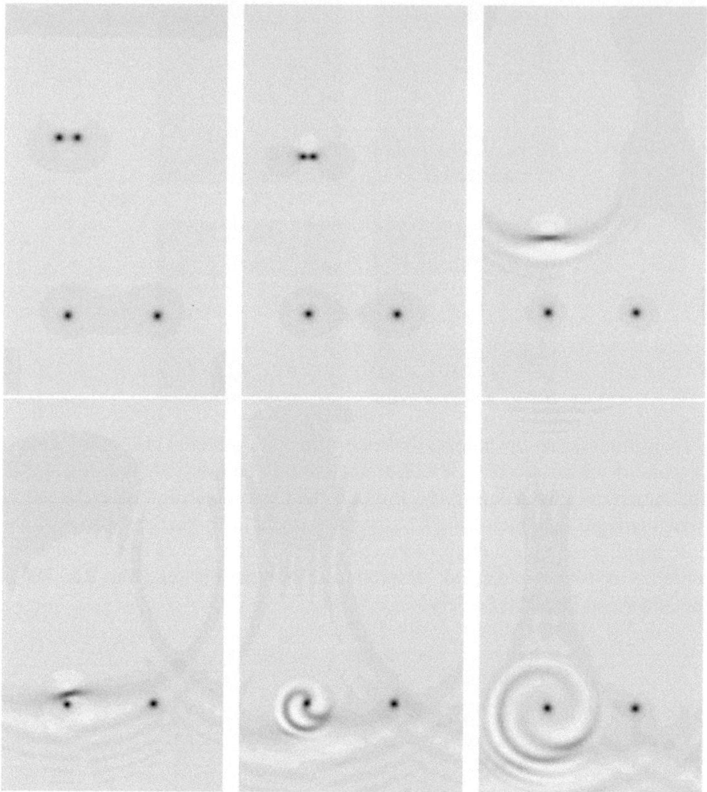

Fig. 10.26 Generation and disintegration sequence of a Jones-Roberts soliton computed with the damped GPE. A vortex-antivortex pair travels towards two stationary vortices in the presence of dissipation (top, left). Because of the dissipation, the pair shrinks in size (top, middle), until it annihilates (top right), leaving behind a Jones-Roberts soliton. The circulation vanishes. The soliton keeps travelling towards a vortex (bottom left), until it collides with it, disintegrating in a burst of sound waves (bottom, middle and right). Courtesy of G.W. Stagg

spatial extension which keeps it stable. In contrast, the dark soliton solution of the 1D GPE which we have seen in Sect. 5.2 becomes unstable if extended to 2D or 3D because the nodal plane where the phase jumps bends transversally, breaking up into line or ring vortices (*snake instability*), as predicted [67] and then observed in both nonlinear optical media [68] and in atomic condensates [69].

10.10 Clustering and Negative Temperatures

In 1949, Lars Onsager [70] proposed that the explanation for the observation of isolated large scale vortex structures created by the inverse energy cascade in 2D flows should be sought on statistical grounds based on the dynamics of point vortices in the absence of forcing and dissipation [71]. In a footnote to his paper he prophetically remarked that quantised vortices would provide the best physical context for his explanation.

Onsager recalled that the governing equations of motion (8.16) and (8.17) of a system of N vortices of position $\boldsymbol{r}_k = (x_k, y_k)$ and circulation Γ_k $(k = 1, \cdots N)$ can be written in Hamiltonian form,

$$H = -\frac{1}{4\pi} \sum_k^N \sum_{j \neq k}^N \Gamma_k \Gamma_j \ln(r_{kj}), \tag{10.18}$$

where H is the Hamiltonian and r_{jk} is the distance between the kth and the jth vortex. Hereafter we assume that the net circulation is zero. If the vortex system is confined by boundaries into a finite flow domain, the logarithm term in Eq. (10.18) must be replaced by a suitable Green function of the Laplacian with the appropriate boundary conditions, and terms must be added to account for image vortices. The key points are that the position coordinates x_k and y_k are the canonically conjugate variables, and, for a confined domain, the phase space volume has the finite value A^N, where A is the area of the flow domain.

By applying statistical mechanics, Onsager assigned an absolute temperature T to the vortex configuration in the following way. The volume of phase space $\Phi(E)$ enclosed by the energy surface E is an increasing function of energy E which has values $\Phi(-\infty) = 0$ and $\Phi(\infty) = A^N$. Therefore the number of states with energy between E and $E + dE$, $\Omega(E) = \Phi(E + dE) - \Phi(E)$, is a non-negative function which vanishes at $E = \pm\infty$ and which must have a maximum at some value $E = E_m$, where its derivative $\Omega'(E)$ with respect to E is zero. Thus $\Omega'(E) > 0$ for $E < E_m$ and $\Omega'(E) < 0$ for $E > E_m$. But in statistical mechanics the logarithmic derivative of $\Omega(E)$ defines the absolute temperature via the entropy, S,

$$S(E) = k_B \ln \Omega(E), \qquad \frac{1}{T} = \frac{dS}{dE}, \tag{10.19}$$

where k_B is Boltzmann;s constant.

If $E < E_m$, the lowest energies of the vortex configuration correspond to weak flows and low-entropy, ordered states in which vortices of opposite signs pair up, as in Fig. 10.27a. At larger energy the pairs unbind, increasing the average nearest-neighbor distance, and vortices assume uncorrelated high-entropy configurations, as in Fig. 10.27b. At $E = E_m$ the entropy reaches a maximum $S = S_m$, corresponding to infinite vortex temperature, $T = \infty$. For $E > E_m$, larger energy can only be achieved at lower entropy, hence the vortex configuration must become more ordered: vortices of the same sign will come together to create same-sign clusters and stronger flows, until, at some energy $E = E_c > E_m$ (corresponding to $T = T_c$), the clusters undergo a transition in which only two giant clusters remain, one positive and one negative, as in Fig. 10.27c. The actual shapes of these two giant clusters will depend on the actual shape of the container. This transition can be tracked by the dipole moment $D = (1/N_v) \sum_k sgn(\Gamma_k) x_k$ which acts as order parameter ($D = 0$ for $E < E_c$ and $D \sim (E - E_c)^{1/2}$ for $E > E_c$).

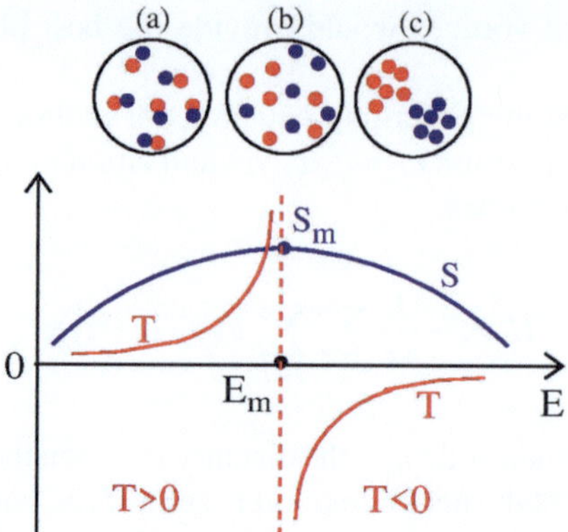

Fig. 10.27 2D quantum turbulence in Onsager's scenario. Plot of the entropy S (blue curve) and the temperature T (red curve) of the vortex configuration as a function of the energy per vortex, E. Notice that, because of the finiteness of the domain, the entropy reaches a maximum value, S_m, at $E = E_m$, then decreases; hence, for $E < E_m$ the vortex temperature T is positive, and for $E > E_m$ it is negative. Typical vortex configurations at small, intermediate and large values of the energy per vortex are shown at the top of the figure (positive and negative vortices are respectively red and blue symbols): (**a**): vortex-antivortex pairs, (**b**): random vortices; (**c**): vortex clusters

Onsager's idea was confirmed in experiments in which vortices were injected into a 2D atomic condensate by dragging obstacles created using laser beams [72,73] without creating too many sound waves. It is interesting to remark that, in the $E < E_m$ regime described above, the annihilation of vortex pairs represents a form of *evaporative heating* [74] that removes the "coldest" vortices and leaves behind the "hottest" vortices: after redistribution of the remaining energy, the energy per vortex will increase. This process is analogous to the mechanism of *evaporative cooling* used to create the BEC in the first place, whereby the hottest particles are lost, resulting in a cooler cloud of atoms.

Problems

10.1 *Navier-Stokes dissipation*:

Using the incompressible Navier-Stokes Eq. (10.4), the incompressibility condition $\nabla \cdot \boldsymbol{u} = 0$ and assuming that there is no forcing, show that the rate of kinetic energy dissipation (per unit mass), defined by Eq. (10.13), obeym

$$\epsilon = \frac{\nu}{V} \int_V \omega^2 dV, \tag{10.20}$$

where V is the volume of the fluid. Assume that $\boldsymbol{u} = 0$ at the surface S of V (no-slip boundary conditions).

10.2 *Vorticity decay*:

Show that in the towed grid experiment, once the peak of the energy spectrum has shifted to the channel size D, energy $E(t)$ and dissipation $\epsilon(t)$ are related by the relation,

$$E \approx \frac{3C}{2}\epsilon^{2/3}k_D^{-2/3},$$

which is found by integrating in k Eq. (10.12) from $k_D = 2\pi/D$ to $k_\eta = 2\pi/\eta \gg k_D$. Then, from $V = D^3$, Eqs. (10.10) and (10.13), show that, for large times, vorticity and energy decay respectively as,

$$\omega(t) \approx \frac{(3C)^{3/2}D}{2\pi\sqrt{\nu}}t^{-3/2},$$

and,

$$E(t) \approx \frac{(3C)^3 D^2}{8\pi^2}t^{-2}.$$

10.3 *Decay of counterflow turbulence*:

Consider Vinen's equation (10.2) in the absence of drive ($V_{ns} = 0$) and show that the vortex line density decays as $L \sim t^{-1}$.

10.4 *Energy of Vinen turbulence and its decay*:

Quantum turbulence of the Vinen type with vortex line density L is contained in a cubic volume of size D. Show that the energy is approximately,

$$E = \frac{\rho\kappa^2 L D^3}{4\pi}\ln(\ell/a_0).$$

where $\ell = L^{-1/2}$. Finally show that the vortex line density decays as $L \sim t^{-1}$.

References

1. W.F. Vinen, Proc. Roy. Soc. A **240**, 114 (1957); **240**, 128 (1957); **242**, 493 (1957); **243**, 400 (1957)
2. K.W. Schwarz, Phys. Rev. B **38**, 2398 (1988)
3. L.K. Sherwin-Robson, C.F, Barenghi, A.W. Baggaley, Phys. Rev. B **91**, 104517 (2015)
4. W. de Haas, H. van Beelen, Physica B **83B**, 125 (1976)
5. C.F. Barenghi, K. Park, R.J. Donnelly, Phys. Lett. **84**, 435 (1981)
6. J.T. Tough, in *Progress in Low Temperature Physics*, vol. VIII, chapt. 3, edited by D.F. Brewer (North-Holland Publishing Co., 1982)

7. E. Varga, S. Babuin, L. Skrbek, Phys. Fluids **27**, 065101 (2015)
8. B. Mastracci, W. Guo, Phys. Rev. Fluids **3**, 063304 (2018)
9. L.B. Opatowsky, J.T. Tough, Phys. Rev. B **24**, 5420(R) (1981)
10. M.R. Smith, R.J. Donnelly, N. Goldenfeld, W.F. Vinen, Phys. Rev. Lett. **71**, 2583 (1993)
11. S.R. Stalp, L. Skrbek, R.J. Donnelly, Phys. Rev. Lett. **82**, 4831 (1999)
12. J. Maurer, P. Tabeling, Europhys. Lett. **43**, 29–34 (1998)
13. O. Reynolds, Proc. Roy. Soc. London **35**, 84 (1883)
14. A.A. Ghira, G.E. Elsinga, C.B. da Silva, Phys. Rev. Fluids **7**, 104605 (2022)
15. J.G. Chen, J.C. Vassilicos, J. Fluid Mech. **938**, A7-1 (2022)
16. T. Ishihara, T. Gotoh, Y. Kaneda, Annu. Rev. Fluid Mech. **41**, 165 (2009)
17. C. Nore, M. Abid, M.E. Brachet, Phys. Rev. Lett. **78**, 3896 (1997)
18. P.M. Walmsley, A.I. Golov, H.E. Hall, A.A. Levchenko, W.F. Vinen, Phys. Rev. Lett. **99**, 265302 (2007)
19. T. Araki, M. Tsubota, S.K. Nemirovskii, Phys. Rev. Lett. **89**, 145301 (2002)
20. J. Salort, C. Baudet, B. Castaing, B. Chabaud, F. Daviaud, T. Didelot, P. Diribarne, B. Dubrulle, Y. Gagne, F. Gauthier, A. Girard, B. Hebral, B. Rousset, P. Thibault, P.E. Roche, Phys. Fluids **22**, 125102 (2010)
21. J. Salort, B. Chabaud, E. Leveque, P.-E. Roche, Europhys. Lett. **97**, 34006 (2012)
22. C.F. Barenghi, V. L'vov, P.-E. Roche, Proc. Nat. Acad. Sci. USA **111** (Suppl. 1) 4683 (2014)
23. P.A. Davidson, *Turbulence* (Oxford University Press, An introduction for scientists and engineers, 2004)
24. A.W. Baggaley, J. Laurie, C.F. Barenghi, Phys. Rev. Lett. **109**, 205304 (2012)
25. A.W. Baggaley, C.F. Barenghi, A. Shukurov, Y. Sergeev, Europhys. Lett. **90**, 260002 (2012)
26. P.-E. Roche, C.F. Barenghi, Europhys. Lett. **81**, 36002 (2008)
27. P.-E. Roche, P. Diribarne, T. Didelot, O. Français, L. Rousseau, H. Willaime, Europhys. Lett. **77**, 66002 (2007)
28. G.E. Volovik, J. Low Temp. Phys. **136**, 309 (2004)
29. P.M. Walmsley, A.I. Golov, Phys. Rev. Lett. **100**, 245301 (2008)
30. A.W. Baggaley, C.F. Barenghi, Y.A. Sergeev, Phys. Rev. B **85**, 060501(R) (2012)
31. N.G. Berloff, B.V. Svistunov, Phys. Rev. A **66**, 013603 (2002)
32. G.W. Stagg, N.G. Parker, C.F. Barenghi, Phys. Rev. A **94**, 053632 (2016)
33. M. Tsubota, T. Araki, S.K. Nemirovskii, Phys. Rev. B **62**, 11751 (2000)
34. D. Kivotides, J.C. Vassilicos, D.C. Samuels, C.F. Barenghi, Phys. Rev. Lett. **86**, 3080 (2001)
35. B. Svistunov, Phys. Rev. B **52**, 3647 (1995)
36. W.F. Vinen, Phys. Rev. B **61**, 1410 (2000)
37. P.C. di Leoni, P.D. Mininni, M.E. Brachet, Phys. Rev. A **95**, 053636 (2017)
38. G. Krstulovic, Phys. Rev. E **86**, 055301(R) (2012)
39. V.S. L'vov, S. Nazarenko, JETP Lett. **91**, 428 (2010)
40. J.T. Mäkinen, S. Autti, V.S. L'vov, P.M. Walmsley, P.J. Heikkinen, J.J. Hosio, R. Hänninen, V.V. Zavjalov, V.B. Eltsov, Nature Phys. **19**, 898 (2023)
41. A.P. Finne, T. Araki, R. Blaauwgeers, V.B. Eltsov, N.B. Kopnin, M. Krusius, L. Skrbek, M. Tsubota, G.E. Volovik, Nature **424**, 1022 (2003)
42. L. Galantucci, E. Rickinson, A.W. Baggaley, N.G. Parker, C.F. Barenghi, Phys. Rev. Fluids **8**, 034605 (2023)
43. M.C. Tsatsos, P.E.S. Tavares, A. Cidrim, A.R. Fritsch, M.A. Caracanhas, F.E.A. dos Santos, C.F. Barenghi, V.S. Bagnato, Phys. Rep. **622**, 1 (2016)
44. E.A.L. Henn, J.A. Seman, G. Roati, K.M.F. Magalhães, V.S. Bagnato, Phys. Rev. Lett. **103**, 145301 (2009)
45. J.H.V. Nguyen, M.C. Tsatsos, D. Luo, A.U.J. Lode, G.D. Telles, V.S. Bagnato, R.G. Hulet, Phys. Rev. X **9**, 011052 (2019)
46. A.D. Garcia-Orozco, L. Madeira, M.A. Moreno-Armijos, A.R. Fritsch, P.E.S. Tavares, P.C.M. Castilho, A. Cidrim, G. Roati, V.S. Bagnato, Phys. Rev. A **106**, 023314 (2022)
47. N. Navon, A.L. Gaunt, R.P. Smith, Z. Hadzibabic, Nature **539**, 72 (2016)

48. N. Navon, C. Eigen, J. Zhang, R. Lopes, A.L. Gaunt, K. Fujimoto, M. Tsubota, R.P. Smith, Z. Hadzibabic, Science **366**, 382 (2019)

49. L.H. Dogra, G. Martirosyan, T.A. Hilker, J.A.P. Glidden, J. Etrych, A. Cao, C. Eigen, R.P. Smith, Z. Hadzibabic, Nature **620**, 521 (2023)

50. H.A.J. Middleton-Spencer, A.D.G. Orozco, L. Galantucci, M. Moreno, N.G. Parker, L.A. Machado, V.S. Bagnato, C.F. Barenghi, Phys. Rev. Res. **5**, 043081 (2023)

51. C.F. Barenghi, H.A.J. Middleton-Spencer, L. Galantucci, N.G. Parker, AVS Quantum Sci. **5**, 025601 (2023)

52. B. Saint-Michel, E. Herbert, J. Salort, C. Baudet, M. Bon Mardion, P. Bonnay, M. Bourgoin, B. Castaing, L. Chevillard, F. Daviaud, P. Diribarne, B. Dubrulle, Y. Gagne, M. Gibert, A. Girard, B. Hébral, Th. Lehner, B. Rousset, and SHREK Collaboration. Phys. Fluids **26**, 125109 (2014)

53. H.L. Grant, R.W. Steward, A. Moilliet, J. Fluid Mech. **12**, 241 (1962)

54. V.E. Zakharov, V.S. L'vov, G. Falkovich, *Kolmogorov Spectra of Turbulence I: Wave Turbulence* (Springer, Berlin, 2012)

55. M.A. Rutgers, Phys. Rev. Lett. **81**, 2244 (1998)

56. M. Chertkov, C. Connaughton, I. Kolokolov, V. Lebedev, Phys. Rev. Lett. **99**, 084501 (2007)

57. J. Paret J, P. Tabeling, Phys. Rev. Lett. **79**, 4162 (1997)

58. W.J. Kwon, G. Moon, J. Choi, S.W. Seo, Y. Shin, Phys. Rev. A **90**, 063627 (2014)

59. G.W. Stagg, A.J. Allen, N.G. Parker, C.F. Barenghi, Phys. Rev. A **91**, 013612 (2015)

60. A. Cidrim, F.E.A. dos Santos, L. Galantucci, V.S. Bagnato, C.F. Barenghi, Phys. Rev. A **93**, 033651 (2016)

61. A.J. Groszek, T.P. Simula, D.M. Paganin, K. Helmerson, Phys. Rev. A **93**, 043614 (2016)

62. A.W. Baggaley, C.F. Barenghi, Phys. Rev. A **97**, 033601 (2018)

63. S. Nazarenko, M. Onorato, J. Low Temp. Phys. **146**, 31 (2007)

64. C.A. Jones, P.H. Roberts, J. Phys. A: Math. Gen. **15**, 2599 (1982)

65. C.A. Jones, S.J. Putterman, P.H. Roberts, J. Phys. A: Math. Gen. **19**, 2991 (1986)

66. N. Meyer, H. Proud, M. Perea-Ortiz, C. O'Neale, M. Baumert, M. Holynski, J. Kronjger, G. Barontini, K. Bongs, Phys. Rev. Lett. **119**, 150403 (2017)

67. C. Josserand, Y. Pomeau, Europhys. Lett. **30**, 43 (1995)

68. A.V. Mamaev, M. Saffman, A.A. Zozulya, Phys. Rev. Lett. **76**, 2262 (1996)

69. B.P. Anderson, P.C. Haljan, C.A. Regal, D.L. Feder, L.A. Collins, C.W. Clark, E.A. Cornell, Phys. Rev. Lett. **86**, 2926 (2001)

70. L. Onsager, Nuovo Cim. **6**(S2), 279 (1949)

71. G.L. Eyink, K.R. Sreenivasan, Rev. Mod. Phys. **78**, 87 (2006)

72. S.P. Johnstone, A.J. Groszek1âŁ , P. Starkey, C.J. Billington, T.P. Simula, K. Helmerson, Science **364**, 1267 (2019)

73. G. Gauthier, M.T. Reeves, X. Yu, A.S. Bradley, M.A. Baker, T.A. Bell, H. Rubinsztein-Dunlop, M.J. Davis, T.W. Neely, Science **364**, 1264 (2019)

74. T. Simula, M.J. Davis, K. Helmerson, Phys. Rev. Lett. **113**, 165302 (2014)

Two-Component Bose-Einstein Condensates

11

Abstract

In this chapter we consider quantum fluid systems consisting of two interacting condensates. There are two key regimes: the miscible regime, in which the two condensates overlap spatially, and the immiscible regime, in which they segregate. We explore how these regimes enrich the stationary solutions and dynamics of the system, including soliton and vortex states.

11.1 Two-Component Gases

In Sect. 3.1 we discussed the atomic species used to generate Bose-Einstein condensates. It is also possible to create systems consisting of two interacting Bose-Einstein condensates, the so-called *two-component condensates*. These systems are either binary mixtures of atoms of the same element but in different hyperfine levels (for example ^{87}Rb), mixtures of different isotopes (for example ^{85}Rb - ^{87}Rb) or mixtures of different elements (for example ^{87}Rb - ^{41}K and ^{87}Rb - ^{133}Cs).

As we know from single-component condensates, an atom within a condensate can interact with another atom in the condensate; this is an *intra-species interaction*. In two-component condensates, an atom in one condensate can additionally interact with an atom in the other condensate; this is an *inter-species interactions*. Thus we must consider the intra-species interactions within each condensate and the inter-species interactions between the condensates. At the extreme diluteness of condensates, each of these interactions can be approximated as a contact interaction characterised by a distinct s-wave scattering length. Feshbach resonances can be used to tune the atomic interactions in condensate mixtures, typically tuning one of the scattering lengths while leaving the others at their natural values. Table 11.1 presents some example combinations of scattering lengths considered in experiments. By using a combination of trapping protocols, such as both magnetic and optical methods, it is possible to exert independent potentials on each component. Mixtures with three or more components can also be created using spinor BECs in which the atoms

© The Author(s), under exclusive license to Springer Nature Switzerland AG 2026   215
C. F. Barenghi et al., *Quantum Fluids, Solitons, and Vortices*, Lecture Notes in Physics 1050, https://doi.org/10.1007/978-3-032-20171-3_11

Table 11.1 Example two-component systems created experimentally and their values of the intra-species and inter-species scattering lengths (in terms of the Bohr radius a_B. In all cases, a Feshbach resonance is used to controllably tune one of the scattering lengths away from its natural value

Mixture	$a_{11}(a_B)$	$a_{22}(a_B)$	$a_{12}(a_B)$
^{87}Rb + ^{85}Rb [2]	99	Tunable	213
^{133}Cs + ^{170}Yb [3]	Tunable	63.9	96.2
^{133}Cs + ^{174}Yb [3]	Tunable	104.9	−74.8
^{41}K + ^{87}Rb [4]	65	100.4	Tunable
^{23}Na + ^{87}Rb [5]	60.1	100.1	Tunable

posses a spin degree of freedom [1]; however here will only consider two-component systems.

11.1.1 Coupled Gross-Pitaevskii Equations

At sufficiently low temperatures and densities these two-component condensates can be modelled by the following coupled Gross-Pitaevskii equations,

$$i\hbar\frac{\partial \Psi_1}{\partial t} = -\frac{\hbar^2}{2m_1}\nabla^2\Psi_1 + V_1\Psi_1 + g_{11}|\Psi_1|^2\Psi_1 + g_{12}|\Psi_2|^2\Psi_1, \qquad (11.1)$$

$$i\hbar\frac{\partial \Psi_2}{\partial t} = -\frac{\hbar^2}{2m_2}\nabla^2\Psi_2 + V_2\Psi_2 + g_{22}|\Psi_2|^2\Psi_2 + g_{21}|\Psi_1|^2\Psi_2, \qquad (11.2)$$

where $\Psi_j(\boldsymbol{r}, t)$, m_j, g_{jj}, and $V_j(\boldsymbol{r})$ for $j = 1, 2$ are respectively the wavefunctions, the atom masses, the intra-species interaction parameters and the trapping potentials of the two condensate species (V_1 is not necessarily the same as V_2). The inter-species interactions are quantified by the parameter $g_{12} = g_{21}$. Note that the two condensates interact directly only via their densities $n_1 = |\Psi_1|^2$ and $n_2 = |\Psi_2|^2$. The interaction parameters are defined as,

$$g_{jj} = \frac{4\pi\hbar^2 a_{jj}}{2m_j} \quad (j = 1, 2), \qquad g_{12} = g_{21} = \frac{2\pi\hbar^2(m_1 + m_2)a_{12}}{m_1 m_2}, \qquad (11.3)$$

where a_{jj} ($j = 1, 2$) and a_{12} are the intra-species and inter-species scattering lengths respectively.

The normalization conditions are,

$$N_j = \int |\Psi_j|^2 \mathrm{d}^3\boldsymbol{r} \quad (j = 1, 2), \qquad (11.4)$$

where N_1 and N_2 are the numbers of atoms of condensate 1 and 2 respectively, and the energy of the entire system is,

$$E = \int \left(\frac{\hbar^2}{2m_1}|\nabla\Psi_1|^2 + \frac{\hbar^2}{2m_2}|\nabla\Psi_2|^2 \right) \mathrm{d}^3 r + \int \left(V_1|\Psi_1|^2 + V_2|\Psi_2|^2 \right) \mathrm{d}^3 r +$$

$$+ \int \left(\frac{1}{2}g_{11}|\Psi_1|^4 + \frac{1}{2}g_{22}|\Psi_2|^4 + g_{12}|\Psi_1|^2|\Psi_2|^2 \right) \mathrm{d}^3 r.$$

$$(11.5)$$

The corresponding time-independent GPEs, which are needed to find the ground state and other stationary states, are obtained by substituting $\Psi_1(r, t)=\psi_1(r)e^{-i\mu_1 t/\hbar}$ and $\Psi_2(r, t) = \psi_2(r)e^{-i\mu_2 t/\hbar}$ into Eqs. (11.1) and (11.2.) This leads to,

$$\mu_1\psi_1 = \left(-\frac{\hbar^2}{2m_1}\nabla^2\psi_1 + V_1 + g_{11}|\psi_1|^2 + g_{12}|\psi_2|^2 \right)\psi_1, \qquad (11.6)$$

$$\mu_2\psi_2 = \left(-\frac{\hbar^2}{2m_2}\nabla^2\psi_2 + V_2 + g_{22}|\psi_2|^2 + g_{12}|\psi_1|^2 \right)\psi_2. \qquad (11.7)$$

The coupled GPE model has been shown to provide an accurate description of two-component condensates and capable of closely reproducing experimental observations, including their ground states and a range of dynamical phenomena, providing the system is sufficiently cold and dilute. At higher atomic densities, such as in two-component quantum droplets, it is important to extend the above model to describe the additional effects of beyond-mean-field physics; this will be introduced in Chap. 13.

11.2 Miscible-Immiscible Transition

An important property of two-component condensates is *miscibility* [6,7]. Under certain conditions, the two condensates segregate spatially rather than remaining mixed. If the system is homogeneous, this immiscible-miscible transition depends on the single parameter Δ, defined as,

$$\Delta = \frac{g_{11}g_{22}}{g_{12}^2} - 1. \qquad (11.8)$$

If $\Delta > 0$ the two species overlap in space (miscible state), whereas if $\Delta < 0$ the two species phase-separate like oil and water (immiscible state).

To see this, we will consider the ground state of the system and compare which state is energetically favoured, following the approach of Ao and Chui [7]. We consider the intra-species interactions to be repulsive ($g_{11}, g_{22} > 0$) to ensure inherent stability to collapse of each component (although we will later that collapse can still be induced for a sufficiently attractive inter-species interaction). For simplicity that $V_1 = V_2 = 0$, such as within a box trap. If we neglect the spatial variations of ψ_1 and ψ_2 at the edges of this box trap and in the bulk (i.e. we assume that ψ_1 and ψ_2 are constant), Eqs. (11.6) and (11.7) become

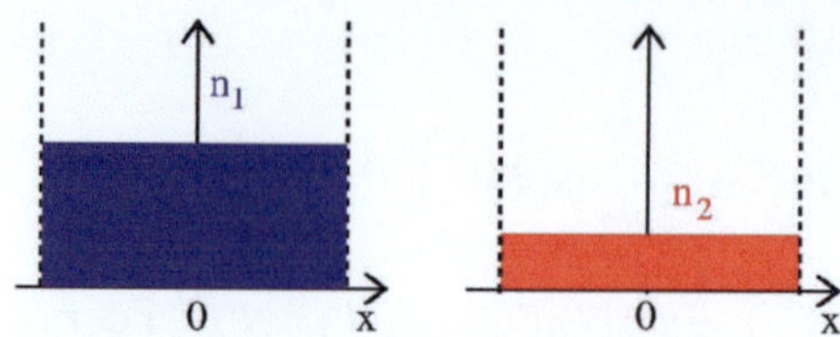

Fig. 11.1 Miscible state: Schematic one-dimensional ground state density profiles n_1 (left, blue) and n_2 (right, red) of a two-component condensate in the miscible state in the same box trap. The vertical dashed lines represent the box trap. Notice that the two species overlap spatially

$$\mu_1 \psi_1 = (g_{11}n_1 + g_{12}n_2)\,\psi_1, \tag{11.9}$$

$$\mu_2 \psi_2 = (g_{22}n_2 + g_{12}n_1)\,\psi_2. \tag{11.10}$$

There are two cases:

(i) Miscible state: We assume that the two condensates overlap spatially as illustrated in Fig. 11.1. The condensate densities can be written as $n_1 = N_1/\mathcal{V}$ and $n_2 = N_2/\mathcal{V}$ where $\mathcal{V}$ is the volume of the box trap. Since $\psi_1, \psi_2 \neq 0$ in $\mathcal{V}$, Eqs. (11.9) and (11.10) reduce to,

$$\mu_1 = g_{11}n_1 + g_{12}n_2, \tag{11.11}$$
$$\mu_2 = g_{22}n_2 + g_{12}n_1, \tag{11.12}$$

Thus the densities n_1 and n_2 of the ground state are,

$$n_1 = \frac{\mu_1 g_{22} - \mu_2 g_{12}}{(g_{11}g_{22} - g_{12}^2)}, \tag{11.13}$$

$$n_2 = \frac{\mu_2 g_{11} - \mu_1 g_{12}}{(g_{11}g_{22} - g_{12}^2)}. \tag{11.14}$$

The energy of the ground state is obtained from Eq. (11.5). We find

$$E_{\mathrm{mis}} = \frac{1}{2}\left(g_{11}\frac{N_1^2}{\mathcal{V}} + g_{22}\frac{N_2^2}{\mathcal{V}} + 2g_{12}\frac{N_1 N_2}{\mathcal{V}}\right). \tag{11.15}$$

Increasing the interaction strength g_{12} thus increases the system's energy.

(ii) Immiscible state: Consider the phase-segregated, immiscible state. Condensate 1 occupies the volume $\mathcal{V}_1$ and is absent in volume $\mathcal{V}_2$, while condensate 2 occupies volume $\mathcal{V}_2$ and is absent in volume $\mathcal{V}_1$, where $\mathcal{V} = \mathcal{V}_2 + \mathcal{V}_2$. This is illustrated in Fig. 11.2. Consider Eq. (11.9) in the region $\mathcal{V}_1$: we can divide by $\phi_1 \neq 0$, and, since

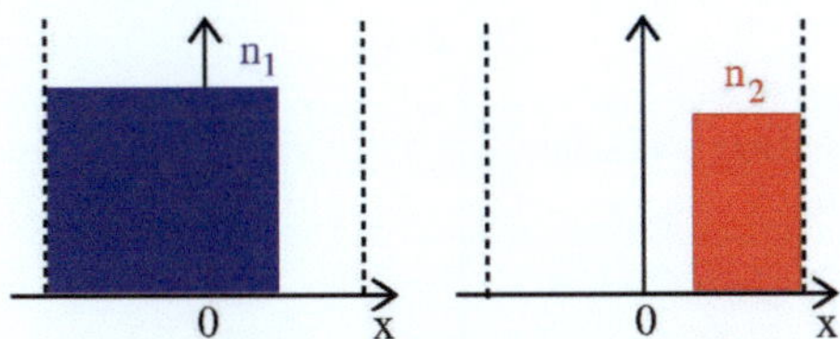

Fig. 11.2 Immiscible state: Schematic one-dimensional ground state density profiles n_1 (left, blue) and n_2 (right, red) of a two-component condensate in the immiscible state confined in the same box trap. The vertical dashed lines represent the box trap. Note that the two species do not overlap

in this region $n_2 = 0$, we conclude that $n_1 = \mu_1/g_{11}$. Similary, from Eq. (11.10) we conclude that $n_2 = \mu_2/g_{22}$. The energy of the system is therefore,

$$E_{\text{imm}} = \frac{1}{2}\left(g_{11}\frac{N_1^2}{\mathcal{V}_1} + g_{22}\frac{N_2^2}{\mathcal{V}_2}\right). \tag{11.16}$$

If we minimize the energy by setting $dE_{\text{imm}}/d\mathcal{V}_1 = 0$, we find that,

$$\frac{\mathcal{V}_1}{\mathcal{V}} = \frac{1}{\left(1 + \sqrt{g_{22}/g_{11}}\,N_2/N_1\right)}, \tag{11.17}$$

$$\frac{\mathcal{V}_2}{\mathcal{V}} = \frac{1}{\left(1 + \sqrt{g_{11}/g_{22}}\,N_1/N_2\right)}, \tag{11.18}$$

such that the minimized energy is

$$E_{\text{imm}} = \frac{1}{2}\left(g_{11}\frac{N_1^2}{\mathcal{V}} + g_{22}\frac{N_2^2}{\mathcal{V}} + 2g_{12}\frac{N_1 N_2}{\mathcal{V}}\right). \tag{11.19}$$

Combining Eqs. (11.15) and (11.19) we have

$$E_{\text{mis}} - E_{\text{imm}} = \frac{N_1 N_2}{\mathcal{V}}\left(g_{12} - \sqrt{g_{11}g_{22}}\right). \tag{11.20}$$

Thus, if the inter-species interactions g_{12} are sufficiently large according to,

$$g_{12}^2 > g_{11}g_{22}, \tag{11.21}$$

then the immiscible state is lower in energy and is favoured over the miscible state. The miscible state is favoured for $g_{12} < \sqrt{g_{11}g_{22}}$. The same result is obtained if one performs a stability analysis of the system [6].

There is an additional regime which arises when $g_{12} < -\sqrt{g_{11}g_{22}}$. Here the attractive inter-species interactions overwhelm the repulsive intra-species interactions such that the system is unstable to collapse [6]. The system behaves qualitatively like a homogeneous attractive single-component condensate (discussed in Chap. 3.1). Figure 11.3 illustrates the miscible:immiscible:collapse regimes of the system as a function of the interaction parameters.

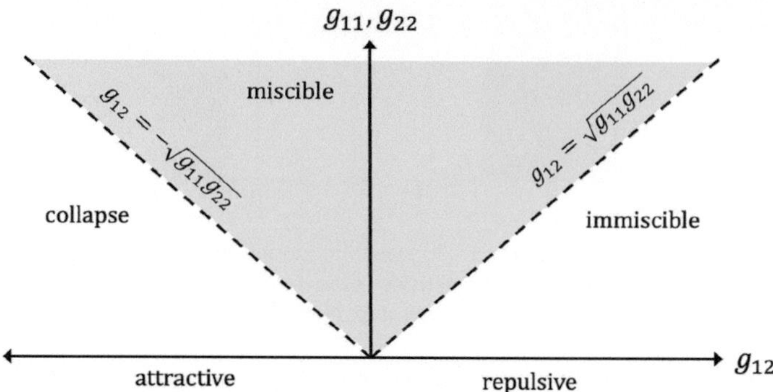

Fig. 11.3 Phase diagram of the homogeneous two-component condensate for repulsive intra-species interactions and either repulsive or attractive inter-species interactions. Shown are the regimes of miscibility, immiscibility and collapse of the mixture

The above results apply to a homogeneous system. The condition for immiscibility becomes slightly modified in the presence of a trap [8] or a significant thermal cloud [9].

11.3   Trapped Solutions

We now consider how the ground states become modified by the presence of harmonic trapping on both components. Figure 11.4 presents some illustrative examples based on the 3D ground state solutions of the coupled GPEs and for repulsive interactions. For simplicity, the system is taken to be spherically-symmetric. In the miscible regime $g_{12} < \sqrt{g_{11}g_{22}}$ (row (a)), the components overlap spatially as expected. The condensates are wider than they would be in isolation—this is because the inter-species repulsion causes additional broadening of the profiles.

In the immiscible regime $g_{12} > \sqrt{g_{11}g_{22}}$ (row (b)) the components segregate, with one component located in the centre of the trap and the other component forming an outer shell. Here it is the Rb component which is located at the trap centre - this is because the greater atomic mass favours Rb being preferentially located at the potential minimum. As well as experiencing the trapping potential, the Rb component features an additional squeezing from the outer Na component, making its profile narrower and more peaked. This becomes more evident as the number of Na atoms is increased. Meanwhile, for a larger number of Rb atoms the Na condensate becomes pushed further from the trap centre.

Although not shown, if the inter-species interactions were weakly attractive we would see the condensates spatially overlap centred on the trap centre, but with narrower and more peaked profiles than for the $g_{12} > 0$ miscible case. This is because the condensates seek to maximise their overlap to take advantage of the attractive interactions. As the atom number is increased or the interactions are made more

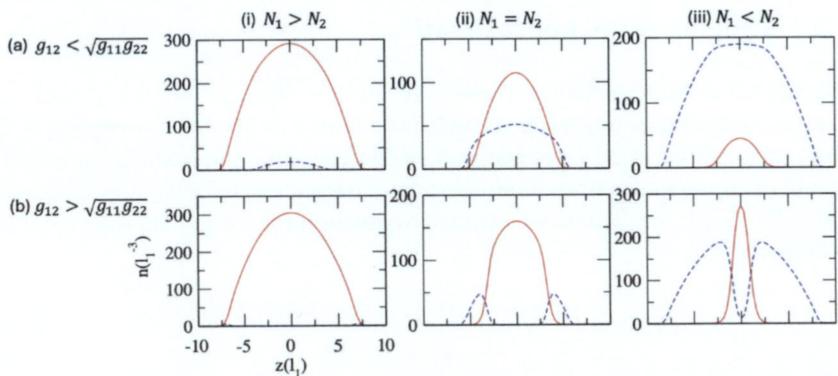

Fig. 11.4 Ground state density profiles for a two-component system under spherically-symmetric harmonic trapping. This is modellied on a mixture of ^{87}Rb (solid red lines, component 1) and ^{23}Na (dashed blue lines, component 2) with $a_{11} = 6$ nm, $a_{22} = 3$ nm, $\omega_1 = 2\pi \times 160$ Hz and $\omega_2 = 2\pi \times 310$ Hz [14]. Row (a) considers weak inter-species interactions $a_{12} = 1.8$nm (miscible regime) while row (b) considers strong inter-species interactions $a_{12} = 3.6$ nm (immiscible regime). Column (i) has an excess of Rb atoms ($N_1 = 2 \times 10^5$, $N_2 = 2 \times 10^3$), (ii) has equal atoms numbers ($N_1 = N_2 = 2 \times 10^4$) and (iii) has an excess of Na atoms ($N_1 = 2 \times 10^3$, $N_2 = 2 \times 10^5$). Adapted with permission from [11]. Copyright 2014, The Author(s). All rights reserved

attractive, the condensates become more peaked until the system becomes unstable to collapse and no stable ground state exists; this is qualitatively similar to the stability of trapped attractive condensates discussed in Chap. 3.

In the above examples for the immiscible regime, the ground state is a symmetric configuration in which one condensate sits in the centre of the system and is surrounded by the other condensate. For certain parameters, an asymmetric configuration can be favoured, in which the ground state consists of the two condensates sitting side-by-side in the trap [10]. This is because interfaces between the components are energetically-costly and, depending on the play-off with the remaining energy contributions, this can favoured configurations with minimised interfacial area; we will return to this in Sect. 11.4.

For the immiscible regime there exists a range of metastable stationary states of slightly higher energy than the ground state in which there are multiple domains of each condensate(either in symmetric and asymmetric configurations), as well as the state of alternating symmetry to the ground. This becomes an important consideration for numerically identifying the ground states; depending on the initial condition and minimisation approach, it is possible to converge to a metastable state rather than the true ground state [11]. It also becomes an important consideration in experiments, where small discrepancies in trap geometry can lead to significantly different ground state solutions [12].

11.3.1 Thomas-Fermi Approximation

In Sect. 3.5.2 we introduced the Thomas-Fermi approximation for the ground state of a repulsively-interacting trapped condensate.; this was obtained by neglecting the kinetic energy terms in the time-independent GPE, an assumption which is possible in the limit of strong interactions. Following the same approach for the two-component system[10, 13, 14], the Thomas-Fermi approximation for the ground state solutions can be written as,

$$n_1(\mathbf{r}) = \frac{[\mu_1 g_{11} - \mu_2 g_{12}] - [g_{22} - g_{12}]V_1(\mathbf{r})}{g_{11}g_{22} - g_{12}^2},$$

(11.22)

$$n_2(\mathbf{r}) = \frac{[\mu_2 g_{11} - \mu_1 g_{12}] - [g_{11} - g_{12}]V_2(\mathbf{r})}{g_{11}g_{22} - g_{12}^2},$$

(11.23)

if the numerators are greater than 0, otherwise the densities are zero. This is valid in the limit of large N_1 and N_2 and repulsive intra-species interactions ($g_{11} > 0, g_{22} > 0$).

In the immiscible regime, the kinetic energy associated with interfaces between the components, which is not accounted for within the Thomas-Fermi approximation, can have a significant influence on the form of the ground state. For example, the Thomas-Fermi approximation always predicts a symmetric ground state under harmonic trapping whereas, as discussed above, the true ground state can favour asymmetric states in certain regimes.

11.4   Interfaces and Domains

The spatial segregation of two immisible condensates leads to an *interface*, or *domain wall*, between the components, in which the densities relax from zero to their background values. The exact ground state density profiles at the interface, including the degree of overlap, is dictated by the interplay between the interaction energies and kinetic energies, and so depends on the interaction strengths, atom numbers and atomic masses. There is no general solution for the profiles, although they resemble the tanh-profiles familiar from the healing profile of a single component condensate at a wall; indeed for certain parameter values the profiles exactly follow a tanh-profile [15].

We discern two main regimes of interface, illustrated in Fig. 11.5:

- **Strongly segregated regime:** In the limit of strong segregation (that is $g_{12}^2/g_{11}g_{22}$ $\rightarrow \infty$,) the overlap between the components is strongly suppressed. The interspecies density term in the coupled GPEs become negligible such that each component essentially behaves independently under the influence of a sharp potential wall generated by interface of the other component. The boundary of each condensate thus follows the tanh-shaped healing solution we met in Sect. 3.4.2,

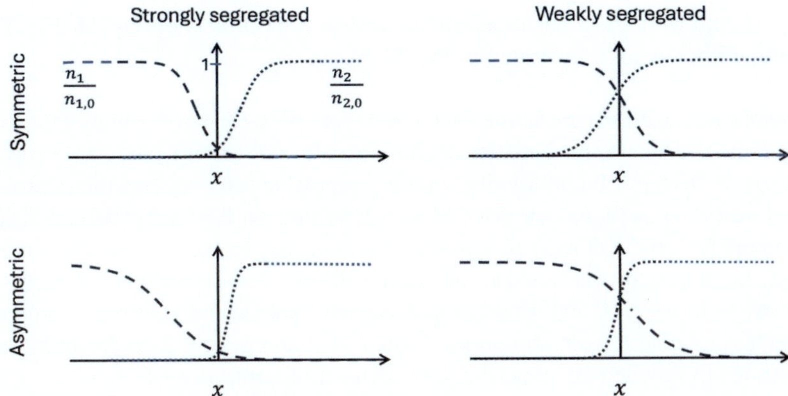

Fig. 11.5 Regimes of interface between adjacent immiscible condensates. Left column: In the strongly segregated regime, overlap is minimised with each component "healing" against the sharp potential created by the adjacent component. Right column: In the weakly segregated regime, each component penetrates into the other on a lengthscale characterised by the penetration length Λ. If the healing lengths are equal (unequal) then the interface is symmetric (asymmetric)

with width characterised by the component's healing length and with only low-amplitude exponentially-decaying tails penetrating the other component.

- **Weakly segregated regime:** For less strong segregation, each component has more significant penetration into the other component. Using dimensional analysis it is possible to show that the penetration length of component 1 into component 2, denoted Λ_1, and vice-versa, is given by [7],

$$\Lambda_{1,2} = \frac{\xi_{1,2}}{\sqrt{\frac{g_{12}}{\sqrt{g_{11}g_{22}}} - 1}}. \tag{11.24}$$

As $g_{12}^2/g_{11}g_{22}$ approaches unity, the penetration length diverges, consistent with entering the miscible regime.

Note that if the healing lengths of equal (unequal) then the interface is symmetric (asymmetric).

The interface has a finite energy, i.e. a surface tension. This has important consequences for the system. The energetic cost of the interface means that the system will tend to minimise the amount of interface, which favours ground states with minimal number of domains or topologies which minimise the interfacial area. The presence of a surface tension also supports modes of the interface, including stable capillary waves and unstable modes which radiate sound waves into the bulk [16].

11.5 Two-Component Solitons

Two-component condensates support a variety of coupled 1D soliton structures, depending on the sign and amplitudes of the inter-species and intra-species interactions [17], and were first considered in the context of nonlinear optics [18, 19]. These are outlined below and illustrated in Fig. 11.6:

- *Domain wall solitons* are formed at the interface between two strongly immiscible condensates, where the densities asymmetrically fall to zero. They are supported for $a_{11} > 0$, $a_{22} > 0$ and relatively strong repulsive inter-species interactions.
- *Dark-antidark solitons* comprise of a dark soliton on the background density of component 1 and a density prominence on the background density of component 2; the latter structure is called a "anti-dark soliton". These structures are supported for $a_{11} > 0$, $a_{22} > 0$ and weakly repulsive inter-species interactions. Component 2 feels a repulsion from component 1 such that component 2 preferentially concentrates in low density area (the dark soliton) of component 1.
- *Dark-grey solitons* feature a dark soliton in component 1 and a low-amplitude dark soliton, a "grey soliton", in component, 2. They supported for $a_{11} > 0$, $a_{22} > 0$ and $a_{12} < 0$. The atoms in component 2 are attracted to component 1 and thus repelled from the low density dark soliton, leading to a density dip being stabilised in component 2.
- *Bright-antidark solitons* comprise of a bright soliton of component 1 coupled to an anti-dark soliton on the background of component 2. They are supported for $a_{11} < 0$, $a_{22} > 0$ and $a_{12} < 0$. The attractive inter-species interactions means that the bright soliton attracts the atoms of the other component, hence supporting the anti-dark soliton.
- *Bright-dark solitons* consist of a bright soliton-like wavepacket of component 1 which sits within a dark soliton in component 2, and is supported for $a_{22} > 0$, $a_{12} > 0$ and both attractive and repulsive a_{11}. They are preserved by the mutual repulsion of the components, which favours the atoms of component 1 sitting in the low density region of component 2.

11.5.1 Bright-Dark Solitons

The bright-dark solitons are particularly robust forms of two-component solitons and have been observed in multiple condensate experiments, and so we present in further detail. These solitons were first predicted in the context of nonlinear optics [20, 21] and later observed [22]. Although bright-dark solitons can be formed for both repulsive and attractive interactions within the bright component, here we focus on the repulsive case for which exact solutions are established.

Consider the 1D versions of the coupled GPEs, a homogeneous system ($V_1 = V_2 = 0$),

$$i\hbar \frac{\partial \psi_1}{\partial t} = -\frac{\hbar^2}{2m_1} \frac{\partial^2 \psi_1}{\partial x^2} + g_{11}|\psi_1|^2 \psi_1 + g_{12}|\psi_2|^2 \psi_1 - \mu_1 \psi_1, \qquad (11.25)$$

$$i\hbar \frac{\partial \psi_2}{\partial t} = -\frac{\hbar^2}{2m_2} \frac{\partial^2 \psi_2}{\partial x^2} + g_{22}|\psi_2|^2 \psi_2 + g_{12}|\psi_1|^2 \psi_2 - \mu_2 \psi_2, \qquad (11.26)$$

Fig. 11.6 Illustration of the
types of coupled solitons
supported in two-component
condensates. The solid
(dotted) line represents the
density profile of component
1 (2)

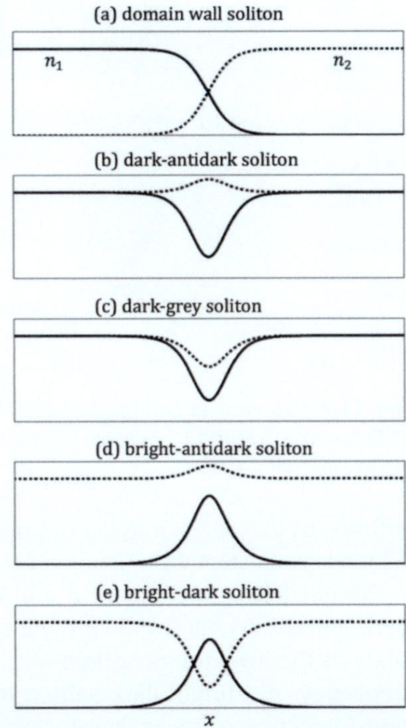

where the macroscopic wavefunctions and nonlinear coefficients take their 1D forms.
Writing in the homogeneous system units of component 1, the equations become,

$$i\frac{\partial \psi_1'}{\partial t'} = -\frac{1}{2}\frac{\partial^2 \psi_1'}{\partial x'^2} + |\psi_1'|^2\psi_1' + \frac{g_{12}}{g_{11}}|\psi_2'|^2\psi_1' - \psi_1', \tag{11.27}$$

$$i\frac{\partial \psi_2'}{\partial t'} = -\frac{1}{2}\frac{m_1}{m_2}\frac{\partial^2 \psi_2'}{\partial x'^2} + \frac{g_{22}}{g_{11}}|\psi_2'|^2\psi_2' + \frac{g_{12}}{g_{11}}|\psi_1'|^2\psi_2' - \mu_2'\psi_2'. \tag{11.28}$$

In the special case of equal masses ($m_1 = m_2$) and equal nonlinearities ($g_{11} = g_{22} = g_{12}$), this reduces to the *Manakov equation*, which is known to be fully nte-grable. The equation has the exact bright-dark soliton solution [23],

$$\psi_1(x) = i\sin\alpha + \cos\alpha\tanh[\kappa(x - q(t), \tag{11.29}$$

$$\psi_2(x) = i\sqrt{\frac{N_2'\kappa}{2}}\exp(i\Omega_2 t)\exp(ix\kappa\tan\alpha)\operatorname{sech}[\kappa(x - q(t)], \tag{11.30}$$

where $N_2' = \int |\psi_2'|^2\,dx$ is the rescaled number of atoms of component 2, $\kappa = \sqrt{\cos^2\alpha + (N_2'/4)^2} - N_2'/4$ is the inverse soliton width, $\Omega_2 = \kappa^2(1 - \tan\alpha)/2 + 1 - \mu_2'$ is a frequency shift of the bright component, α is the velocity-angle of the dark soliton, and $q(t) = q(0) + t\kappa\tan\alpha$ is the soliton position. The expression for κ

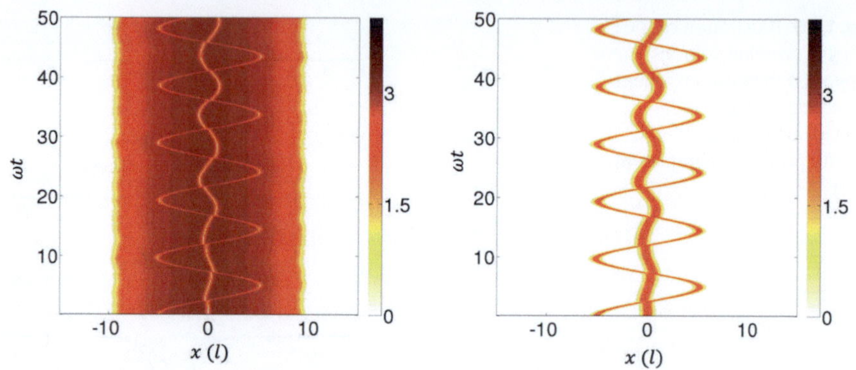

Fig. 11.7 Evolution of two bright-dark solitons within a harmonic trap according to simulations of the coupled 1D GPEs. Adapted with permission from [11]. Copyright 2014, The Author(s). All rights reserved

informs us that as the number of bright atoms increases, the soliton width increases, which is consistent with the increase repulsion within the bright component.

Similar to single-component solitons, a particle model can be developed for bright-dark solitons, which is valid providing the axial potential varies slowly on the length-scale of the bright-dark soliton κ^{-1}. This can be used to determine the oscillation frequency of a bright-dark soliton in a trap, where it is predicted that the oscillation frequency is lower than for a dark soliton and become reduced as the bright component is enlarged [23]. Figure 11.7 illustrates the oscillation of two bright-dark solitons where each component experiences a harmonic potential. Note how the solitons undergo stable oscillations and also how they pass through each other unscathed, characteristic of soliton dynamics.

Bright-dark solitons have been created experimentally with two-component condensates. Becker et al. [24] generated a dark soliton within a single ^{87}Rb condensate, then transfered approximately 10% of the atoms in the local vicinity of the dark soliton to a different spin state, creating a localised second component. This then evolved into a bright-dark soliton and oscillated in the trap in excellent agreement with GPE predictions. The bright soliton contained around 4000 atoms and had a width of approximately 6.7 μ, significantly broader than the healing length which was approximately 0.8 μ. Hoefer et al. [25] generated a train of bright-dark solitons by inducing two components to move relative to each other; the ensuing counterflow instability led to the formation of dark soliton phase defects in one component and the second component then filling each dark soliton to create bright-dark solitons.

11.6 Filled Vortex Cores

An interesting situation occurs when the core of a vortex line in the first condensate is filled of atoms of a second condensate. This case is also referred to as the problem of

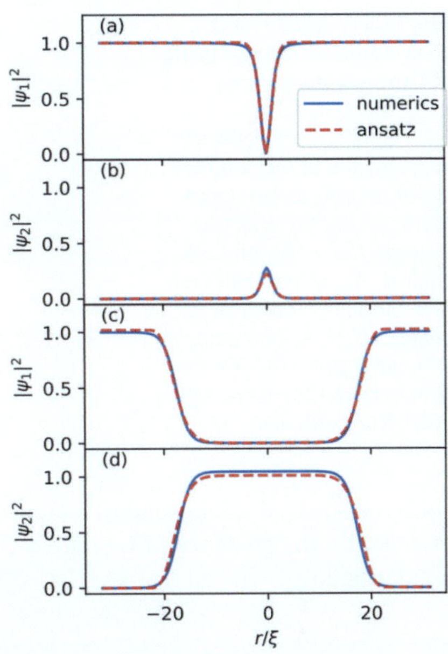

Fig. 11.8 Infilled vortex. Radial density profiles of the majority component, $n_1 = |\psi_1|^2$ (**a, c**) and of the minority component, $n_2 = |\psi_2|^2$ (**b, d**) for $N_2 = 2$ (**a, b**) and $N_2 = 1000$ (**c, d**) atoms. Solid blue line: numerical solutions. Dashed red lines: variational method solutions. Notice how the core of the vortex in the first condensate becomes larger if atoms of the second condensate are added. Reprinted under CC-BY-4.0 license from [26]. Copyright 2024, The Author(s)

the *massive vortex* [26].[1] The natural questions are whether the vortex core changes size, and whether, if the vortex moves, its trajectory is affected by the infilling atoms. Precise answers to these and related questions open the way to the use of a second condensate to track the trajectories of vortex lines in the first condensate, in the same spirit of PTV (Particle Tracking Velocimetry) based on micron-size frozen hydrogen snowballs in liquid helium. The general formulation of the problem of the motion of massive vortices in a condensate was tackled by Richaud and collaborators [27] in the approximation of the vortex point model (VPM).

The solution to the first question was found numerically and analytically (using a variational method) [26]. The authors solved Eqs. (11.6) and (11.7) in 2D on the x, y plane (the vortex being aligned along the z-direction) for a system of uniform density with no confining trap ($V_1 = V_2 = 0$), $m_2/m_1 = 1$, $g_{11} = g_{22} = 1$ and $g_{12} = 1.1$. Calling species 1 the majority component with the vortex, and species 2 the minority component which infills the vortex, they found that the vortex core becomes larger— see Fig. 11.8—for increasing number of infilled atoms of the minority component.

To answer the second question, the same authors studied numerically the scattering of two colliding vortex-antivortex pairs at increasing number of infilling vortices of the second species. Initially the vortices of the majority component were placed at $(x_k, y_k) = (\pm 64\xi_1, \pm 16\xi_1)$ where ξ_1 is the healing length of the majority species. In the absence of the minority species ($N_2 = 0$) the deflected trajectories of the two pairs are represented by the open black circles of Fig. 11.9. The blue crosses and the

---

[1] We must keep in mind that even, if the vortex core is hollow, when it moves in the superfluid it has a small nonzero hydrodynamical mass; in many cases (but not all) this hydrodynamic mass can be neglected.

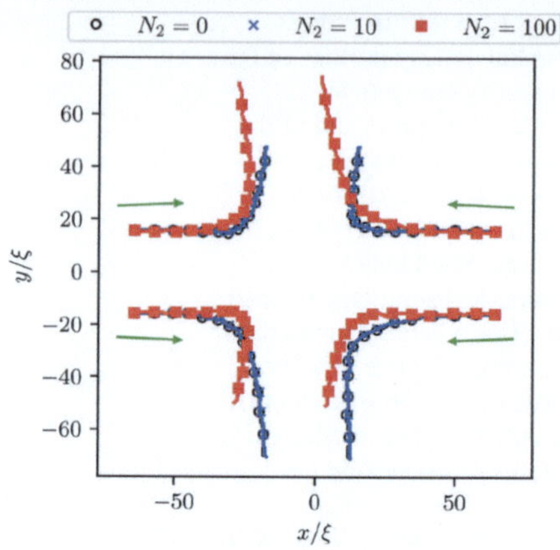

Fig. 11.9 Infilled vortices. Trajectories on the x, y plane of two vortex-antivortex pairs which move toward each other and are deflected as a function of the number N_2 of infilling atoms. Open black circles: $N_2 = 0$; blue crosses: $N_2 = 10$; solid red squares: $N_2 = 100$. Notice how the angle of deflection changes if N_2 is increased. Reprinted under CC-BY-4.0 license from [26]. Copyright 2024, The Author(s)

solid red squared correspond to increasing N_2. Notice how the vortex trajectories are affected by the presence of a relatively small number of infilling atoms of a second condensate.

11.7   Dynamical Instabilities of Two-Component Condensates

Here we introduce several key dynamical phenomena which are supported by two-component condensates.

11.7.1 Counterflow Instability

Consider a homogeneous system of two miscible condensates which have a relative velocity difference between them. Stability analysis reveals that modes become unstable when the relative speed exceeds a critical value., with the most unstable modes being those with wavevectors parallel to the relative velocity [28,29]. This modulational instability becomes reinforced by nonlinear effects leading to the generation of large-amplitude defects in the form of solitons in effectively 1D systems and vortices in 2D and 3D systems [30]. The counterflow instability has been demonstrated experimentally [32] for two elongated and overlapping condensates of ^{87}Rb in different magnetic states. An external magnetic field gradient was used to drive the condensates in different directions to realise the counterflow. For high counterflow speeds, it was seen that trains of coupled dark solitons were formed in the condensate; simulations confirmed that these structures were the product of the counterflow-driven modulational instability. This is depicted in Fig. 11.10. Due to the three-dimensional nature of the condensates, the dark solitons were seen to decay into vortices via the snake instability we met in Chap. 5. For slower counterflow, a state of many dark-bright solitons was created in the same experimental set-up, seeded at the condensate interfaces [25].

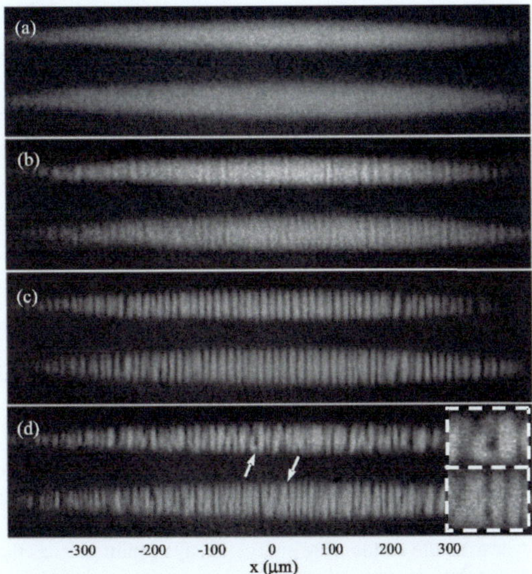

Fig. 11.10 In the experiment of Ref. [32], two overlapping elongated condensates were driven into relative motion. For sufficiently high relative speed, the initial smooth condensates (i) underway a modulational instability to develop regular linear corrugations (ii) which grew into a state of many dark solitons in each component (iii). At later times the patterning become more irregular and some of the density defects appear as vortices, highlighted by the arrows/insets. Reprinted figure with permission from [32]. Copyright 2011 by the American Physical Society

11.7.2 Formation and Relaxation of Domains

Consider a homogeneous miscible system which is suddenly quenched into the immiscible regime, for example, via rapid Feshbach tuning of the interactions [31,33]. The background densities undergo a modulational instability, with oscillatory perturbations grow exponentially in time. The mode with the fastest growth rate has the wavenumber,

$$Q^2 = \frac{m}{\hbar}\left[\sqrt{n_1 g_{11} + n_2 g_{22} + 4n_1 n_2 g_{11} g_{22}\left(\frac{g_{12}^2}{g_{11}g_{22}-1}\right)} - n_1 g_{11} - n_2 g_{22}\right].$$

(11.31)

This instability leads to the rapid bunching of each component into domains of characteristic lengthscale $1/Q$ and the inter-species repulsion drives the components apart, such that the system forms domains that alternate from one component to another. The domain walls can propagate and have soliton-like properties [34]. The presence of multiple domains and interfaces means that the system is in a state of high energy. Over time there a slow relaxation towards the ground state with a minimal amount of domains and interface, driven by tunnelling between different domains of the same component, collisions between domains and changes in interface topology. During this process, transient dark-bright solitons can be formed when a domain

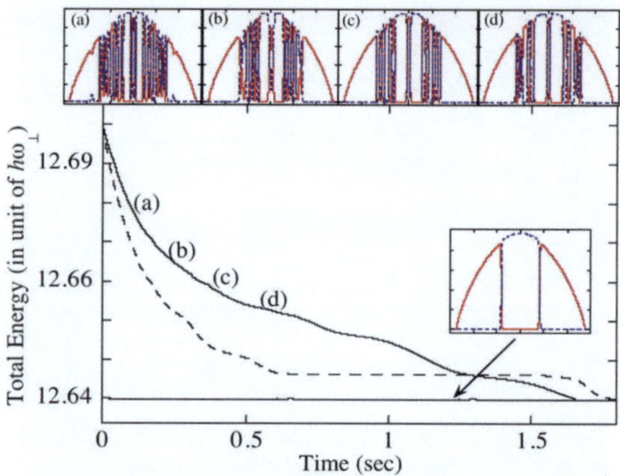

Fig. 11.11 Starting the non-equilibrium state of two overlapping, immiscible, elongated condensates, a modulational instability drives the formation of many alternating domains in each component. The top panels show the axial density profiles of each component and the presence of alternating domains. Over time the condensate energy decays (main plot) as the number of domains reduces. The system approaches the ground state solution, with a central domain of one component surrounded by the other component. The results are based on simulations of the 3D GPE with phenomenlogical damping term γ to mimic thermal dissipation of the system; results are for a weak dissipation ($\gamma = 0.025$, solid line) and a stronger value ($\gamma = 0.05$, dashed line). Reprinted figure with permission from [33]. Copyright 2004 by the American Physical Society

becomes sufficiently small to be of the order of the dark-bright soliton lengthscale. The presence of domains and their relaxation over time is shown in Fig. 11.11, based on the computational simulations of the GPE [33].

11.7.3 Kelvin-Helmholtz Instability

The Kelvin-Helmholtz instability, well known from classical fluid dynamics, describes the instability at the interface between two parallel laminar fluid streams with differing velocities. The interface develops corrugations and rolls up, disrupting the laminar flow and usually triggering a transition to turbulent flow. The quantum analog of the Kelvin-Helmholtz instability has been considered for an immiscible two-component system where the components are set into relative motion [35,36]. Figure 11.12 illustrates the ensuing instability. For sufficiently high relative speed, the interface develops a sinusoidal corrugation, concurrent with the generation of regularly-spaced vortices at the interface.; the line density of vortices grows with the relative velocity. These vortices then begin to interact with each other, driving the roll up of the interface, and the clustering of associated clustering of vortices; over time progressively larger clusters form. The quantum fluid system thus mimics the roll-up iof classical fluids. An effective shear layer is formed from the collective motion of the vortices in the vicinity of the original interface. When the condensates are weakly immiscible, with significant penetration into each other, then the effects of counterflow in the penetration region can become significant and complicate the dynamics.

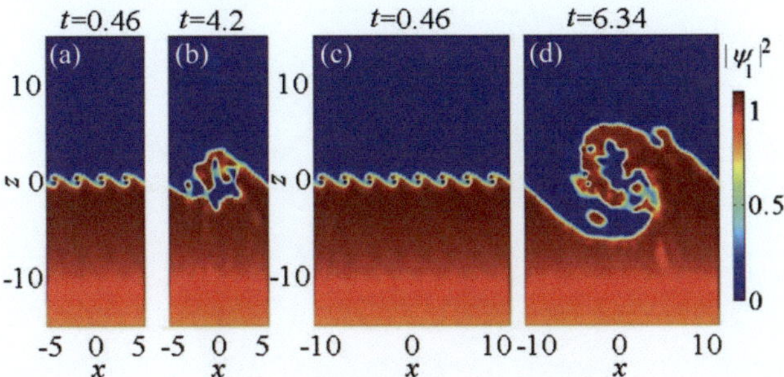

Fig. 11.12 The Kelvin-Helmholtz instability between two adjacent immiscible condensates. Relative flow between the condensates causes the visible undulation of the interface and creation of vortices, which subsequently roll-up into larger structure of many vortices. The behaviour is consistent across two different widths of simulation box. Reprinted figure with permission from [35]. Copyright 2004 by the American Physical Society

The quantum Kelvin-Helmholtz instability can also be generated in a single component condensate by bring together two initially-separated flowing parts of the condensate [37]; indeed this was recently demonstrated experimentally using a single atomic superfluid confined to a ring-trap [38].

11.7.4 Rayleigh-Taylor Instability

In classical fluids, the Rayleigh-Taylor instability arises when a lighter fluid lies underneath a heavier fluid. Buoyancy drives the instability of the interface, resulting in the lighter fluid punching through the interface on a characteristic lengthscale, creating mushroom-like protrusions into the heavier fluid. An analogous scenario has been considered for immiscible two-component condensates [35, 39, 40]. Here it was considered that the condensate atoms are in two opposite spin states and a magnetic field gradient is applied such that the condensate experience opposing forces, thereby mimicking differences in potential energy and thus buoyancy. The instability is seen to progress is close analogy with the classical version, with the instability of the interface leading to mushroom-shaped protrusions of the lighter component entering into the heavier component. The key difference to the classical version is the formation of vortex rings under the caps of the mushrooms, driven by the shear flow during the instability. This is illustrated in Fig. 11.13.

Exercises

11.1 *Miscible-immiscible condition*:
Consider a two-component condensate in the immiscible state. Starting from Eq. 11.16, minimize E_{imm} hence derive Eqs. (11.17) and (11.18).

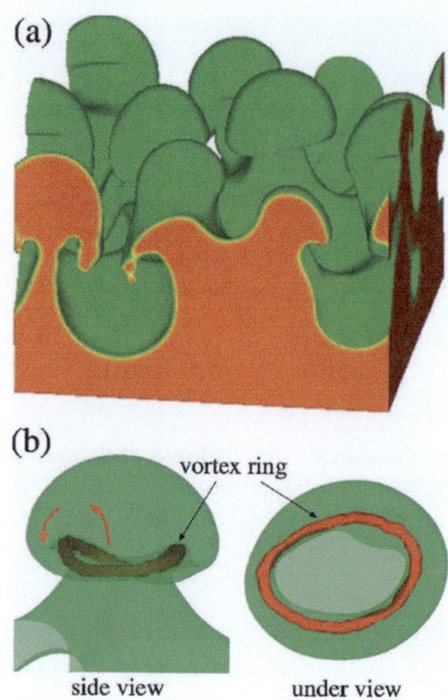

Fig. 11.13 The Rayleigh-Taylor instability between two adjacent immiscible condensates, driven by differential buoyancy forces, leads to the instability of the interface and the lighter component entering the heavier component in localised mushroom-shaped protusions. Vortex rings are formed under the caps of the mushrooms. Shown iare isosurfaces of the density of the lighter component, based on GPE simulations. Reprinted under CC-BY-3.0 license from [39]. Copyright 2009, The Author(s)

11.2 *Thomas-Fermi solution 1*:
Starting from the coupled Gross-Pitaevskii equations (11.1) and (11.2), derive the Thomas-Fermi solution given by Eqs. (11.22) and (11.23).

11.3 *Thomas-Fermi solution 2*:
Starting from the Thomas-Fermi solution given by Eqs. (11.22) and 11.23, and assuming the condensates have the same atomic mass, and experience the same spherically-symmetric harmonic potential ($V_1(\mathbf{r}) = V_2(\mathbf{r}) = m\omega^2 r^2$), obtain an expression for the ratio of the Thomas-Fermi radii R_2/R_1 of the two clouds. Check that for $g_{12} = 0$ you recover the result you would expect for independent condensates.

If the condensates also have the same intra-species scattering lengths ($g_{11} = g_{22}$) and chemical potentials ($\mu_1 = \mu_2$), obtain an expression for the density profile of each component. Hence identify the effective combined nonlinear coefficient experienced by each component.

11.4 *Domain wall solution*:
Consider two immiscible 1D condensates in a uniform system. The condensates are described by the 1D coupled GPEs, Eq. (11.27), written in the homogeneous system units for component 1.

Show that for $g_{12} = 3g_{11} = 3g_{22}$ and $m_1 = m_2$ there exists a domain wall solution:

$$\psi_j(x, t) = Ae^{-i\mu t}\left[1 + (-1)^j \tanh\left(\frac{x}{\sigma}\right)\right], \tag{11.32}$$

for $j = 1, 2$ and where $A, \sigma > 0$. Hence obtain expressions for the dimensionless chemical potential μ and dimensionless lengthscale of the interface σ in terms of A.

References

1. Y. Kawaguchi, M. Ueda, Phys. Rep. **520**, 253 (2012)
2. S.B. Papp, J.M. Pino, C.E. Wieman, Phys. Rev. Lett. **101**, 040402 (2008)
3. A. Guttridge et al., Phys. Rev. A **98**, 022707 (2018)
4. C. D'Errico et al., Phys. Rev. Res. **1**, 033155 (2019)
5. Z. Guo et al., Phys. Rev. Res. **3**, 033247 (2021)
6. C.K. Law et al., Phys. Rev. Lett. **79**, 3105 (1997)
7. P. Ao, S.T. Chui, Phys. Rev. A **58**, 4836 (1998)
8. K.L. Lee, N.B. Jørgensen, I.-K. Liu, L. Wacker, J.J. Arlt, N.P. Proukakis, Phys. Rev. A **94**, 013602 (2016)
9. K.L. Lee et al., New J. Phys. **20**, 053004 (2018)
10. M. Trippenbach et al., J. Phys. B **33**, 4017 (2000)
11. R. W. Pattinson, *Two-Component Bose-Einstein Condensates: Equilibria and Dynamics at Zero Temperature and Beyond* (Newcastle University, 2014)
12. R.W. Pattinson et al., Phys. Rev. A **87**, 013625 (2013)
13. T-L. Ho, V.B. Shenoy, Phys. Rev. Lett. **77**, 3276 (1996)
14. H. Pu, N.P. Bigelow, Phys. Rev. Lett. **80**, 1130 (1998)
15. O. Indekeu et al., Phys. Rev. A **91**, 033615 (2015)
16. I.E. Mazets, Phys. Rev. A **65**, 033618 (2002)
17. P.G. Kevrekidis et al., Eur. Phys. J. D **28**, 181 (2004)
18. A.P. Sheppard, Y.S. Kivshar, Phys. Rev. E **55**, 4773 (1997)
19. Y.S. Kivshar, B. Luther-Davies, Phys. Rep. **298**, 81 (1998)
20. S. Trillo, S. Wabnitz, E.M. Wright, G.I. Stegeman, Opt. Lett. **13**, 871 (1988)
21. D.N. Christodoulides, Phys. Lett. A **132**, 451 (1988)
22. M. Shalaby, A.J. Barthelemy, IEEE J. Quantum Electron. **28**, 2736 (1992)
23. T. Busch, J.R. Anglin, Phys. Rev. Lett. **87**, 010401 (2001)
24. C. Becker et al., Nat. Phys. **4**, 496 (2008)
25. C. Hamner, J.J. Chang, P. Engels, M.A. Hoefer, Phys. Rev. Lett. **106**, 065302 (2011)
26. R. Doran, A.W. Baggaley, N.G. Parker, Phys. Rev. A **109**, 023318 (2024)
27. A. Richaud, V. Penna, A.L. Fetter, Phys. Rev. A **103**, 023311 (2021)
28. C.K. Law, C.M. Chan, P.T. Leung, M.-C. Chu, Phys. Rev. A **63**, 063612 (2001)
29. M. Abad, A. Recati, S. Stringari, F. Chevy, Eur. Phys. J. D **69**, 126 (2015)
30. H. Takeuchi, S. Ishino, M. Tsubota, Phys. Rev. Lett. **105**, 205301 (2010)
31. P. Ao, S.T. Chui, J. Phys. B **33**, 535 (2000)
32. M.A. Hoefer, J.J. Chang, C. Hamner, P. Engels, Phys. Rev. A **84**, 041605(R) (2011)
33. K. Kasamatsu, M. Tsubota, Phys. Rev. Lett. **93**, 100402 (2004)
34. L.-Z. Men, Y.-H. Qin, L.-C. Zhao, Z.-Y. Yang, Chin. Phys. B **28**, 060502 (2019)
35. D. Kobyakov et al., Phys. Rev. A **89**, 013631 (2014)
36. N. Suzuki, H. Takeuchi, K. Kasamatsu, M. Tsubota, H. Saito, Phys. Rev. A **82**, 063604 (2010)
37. A.W. Baggaley, N.G. Parker, Phys. Rev. A **97**, 053608 (2018)

38. D. Hernandez-Rajkov et al., Nat. Phys. **20**, 939 (2024)
39. K. Sasaki, N. Suzuki, D. Akamatsu, H. Saito, Phys. Rev. A **80**, 063611 (2009)
40. S. Gautam, D. Angom, Phys. Rev. A **81**, 053616 (2010)
41. T. Kadokura, T. Aioi, K. Sasaki, T. Kishimoto, H. Saito, Phys. Rev. A **85**, 013602 (2012)

Dipolar Bose-Einstein Condensates 12

Abstract

The aforementioned experiments and results were limited to BECs comprising of local effective interactions. However, a range of exotic long-range interactions are possible. In particular, the creation of condensates with atoms possessing significant magnetic dipole moments affords a new opportunity to explore the interplay of magnetic effects with short-range interactions. In this chapter, we explore how dipolar interactions modify the elementary excitation spectrum and shape the properties of solitons and vortices in quantum gases, while also uncovering their role in the emergence of novel ultracold states of matter.

12.1 Introduction

Generally, interactions between particles in an atomic BEC are not as simple as the billiard-ball picture that has been presented so far. The alkali atoms that were first Bose condensed also exhibit a negligible magnetic dipole-dipole interaction with an effective range of femtometres (10^{-15}m). However, it wasn't until chromium was condensed in 2006 [1], followed by dysprosium (2011) [2] and erbium (2012) [3], that these long-range interactions affected the observed physics, requiring new theoretical tools.

In this Chapter, we extend the Gross-Pitaevskii theory to include the dipole-dipole interaction, and show how this modifies the known properties of quantum gases, setting the scene to study an entirely new phase of matter, the *supersolid*, which will be explored in Chap. 13.

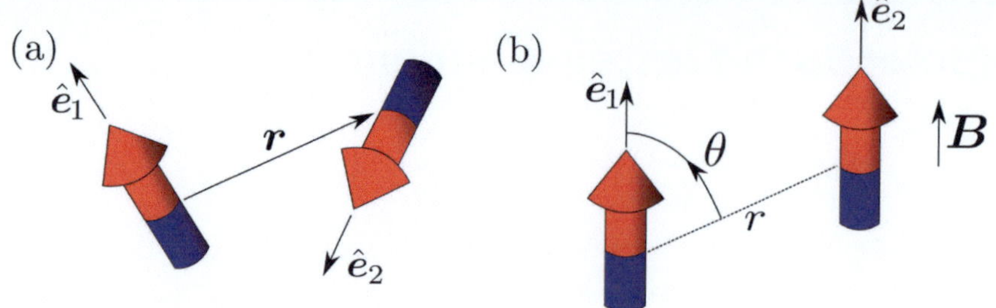

Fig. 12.1 Schematic of the dipole-dipole interaction (DDI). **a** General case with arbitrary polarisation. **b** Polarised case, reducing the number of parameters to two: the distance between dipoles r and the angle between them θ

12.2   The Dipole-Dipole Interaction

The general dipole-dipole interaction (DDI) potential between two dipoles with arbitrary polarisation takes the form [4]

$$U_{\mathrm{dd}}(\boldsymbol{r}) = \frac{C_{\mathrm{dd}}}{4\pi} \left(\frac{(\hat{\boldsymbol{e}}_1 \cdot \hat{\boldsymbol{e}}_2)r^2 - 3(\hat{\boldsymbol{e}}_1 \cdot \boldsymbol{r})(\hat{\boldsymbol{e}}_2 \cdot \boldsymbol{r})}{r^5} \right), \tag{12.1}$$

where C_{dd} is $\mu_0 \mu_m^2$ (the permeability of free space multiplied by the dipole moment squared) for magnetic dipoles and d^2/ϵ_0 (electric dipole moment squared over the permittivity of free space) for electric dipoles, $\boldsymbol{r}$ is the vector joining the dipoles, while $r = |\boldsymbol{r}|$, and $\hat{\boldsymbol{e}}_j$ is the orientation of dipole j. See Fig. 12.1a for a schematic of this setup.

As we are describing an ensemble of particles with a mean-field description, it is convenient to limit this to the polarised case, such that $\hat{\boldsymbol{e}}_1 = \hat{\boldsymbol{e}}_2$, implemented by a homogeneous external magnetic or electric field. In spherical polar coordinates the DDI then takes the form

$$U_{\mathrm{dd}}(\boldsymbol{r}) = \frac{C_{\mathrm{dd}}}{4\pi} \left(\frac{1 - 3\cos^2 \theta}{r^3} \right), \tag{12.2}$$

with θ being the angle between the polarising direction and $\boldsymbol{r}$. In this form it is clear that the DDI is both long-ranged[1] ($\propto 1/r^3$) and anisotropic (angle dependent).

The behaviour of the DDI is analogous to two bar magnets. When $\theta = 0$ the dipoles are in an end-to-end configuration and the interaction energy is negative causing mutual attraction, whereas for $\theta = \pi/2$ dipoles are in a side-by-side configuration and the interaction energy is positive causing mutual repulsion.

---

[1] A potential of the form r^{-m} ($m > 0$) is *long-range* when the dimension D of the system satisfies $D \geq m$ and *non-local* otherwise.

12.3   Ultra-Cold Dipolar Systems

Dipolar interactions fall into two categories: electric or magnetic dipoles. In the following section we explore these categories in the context of quantum gases, and discuss the experimental progress that has been made on both. We note that a recent comprehensive review on experiments with dipolar atoms can be found in Ref. [5].

12.3.1 Electric Dipoles

Dipolar Excitons

When an electron is promoted from the valence band to a higher energy state, it leaves behind a positively charged hole. The Coulomb attraction between this electron-hole pair forms a bound quasi-particle known as an exciton. Unlike ultracold atomic gases, exciton systems are inherently dynamic: light creates excitons, which eventually recombine and emit photons. In materials like two-dimensional GaAs quantum wells, continual optical pumping leads to a steady-state population determined by a balance of excitation and decay processes.

If these excitons possess a built-in dipole moment—commonly achieved in bilayer or spatially indirect setups—they become dipolar excitons, which interact via strong, long-range dipole-dipole forces. This interaction manifests as a pronounced blue-shift in their emission spectrum and substantial spectral broadening due to collisions [6–8].

The spin configuration of excitons plays a pivotal role in determining their macroscopic behavior. Excitons are categorized into optically active (bright) and inactive (dark) spin states. While non-interacting theories predict that condensation should favor dark states [9], real systems experience destabilizing exchange interactions. These can drive a transition from dark to bright states above a critical density—a process known as the brightening instability. Applying a vertical electric field enhances repulsion between excitons and suppresses close-range interactions, mitigating this effect. Experiments have confirmed this stabilization mechanism [7, 8, 10], and theoretical modeling has further clarified the dynamics [11]. Additionally, signatures of superfluid transport have emerged in bilayer systems of dipolar excitons [12, 13], raising intriguing possibilities for analogies to high-temperature superconductivity in bilayer graphene.

Dipolar Molecules

Ultracold gases of polar molecules have gained intense interest over the past decade, culminating in the first realization of a ground-state bosonic molecular BEC in 2024 [14]. These systems inherit many characteristics from atomic BECs but offer much stronger interparticle interactions due to their permanent electric dipole moments. However, producing such molecular condensates poses significant challenges. Typically, two distinct atomic species must be simultaneously cooled into degenerate states while kept spatially separated. After achieving individual condensation, the atomic clouds are merged, and molecular formation is initiated via Feshbach association, achieved by scanning the magnetic field across a Feshbach resonance (see Sec. 3.1). To qualify as a BEC, the molecules must occupy their rovibrational ground state, i.e. with no internal motion from rotation or vibration.

A key breakthrough enabling stability in these systems was the application of *microwave shielding*, where two internal molecular states are coupled via a microwave field [15]. This creates a repulsive core at short range, effectively preventing collapse due to three-body recombination and allowing the long-range dipolar interaction to dominate. The strength of these losses also depends on whether the molecules are chemically reactive. For instance, reactive loss processes like atom exchange (*e.g.*, $KRb + KRb \rightarrow K_2 + Rb_2$) can severely limit lifetime. Choosing chemically stable species such as NaK or CsRb circumvents this issue and provides a more robust platform for exploring dipolar quantum gases.

Rydberg Atoms

Rydberg atoms—those in which one electron occupies a highly excited orbital— exhibit the strongest dipole-dipole interactions among known cold atom systems. The size of a Rydberg atom scales with the square of the principal quantum number, n^2, reaching up to micrometer scales. This leads to interaction strengths that grow as n^4. Upon excitation to a Rydberg state, the large spatial extent of the electron cloud prevents further excitation within a surrounding region known as the *blockade radius*, a phenomenon measurable through suppressed second-order photon correlations (g_2).

However, the practical use of Rydberg atoms is limited by their short lifetimes, which scale unfavorably as n^{-3}, and deteriorate further at higher densities [16]. In many cases, Rydberg systems do not possess permanent dipole moments unless externally polarized. An exception arises when a Rydberg atom binds weakly to a nearby ground-state atom, forming a Rydberg molecule with a permanent electric dipole moment [17]. These exotic bound states provide a novel platform for studying dipolar effects in cold gases.

12.3.2 Magnetic Dipoles

Unlike electric dipoles, where long-range effects require an external electric potential gradient to arise, some species of atoms have a significant permanent dipole moment in their ground state. These systems are called ultracold dipolar atoms, and will be the focus of the rest of this chapter.

Chromium

A significant milestone in dipolar quantum gases was reached in 2005, when the Pfau group successfully created a Bose-Einstein condensate (BEC) of ^{52}Cr atoms [1]. Chromium stands out due to its ground state total spin of $S = 3$, resulting in a relatively large magnetic dipole moment of $6\,\mu_B$, where $\mu_B = 9.274 \times 10^{-24}\,\mathrm{JT}^{-1}$ (Joules per Tesla) is the Bohr magneton. At ultracold temperatures, this non-zero spin introduces a dipolar-specific two-body loss mechanism: spin relaxation, wherein the spin of an atom flips during a collision, redistributing angular momentum between spin and motion. This decay process was mitigated in the experiment by using optical pumping to prepare all atoms in the lowest Zeeman sublevel, while sufficiently strong magnetic fields suppressed spin-changing collisions through energy conservation.

Rare-Earth Elements

Due to their larger magnetic dipole moments, rare-earth atoms have largely supplanted chromium in dipolar quantum gas experiments. The first such strongly magnetic BEC was realized in 2011 by the Lev group at Stanford, using ^{164}Dy atoms [2]. With a magnetic moment of $9.93\,\mu_B$, dysprosium remains one of the most promising atoms for exploring dipolar effects. That same group later produced the first degenerate Fermi gas of ^{161}Dy [18], demonstrating the feasibility of magnetic dipolar fermions.

Shortly thereafter, in 2012, the first erbium BEC was produced by a team in Innsbruck [3]. Erbium, with its substantial magnetic moment of $6.98\,\mu_B$, quickly joined dysprosium as a prime candidate for dipolar studies. The same group went on to realize quantum-degenerate mixtures of erbium and dysprosium, exploiting the multiple stable isotopes of each species. This enabled the creation of five distinct bosonic mixtures and one Bose-Fermi system [19,20].

More recently, thulium and europium have also been cooled to quantum degeneracy [21,22]. Additionally, experiments with holmium atoms have achieved ultracold temperatures, with degeneracy expected to be within reach [23].

12.3.3 Comparison of Dipolar Strengths

It is convenient to define a characteristic lengthscale for dipolar interactions,

$$a_{\mathrm{dd}} = \frac{C_{\mathrm{dd}}m}{12\pi\hbar^2}, \tag{12.3}$$

and subsequently define

$$g_{\mathrm{dd}} = \frac{4\pi\hbar^2 a_{\mathrm{dd}}}{m}. \tag{12.4}$$

Note that Eq. (12.3) is proportional to the atomic mass. So, even though chromium and erbium have similar dipole moments of $6\,\mu_B$ and $7\,\mu_B$, respectively, erbium is over 3 times heavier than chromium, and thus a much larger a_{dd}. We define the quantity

$$\varepsilon_{\mathrm{dd}} = \frac{a_{\mathrm{dd}}}{a_s} \equiv \frac{g_{\mathrm{dd}}}{g}, \tag{12.5}$$

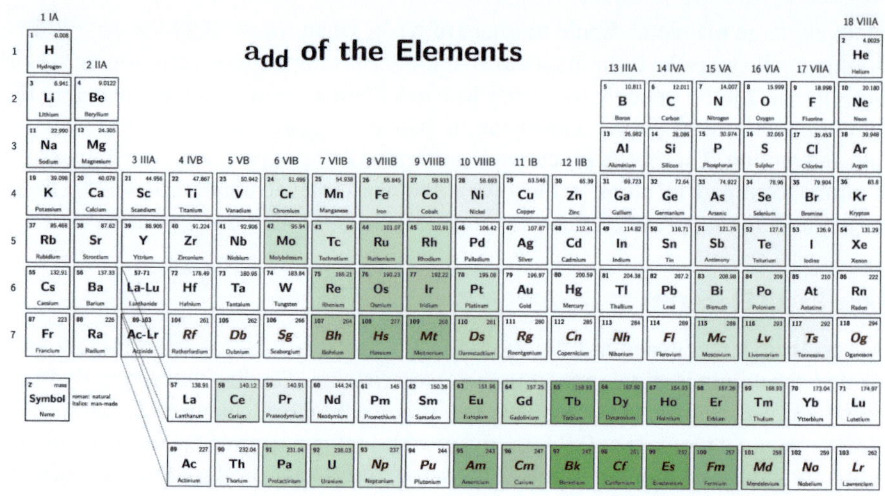

Fig. 12.2 Dipole strengths. A darker shade of green indicates the size of a_{dd}, with the maximum around $200a_0$

as a measure of how dipolar the state is, such that $\varepsilon_{\mathrm{dd}} > 1$ is dominantly dipolar, and $\varepsilon_{\mathrm{dd}} < 1$ as dominantly s-wave.

Figure 12.2 shows the maximum a_{dd} across the periodic table. The middle columns carry the largest dipolar lengths due to their electronic structure. The maximum dipolar length in the naturally occuring non-radioactive elements is dysprosium, with $a_{\mathrm{dd}} = 131a_B$. Molecular gases boast dipolar lengths in the 10^4–10^5 range [14].

12.4   The Dipolar Gross-Pitaevskii Equation

As we have seen, the workhorse for modelling BECs is the GPE. The dipolar interaction simply modifies the potential terms of the GPE, where the pseudopotential Eq. (3.3) now becomes

$$\mathcal{U}(\boldsymbol{r} - \boldsymbol{r}') = g\delta(\boldsymbol{r} - \boldsymbol{r}') + \frac{3g_{\mathrm{dd}}}{4\pi}\left(\frac{1 - 3\cos^2\theta}{r^3}\right). \tag{12.6}$$

Inserting this pseudo-potential into Eq. (3.11) leads to the time-dependent equation of motion

$$i\hbar\frac{\partial\Psi(\boldsymbol{r}, t)}{\partial t} = \left(-\frac{\hbar^2\nabla^2}{2m} + V_{\mathrm{ext}}(\boldsymbol{r}, t) + g|\Psi(\boldsymbol{r}, t)|^2 + \Phi_{\mathrm{dd}}(\boldsymbol{r}, t)\right)\Psi(\boldsymbol{r}, t) \tag{12.7}$$

The dipole-dipole contribution is given by

$$\Phi_{\mathrm{dd}}(\boldsymbol{r}, t) = \int \mathrm{d}^3\boldsymbol{r}'\, U_{\mathrm{dd}}(\boldsymbol{r} - \boldsymbol{r}')|\Psi(\boldsymbol{r}', t)|^2. \tag{12.8}$$

The dipolar GPE was first written in this form by Góral and co-authors in 2000 [24].

12.4.1 Fourier Transform of the Dipole-Dipole Interaction

An important quantity for theoretical modelling of the dipolar interaction is the momentum space representation of the DDI, i.e. $\mathbf{k} = (k_x, k_y, k_z)$ space where $|\mathbf{k}| = k$. The derivation will be an exercise at the end of the chapter. The final result is

$$\tilde{U}_{\mathrm{dd}}(\mathbf{k}) = \frac{3g_{\mathrm{dd}}}{4\pi} \int \mathrm{d}^3 \mathbf{r}\, \frac{1 - 3\cos^2\theta}{r^3} e^{-i\mathbf{k}\cdot\mathbf{r}} = g_{\mathrm{dd}}(3\cos^2\theta_k - 1), \qquad (12.9)$$

where θ_k is the angle between the polarization axis and wave vector in momentum space. For example, if the dipoles are polarized along the z-axis, then $\cos\theta = z/r$ and $\cos\theta_k = k_z/k$.

12.4.2 Dipolar GPE in Lower Dimensions

For the non-dipolar GPE, the dimensional reduction from 3D to 2D or 1D only changes the coefficient of the contact interaction term, as we presented in Sect. 3.5.4. Here, however, the dipolar interaction potential also depends on the spatial dimension and the angle between the tightly trapped axis and the polarization axis, and so we have to derive the lower dimensional GPE with care.

2D Dipolar GPE

Taking an oblate, or pancake-shaped trap ($V_{\mathrm{ext}}(\mathbf{r}) = V_{\mathrm{ext}}(\boldsymbol{\rho}) + 1/2m\omega_z^2 z^2$, where $\boldsymbol{\rho} = (x, y)$ and $\hbar\omega_z \gg \mu$), we can assume that the wavefunction in the tightly confined direction is given by the ground state in the harmonic oscillator, and the total wavefunction can be separated into $\Psi(\mathbf{r}) = \psi(\boldsymbol{\rho}) \exp(-z^2/2l_z^2)/\sqrt{\pi^{1/2}l_z}$, with $l_z = \sqrt{\hbar/m\omega_z}$. Substituting this in to Eq. (12.7) and integrating out z leaves behind the 2D dipolar GPE

$$i\hbar\frac{\partial\psi}{\partial t} = \left(-\frac{\hbar^2\nabla_\rho^2}{2m} + V_{\mathrm{ext}}(\boldsymbol{\rho}, t) + g_{\mathrm{2D}}|\psi|^2 + \int \mathrm{d}^2\boldsymbol{\rho}'\, U_{\mathrm{2D}}(\boldsymbol{\rho} - \boldsymbol{\rho}')|\psi|^2\right)\psi(\boldsymbol{\rho}, t),$$

$$(12.10)$$

where $\nabla_\rho = \left(\frac{\partial}{\partial x}, \frac{\partial}{\partial y}\right)$, $g_{\mathrm{2D}} = g/\sqrt{2\pi}l_z$, and the Fourier transform of the 2D DDI is

$$\tilde{U}_{\mathrm{2D}}(\mathbf{q}) = \frac{4\pi g_{\mathrm{dd}}}{3\sqrt{2\pi}l_z}[F_\parallel(\mathbf{q})\sin^2\alpha + F_\perp(\mathbf{q})\cos^2\alpha], \qquad (12.11)$$

where $\mathbf{q} = (q_x, q_y) = (k_x, k_y)l_z/2$, α is the angle between the polarization axis and the z axis in the $x - z$ plane, and

$$F_\parallel(\mathbf{q}) = -1 + 3\sqrt{\pi}(q_x^2/q)e^{q^2}\mathrm{erfc}(q),$$

$$F_\perp(\mathbf{q}) = 2 - 3\sqrt{\pi}q e^{q^2}\mathrm{erfc}(q). \qquad (12.12)$$

1D Dipolar GPE

Similarly for prolate, or cigar-shaped traps ($V_{\text{ext}}(r) = 1/2m\omega_\rho^2\rho^2 + V_{\text{ext}}(z)$, where $\hbar\omega_\rho \gg \mu$), we can assume that the wavefunction can be separated into $\Psi(r) = \psi(z)\exp(-\rho^2/2l_\rho^2)/\sqrt{\pi^{1/2}l_\rho}$, with $l_\rho = \sqrt{\hbar/m\omega_\rho}$. Substituting this in to Eq. (12.7) and integrating out ρ leaves the 1D dipolar GPE

$$i\hbar\frac{\partial\psi}{\partial t} = \left(-\frac{\hbar^2}{2m}\frac{\partial^2}{\partial z^2} + V_{\text{ext}}(z,t) + g_{1D}|\psi|^2 + \int dz'\, U_{1D}(z-z')|\psi|^2\right)\psi(z,t),$$

$$(12.13)$$

where $g_{1D} = g/2\pi l_\rho^2$, and the Fourier transform of the 1D DDI is

$$\tilde{U}_{1D}(q_z) = \frac{g_{dd}}{4\pi l_\rho^2}(1 - 3\cos^2\phi)\left[1 - 3q_z^2 e^{q_z^2}E_1[q_z^2]\right], \qquad (12.14)$$

where $q_z = k_z l_\rho/\sqrt{2}$, ϕ is the angle between the polarization axis and the z axis, and $E_1[x] = \int_x^\infty dt\, t^{-1}e^{-t}$ is the exponential integral. Note that for $\phi \neq 0$ this equation replicates the correct physics but is not a good predictor for the experimental parameters needed to observe it, as the anisotropy in the DDI means that the approximation of a single azimuthal lengthscale l_ρ does not hold. There are other dimensional reductions that allow for $l_x \neq l_y$ that we do not cover here [25].

12.5　Evaluation of the Dipolar Potential

The dipolar potential is efficiently evaluated in Fourier space through application of the convolution theorem. The convolution of two functions $f(r)$ and $g(r)$ is stated as

$$(f * g)(r) = \int d^3r'\, f(r-r')g(r') \equiv \int d^3r'\, f(r')g(r-r'), \qquad (12.15)$$

and measures the amount of overlap of one function as it is shifted over the other. The convolution theorem states that

$$(f * g)(r) = \mathcal{F}^{-1}\left[\mathcal{F}[f]\mathcal{F}[g]\right], \qquad (12.16)$$

where the Fourier and inverse Fourier transforms are defined as

$$\mathcal{F}[f](\mathbf{k}) = \tilde{f}(\mathbf{k}) = \int_{\mathbb{R}^d} f(\mathbf{r})\,e^{-i\mathbf{k}\cdot\mathbf{r}}\,d\mathbf{r}, \qquad (12.17)$$

$$\mathcal{F}^{-1}[\tilde{f}](\mathbf{r}) = f(\mathbf{r}) = \frac{1}{(2\pi)^d}\int_{\mathbb{R}^d} \tilde{f}(\mathbf{k})\,e^{i\mathbf{k}\cdot\mathbf{r}}\,d\mathbf{k}, \qquad (12.18)$$

for dimension d with corresponding domain $\mathbb{R}^d$. Thus, the convolution is evaluated as the inverse Fourier transform of the product of the Fourier transforms.

Recall the dipolar interaction term $\Phi_{dd}(\boldsymbol{r}, t)$, stated again here as

$$\Phi_{dd}(\boldsymbol{r}, t) = \int d^3 r' \, U_{dd}(\boldsymbol{r} - \boldsymbol{r}')|\Psi(\boldsymbol{r}', t)|^2. \tag{12.19}$$

The convolution theorem allows us to rewrite this as

$$\Phi_{dd}(\boldsymbol{r}, t) = \mathcal{F}^{-1}\left[\tilde{U}_{dd}(\boldsymbol{k})\tilde{n}(\boldsymbol{k}, t)\right], \tag{12.20}$$

where $\tilde{U}_{dd}(\boldsymbol{k})$ is the k-space dipolar pseudo-potential from Eq. (12.9) and $\tilde{n}(\boldsymbol{k}, t) = \mathcal{F}[n(\boldsymbol{r}, t)]$ is the density in k-space.

The numerical evaluation of Eq. (12.20) is handled simply through fast Fourier transforms (FFTs). FFT algorithms are evaluated through the sum of plane waves with varying coefficients. This means they are naturally periodic and introduce alias copies of the condensate wave function that interact with the system of interest through the long-range DDI. In order to offset this effect, there have been truncated variants of the DDI derived. We will only discuss the simplest case, a spherical cut-off. Restricting the range of the DDI potential to a sphere of radius R_c gives us the real-space DDI

$$U_{dd}^{R_c}(\boldsymbol{r}) = \begin{cases} \frac{C_{dd}}{4\pi} \frac{1-3\cos^2\theta}{r^3}, & r < R_c, \\ 0, & \text{otherwise.} \end{cases} \tag{12.21}$$

As long as we choose $R_c > L$, where L is the system size, this potential is physically reasonable. The analytic Fourier transform is [26]

$$\tilde{U}_{dd}^{R_c}(\boldsymbol{k}) = \frac{C_{dd}}{3}\left[1 + 3\frac{\cos(R_c k)}{R_c^2 k^2} - 3\frac{\sin(R_c k)}{R_c^3 k^3}\right](3\cos^2\theta_k - 1), \tag{12.22}$$

which reduces to the full DDI in the limit $R_c \to \infty$. Note that this remains well behaved as $k \to 0$. This form of the potential is the most commonly used for numerical modelling of 3D dipolar gases. Other truncations have been considered, where the cut-off in the different cardinal directions is not the same, particularly useful for modelling cigar-shaped traps, however their Fourier transforms are typically not analytic.

12.6   Tuning the Dipole-Dipole Interaction

We have already discussed how the local interactions may be tuned by Feshbach resonances, however it is also possible to tune the DDI too, including being able to invert the sign [27]. Reducing the amplitude is in principle a simple task. For an atomic species with total spin number J, there are $2m_J + 1$ possible spin states to

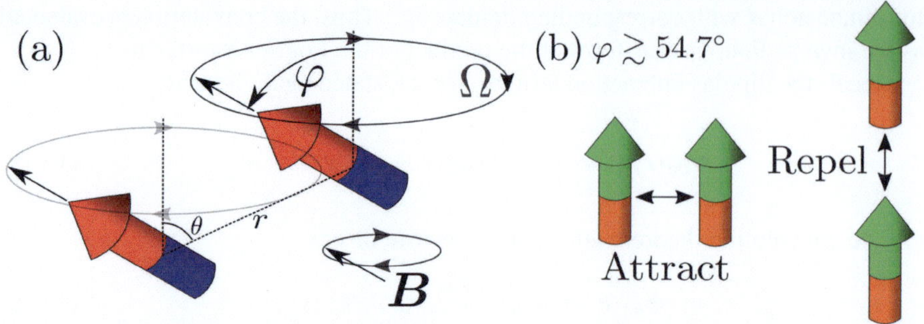

Fig. 12.3 Tuning the dipole-dipole interaction. **a** Using a rotating field at an angle φ from the vertical axis, the dipole-dipole interaction feels a time-averaged interaction for sufficiently large Ω (see Eq. (12.24)). **b** If φ is greater than the magic angle, the rotating dipoles act like static anti-dipoles, where the interaction is reversed from the usual type

choose from, and $a_{\mathrm{dd}} \propto m_J^2$. However, for an m_J not in the ground state there are new open decay channels, such that these other spin states are unstable with limited lifetimes [28].

There is another way to fully tune the dipolar interaction whilst keeping the atoms in their spin ground state. Suppose that the orientation of two dipoles is made to rotate through the application of an external electric or magnetic field, as in Fig. 12.3, such that their time-dependent orientations are

$$\hat{\boldsymbol{e}}_1 = \hat{\boldsymbol{e}}_2 = (\sin\varphi\cos\Omega t, \sin\varphi\sin\Omega t, \cos\varphi). \tag{12.23}$$

Substituting Eq. (12.23) and $\hat{\boldsymbol{r}} = (\sin\theta, 0, \cos\theta)$ into Eq. (12.1) gives

$$\langle U_{\mathrm{dd}}(\boldsymbol{r}, t)\rangle = \frac{C_{\mathrm{dd}}}{4\pi}\left(\frac{1 - 3\cos^2\theta}{r^3}\right)\left[\frac{3\cos^2\varphi - 1}{2}\right], \tag{12.24}$$

where $\langle\cdot\rangle$ denotes an average over the period.

It is clear from Eq. (12.24) that through the choice of φ the magnitude of the dipolar potential is altered by a factor of $(3\cos^2\varphi - 1)/2$. If $\varphi = 0$ then this factor is 1, and we recover the standard DDI, however for $\varphi = \pi/2$ this factor is $-1/2$, leading to repulsion between dipoles for $\theta = 0$, and attraction between dipoles when $\theta = \pi/2$, converse to the standard dipole-dipole interaction. The final interesting case is rotating at the so-called *magic angle*, where the factor in square brackets is 0, and the dipolar interaction can be completely turned off. This can have interesting applications in state preparation, where long-range interactions may need to be removed, for example. Thus, if the oscillation period is made to be shorter than the timescale for internal dynamics of the gas (e.g. phonons or collective modes), then the gas effectively experiences this time-averaged anti dipole-dipole interaction. The Lev group [29] reported the successful application of this method for a dysprosium gas over very short timescales. There, the rotation frequency was 19 times the radial trap frequency, but it is now predicted that a factor of 50 is required in order to see a long-lived stable anti-dipolar gas [30,31]. It is worth noting that microwave shielding in molecular BEC experiments can give rise to this exact anti-dipolar interaction, and results obtained for magnetic atoms with an anti-dipolar interaction will prove useful for understanding this more complex system.

12.7 Elementary Excitations

In 1947 Bogoliubov proposed splitting the mean-field wavefunction into two contributions: the macroscopically occupied condensate described by a complex mean-field $\psi_0(r)$ and weak perturbations describing small-scale excitations to the background $\delta\psi(r, t)$ [32], a theory later applied to superconductivity of metals by de Gennes [33]. This gives a trial wavefunction

$$\psi(r, t) = [\psi_0(r) + \delta\psi(r, t)]\, e^{-i\mu t/\hbar}. \tag{12.25}$$

Substitution of (12.25) into (12.7) and linearising gives the dispersion relation for the low energy elementary excitations. We can link the energy of the excitations to their frequency through $\epsilon(k) = \hbar\omega(k)$. Here, we state the result for the homogeneous BEC,

$$\epsilon(k) = \sqrt{\frac{\hbar^2 k^2}{2m}\left\{\frac{\hbar^2 k^2}{2m} + 2n_0 \tilde{U}(k)\right\}}, \tag{12.26}$$

where n_0 is the density and $\tilde{U}(k)$ is the Fourier transform of any generic two-body interaction pseudo-potential. Using Eq. (12.6) the dispersion relation becomes

$$\epsilon(k) = \sqrt{\frac{\hbar^2 k^2}{2m}\left\{\frac{\hbar^2 k^2}{2m} + 2n_0[g + g_{dd}(3\cos^2\theta_k - 1)]\right\}}, \tag{12.27}$$

which, like the interaction itself, is anisotropic in k-space.

12.7.1 Phonons and Free Particles

Consider two limiting cases of Eq. (12.27), firstly in the limit of large momenta $k \gg 1$ we obtain the free particle dispersion relation

$$\epsilon_{k \gg 1} = \frac{\hbar^2 k^2}{2m}, \tag{12.28}$$

i.e. the effects of particle interactions become negligible.

In the opposite limit, $k \ll 1$ the dispersion relation becomes

$$\epsilon_{k \ll 1} = \hbar k c(k), \tag{12.29}$$

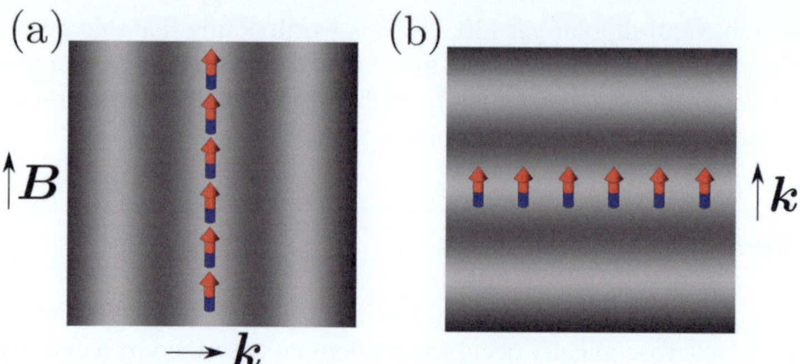

Fig. 12.4 a A phonon with k oriented perpendicularly to the polarisation creates planes of high density where the dipoles are in a end-to-end configuration, causing an instability. **b** If k is parallel to the polarisation direction the dipoles point out of the plane of high density, and remain stable

where these excitations are phonons (sound waves) propagating at speed of sound

$$c(\boldsymbol{k}) = \sqrt{\frac{n_0 g}{m}[1 + \varepsilon_{\mathrm{dd}}(3\cos^2\theta_k - 1)]}. \tag{12.30}$$

This speed of sound is anisotropic, as seen by the θ_k dependence above, a feature that has been experimentally measured in chromium [34] and dysprosium [35]. The nature of this anisotropy is shown in Fig. 12.4. If a phonon is travelling perdendicularly to the polarization axis [(a) $\theta_k = \pi/2$] the atoms in the peak of wave enhance their head-to-tail attraction, and these phonons cost less energy to excite, travelling with speed $c = \sqrt{\frac{n_0 g}{m}[1 - \varepsilon_{\mathrm{dd}}]}$. Conversely, phonons propagating in parallel with the polarization axis [(b) $\theta_k = 0$] cost more energy to excite as they enhance the replusive side-by-side atomic configuration, and $c = \sqrt{\frac{n_0 g}{m}[1 + 2\varepsilon_{\mathrm{dd}}]}$.

A gas will become unstable when ϵ becomes imaginary for any k, signalling a *mode softening*, i.e. a rapid growth of the excitation amplitude. For a gas to become unstable only a single mode has to go soft. The above criteria give the general statement that a homogeneous dipolar (or anti-dipolar) Bose gas in three dimensions is stable in the range $-1/2 \leq \varepsilon_{\mathrm{dd}} \leq 1$. The pure homogeneous dipole gas ($g = 0$, $g_{\mathrm{dd}} \neq 0$) is always unstable, due to there always being a k direction that is dominantly attractive.

12.7.2 Rotons and Maxons

As introduced in Sect. 4.1.1, any minimum in the dispersion relation at finite momenta is known as a roton, and given all superfluid dispersion curves are gapless at $k = 0$ (the so-called "Goldstone" mode associated with the zero energy cost to globally rotate the condensate phase), any minimum must be preceeded by a maximum, known as the maxon. Rotons were also discussed in the context of liquid helium in Sect. 6.10, although the underlying physics differs from that introduced here for dipolar condensates.

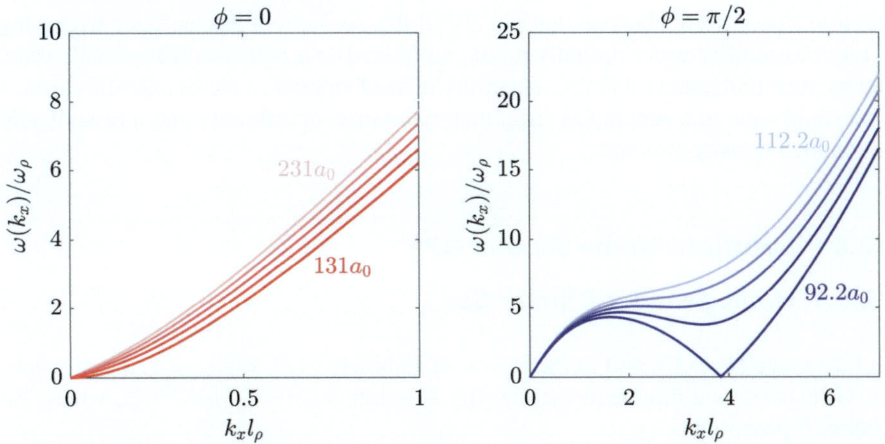

Fig. 12.5 Excitation spectrum of a quasi-1D ^{164}Dy gas with $\omega_\rho = 2\pi \times 300$ Hz and linear density $n_{1D} = 2000\,\mu\text{m}^{-1}$. Left: Dipoles polarized along the tube, the homogeneous gas is phonon unstable when $\varepsilon_{dd} > 1$, or $a_s < 131 a_B$. Each curve is a solution to Eq. (12.31), with jumps of $25 a_B$ in scattering length. Right: Dipoles polarized perpendicular to the tube, the homogeneous gas is roton unstable at $a_s = 92.2 a_B$. Each curve is a jump in $5 a_B$ in scattering length

A roton minimum can occur in any system that has a preferred spatial ordering. The wavenumber at the roton minimum, k_r, indicates that it costs the least energy to excite this mode with wavelength $2\pi/k_r$. In liquid helium, this spatial ordering arises from the strongly correlated interactions, however in dipolar gases it can only arise from an interplay between the dipolar and contact interactions with the external trapping potential [36, 37]—there is no roton in the infinite homogeneous 3D gas.

The roton is best revealed taking the quasi-1D dipolar GPE from Eq. (12.13), where the excitation spectrum of the infinite 1D tube is

$$\omega(k_z) = \sqrt{\frac{\hbar^2 k_z^2}{2m}\left(\frac{\hbar^2 k_z^2}{2m} + 2n_{1D}\left[g + \tilde{U}_{1D}(k_z l_\rho/\sqrt{2})\right]\right)}, \qquad (12.31)$$

for constant 1D density n_{1D}. Solutions to this equation are plotted in Fig. 12.5. When the dipoles are pointing along the long axis of the tube the excitation spectrum shape is similar to the non-dipolar case, hosting phonon and free particle ($\propto k^2$) regimes. Tilting the polarization axis out of the tube completely alters this picture. Instead, there is now a pronounced minimum in the excitation spectrum, that by increasing ε_{dd} can be forced to 0. This means that although the unmodulated infinite tube density is the ground state, it costs a small amount of energy to produce a density modulated state with wavelength $\approx 2\pi/k_r$, where here $k_r \approx 4/l_\rho$. Note, that in an experiment the roton minimum appears at $k_r \approx \sqrt{2}/l_\rho$, and this mismatch is due to the quasi-1D approximation. In 1D condensates, the runaway growth of excitations after triggering a roton instability was observed in an experiment in 2018 [38], followed by a detailed measurement of the full excitation spectrum in 2019 [39]. Rotons in quasi-2D condensates also have two-dimensional stucture, and these *angular rotons* were observed later in 2021 [40]. As we will discuss later, triggering a roton instability was originally predicted to lead to a collapse of the condensate, but we now understand that this instability instead signifies a phase transition from a superfluid state to a supersolid state, that maintains superfluidity and a periodically modulated density structure.

12.8　Solutions to the Dipolar GPE

12.8.1　Homogeneous Dipolar Gas

In free space ($V = 0$), and in the region of stability $-1/2 < \varepsilon_{dd} < 1$, the 3D dipolar GPE permits a homogeneous density solution $\Psi(\boldsymbol{r}) = \sqrt{n_0}e^{-i\mu t/\hbar}$, where the chemical potential is

$$\mu = n_0(g - g_{dd}).\qquad(12.32)$$

The origin of the dipolar contribution is best seen from the momentum space DDI Eq. (12.9), where, in a homogeneous gas, the only wavenumber that contributes is $k = 0$, and $\tilde{U}_{dd}(k = 0) = -g_{dd}$. A similar process can be applied to dipolar systems in lower dimensions, where the chemical potential is found to be dependent on the polarization angle.

12.8.2　Hard-Wall Solution

In a non-dipolar gas, we have seen that the condensate density monotonically increases from 0 at the boundary of a hard wall to a constant determined by the system's chemical potential, at a distance governed by the healing length. In a dipolar gas, this addition of a trap potential allows for a roton minimum to form in the excitation spectrum, and this modifies this simple picture.

In Fig. 12.6 we show the impact of the dipolar interaction on the solution near a hard wall. Away from parameters near the roton instability, here with the dipoles oriented along x, the density exhibits the expected tanh-like distribution, with the now modified healing length $\xi = \hbar/\sqrt{m\mu(\varepsilon_{dd})}$. Turning the polarization perpendicular to x, the excitation spectrum now contains a roton minimum. This modifies the density near to the hard wall, creating a density ripple structure. The wavelength of the ripples is the roton wavelength of the corresponding homogeneous density.

12.8.3　Magnetostriction and the Thomas-Fermi Solution

Consider now the other fundamental trap for ultracold atom research, the harmonic trap. We have seen that in the limit of large N, density gradients can be ignored and thus the kinetic term in the GPE can be thrown away, leading to the Thomas-Fermi

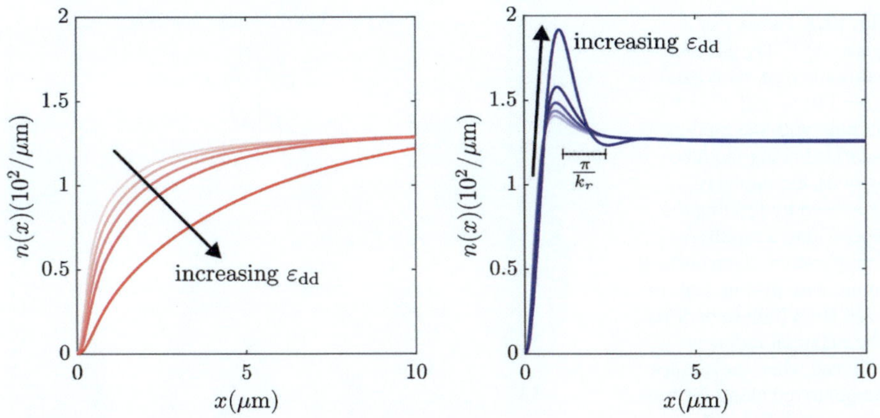

Fig. 12.6 Density of a quasi-1D ^{164}Dy gas with a hard wall at $x = 0$. Left: Dipoles polarized along the tube, the healing length is diverging with increasing $\varepsilon_{dd} \to 1$. Right: Dipoles polarized perpendicular to the tube. The amplitude of the density lumps increases as the roton minimum is lowered, but the wavelength is always the roton wavelength

density solution. This solution is considerably more complex when including the dipolar term, still analytically tractable, but out of the scope of this book. However, the same argument to remove the kinetic term can be applied to the dipolar term, where for negligible density gradients the main contribution to the dipolar term will be the non-oscillating $k = 0$ part, which simply leads to a modified local interaction. This crucially captures the main effect of the dipolar term, *magnetostriction*.

Take the quasi-1D dipolar GPE (Eq. (12.13)) with a harmonic trap $V_{\text{ext}}(z) = 1/2 m\omega_z^2 z^2$, under the approximation that the kinetic term is negligible and only the $k = 0$ (contact part) of the dipolar term contributes to the Hamiltonian. Then, we obtain an equation for the stationary solution

$$\mu\psi = \frac{1}{2} m\omega_z^2 z^2 \psi + g_{1D}\left(1 + \frac{\varepsilon_{dd}}{2}(1 - 3\cos^2\phi)\right)|\psi|^2\psi, \tag{12.33}$$

which, after rearranging for the density $n = |\psi|^2$ gives

$$n(z) = \begin{cases} n_0\left(1 - \dfrac{z^2}{R_z^2}\right), & |z| \le R_z, \\ 0, & \text{otherwise.} \end{cases} \tag{12.34}$$

We have defined

$$n_0 = \frac{\mu}{g_{1D}\left(1 + \frac{\varepsilon_{dd}}{2}(1 - 3\cos^2\phi)\right)}, \quad R_z = \left(\frac{2\mu}{m\omega_z^2}\right)^{1/2}, \tag{12.35}$$

and by using the fact that the total atom number must satisfy $N = \int_{-R_z}^{R_z} dz\, n(z)$, we find that

$$\mu = \frac{1}{2}\left(\frac{3}{2}\left(m\omega_z^2\right)^{1/2} g_{1D}\left(1 + \frac{\varepsilon_{dd}}{2}(1 - 3\cos^2\phi)\right) N\right)^{2/3}. \tag{12.36}$$

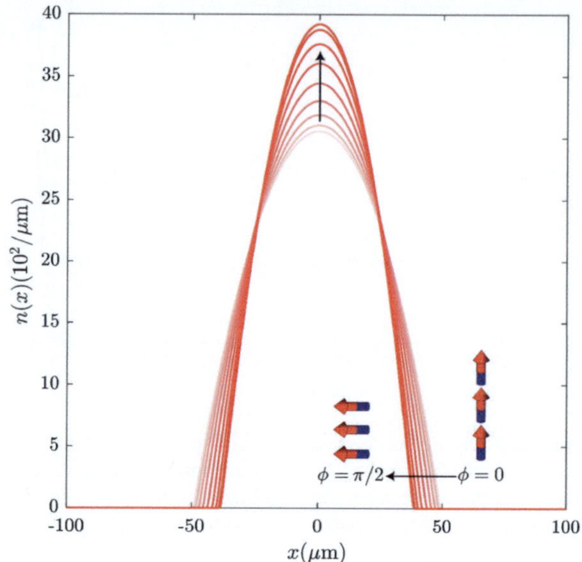

Fig. 12.7 Density of a quasi-1D ^{164}Dy gas in a harmonic trap, with fixed $a_s = 150a_B$ and $N = 10^5$. Starting with the dipoles polarized along the tube ($\phi = 0$), the radius is decreased by rotating the dipoles into a repulsive configuration. Each curve is an increase in ϕ in steps of $\pi/20$ from light to dark red. The maximal radius is achieved when the dipoles are polarized along the tube axis, in an effect called magnetostriction

Numerical solutions to the dipolar GPE in a one-dimensional harmonic trap are shown in Fig. 12.7. This confirms the polarization angle dependency of the Thomas-Fermi radius, and the similar inverse depence on the central density n_0. The stretching of the condensate radius towards the polarization angle is called *magnetostriction*. As the atoms can minimize their energy (or maximize the negative contribution to the energy) by aligning head-to-tail, this happens easiest along the polarization axis. In a 3D system with a spherically symmetric harmonic trap, the Thomas-Fermi radii are not equal. The radius extends parallel to the polarization, and radially contracts perpendicular to it.

12.9 Vortices and Solitons in Dipolar Gases

12.9.1 Vortices

The unique long-range and directional characteristics of the DDI introduce substantial modifications to the well-studied vortex dynamics in superfluid systems. A simple model introduced by Martin et al. [41] can help predict the impact of the dipolar interaction. By writing the total condensate density in the presence of a vortex as $n(\mathbf{r}) = n_0 - n_v(\mathbf{r})$, where n_0 denotes the uniform background density and $n_v(\mathbf{r})$ accounts for the density depletion caused by the vortex and inserting this

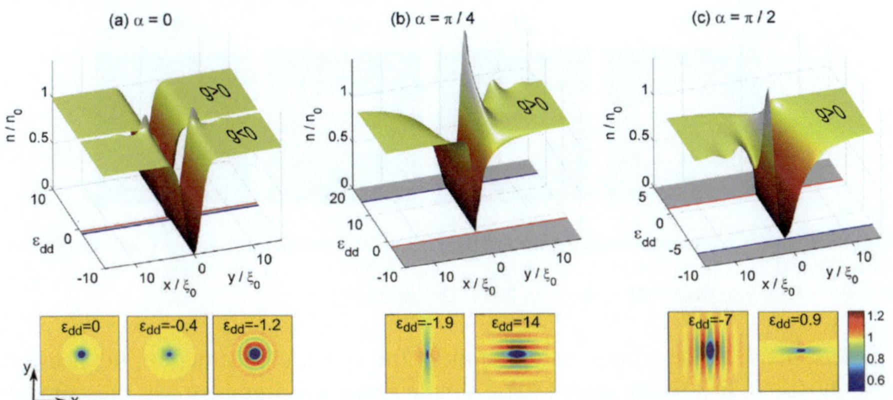

Fig. 12.8 Vortex solutions in an infinite dipolar condensate, as a function of ε_{dd}, in the quasi-two-dimensional regime (Eq. (12.10), $l_z/\xi = 0.5$). Along x (for $y = 0$) the normalised density profile (n/n_0) is shown on the left-hand side of the main plots. Along the right-hand side the normalised density profile is plotted along y (with $x = 0$). **a** Dipoles polarized along z ($\alpha = 0$). **b** Dipoles polarized off-axis at $\alpha = \pi/4$. **c** Dipoles polarized along x ($\alpha = \pi/2$). In each of the main plots, grey bands indicate the unstable regimes of ε_{dd}. Insets: normalised density profile over an area $(40\xi)^2$ for indicated values of ε_{dd}. Reprinted figure with permission from [45]. Copyright (2013) by the American Physical Society

decomposition into the expression for the dipolar interaction energy yields:

$$2E_{dd} = \int d^3\mathbf{r} \int d^3\mathbf{r}'\, n(\mathbf{r}) U_{dd}(\mathbf{r} - \mathbf{r}') n(\mathbf{r}')$$

$$= \int d^3\mathbf{r} \int d^3\mathbf{r}'\, n_0 U_{dd}(\mathbf{r} - \mathbf{r}') n_0$$

$$+ \int d^3\mathbf{r} \int d^3\mathbf{r}'\, n_v(\mathbf{r}) U_{dd}(\mathbf{r} - \mathbf{r}') n_v(\mathbf{r}')$$

$$- 2 \int d^3\mathbf{r} \int d^3\mathbf{r}'\, n_0 U_{dd}(\mathbf{r} - \mathbf{r}') n_v(\mathbf{r}'). \tag{12.37}$$

The first of three lines is a constant energy shift from the dipolar interaction between condensate atoms. The second line reveals that two vortices interact with each other at long-range with the usual DDI. The final term, bearing a negative sign, effectively describes an interaction where the vortex behaves as an "anti-dipole" within the larger dipolar condensate. When the magnetic dipoles are aligned along the vortex line, the resulting interaction pulls atoms toward the vortex core, forming a high-density ringed density profile due to roton-induced effects [42,44–48], see Fig. 12.8a. Conversely, when the dipoles lie within the plane, the repulsive head-to-tail configuration stretches the vortex core into an elliptical shape Fig. 12.8b and c, echoing magnetostrictive behavior [42–46]. Such core deformations have also been observed in superconducting systems, driven by analogous physical mechanisms [49].

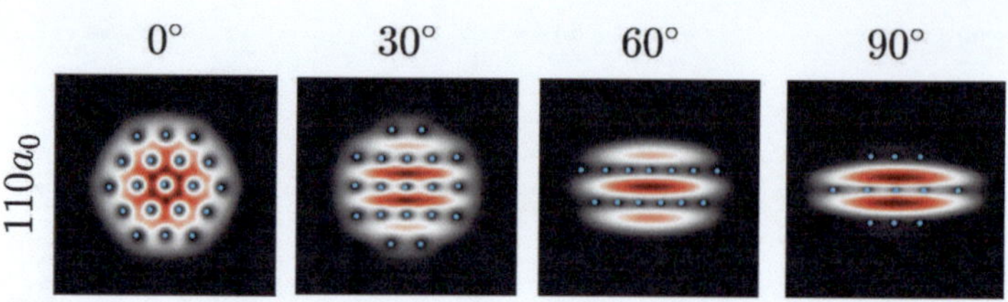

Fig. 12.9 Vortex lattice solutions of a dipolar BEC. Light blue points mark vortex positions. Parameters: $\Omega = 0.7\omega_r$, $N = 15000$, $a_s = 110a_B$, $\omega_r = 2 \times 50\,\text{Hz}$, $\omega_z = 2 \times 130\,\text{Hz}$. Adapted under CC-BY-4.0 license from [54]. Copyright 2023, The Author(s)

In response to the elliptic deformation of the vortex core structure, the phase is also modified. In a recent work [50], a simple geometric argument was presented to find an expression for the phase of a quantum vortex with an elliptic core with aspect ratio λ, given by

$$S(x, y) = q\Lambda \left[(\lambda^2 - 1) F \left[\arctan \left(\frac{y}{x} \right) \,|1 - \lambda^4 \right] \right.$$
$$\left. + \lambda^2 \Pi \left[1 - \lambda^2 \,;\, \arctan \left(\frac{y}{x} \right) \,|1 - \lambda^4 \right] \right] \tag{12.38}$$

where q is the integer wining number, and $F(z|n)$ ($\Pi(m; z|n)$) is the incomplete elliptic integral of the first (third) kind. The integration constant is

$$\Lambda = \pi \left[4 F \left(\frac{\pi}{2} \middle| 1 - \lambda^{-4} \right) - 2\lambda^{-2} \Pi \left(1 - \lambda^{-2}; \frac{\pi}{2} \middle| 1 - \lambda^{-4} \right) \right]^{-1}, \tag{12.39}$$

enforcing that the phase is continuous at the branch cut. This reduces to $S(\theta) = q\theta$ when $\lambda = 1$.

The framework introduced in Eq. (12.37) can be used to understand how pairs of vortices interact with each other. In quasi-2D non-dipolar condensates, vortex-vortex interactions are long-ranged (decaying as $1/r$) but maintain isotropic. These interactions are determined by the sign of each vortex's circulation: like-signed vortices repel and orbit one another, while oppositely signed vortices attract and can annihilate. In fully 3D systems, additional dynamical effects emerge—such as filament recombination or bouncing—depending on relative orientation and velocity [51]. In contrast, dipolar interactions introduce directionality. Two vortices oriented side-by-side relative to the magnetic polarization will repel, whereas head-to-tail alignment leads to attraction. This directional dependence leads to phenomena like reduced annihilation of vortex-antivortex pairs [46], and their directionally dependent velocity [50], and elliptic-shaped orbits of same-sign vortex pairs [46,50,52,53].

Beyond individual vortex pairs, the collective arrangement of vortices is also reshaped by dipolar forces. When dipoles are oriented perpendicular to the two-dimensional plane—resulting in isotropic in-plane interactions—a triangular vortex lattice forms, provided the short-range interaction remains repulsive [42,48,54],

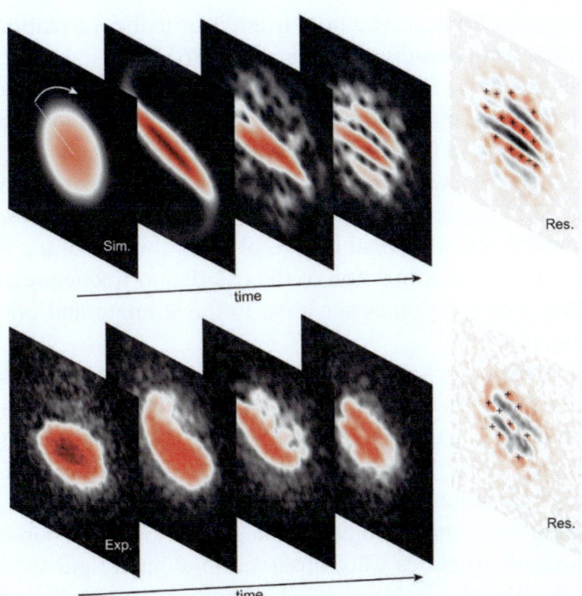

Fig. 12.10 Generation of vortex stripes in a dipolar BEC. Top row: Time series of real-time simulation snapshots at $(16, 164, 542, 1200)$ms using the experimental parameters. Bottom row: Time series of the corresponding experimental snapshots at $(20, 127, 314, 447)$ms. For the last image of each series a vortex detection algorithm is applied [57], showing the calculated residuals, with the detected vortices marked with crosses. Used parameters: $\theta = 35°$, $\Omega = 0.7\,\omega_r$, $a_s = 109a_B$, $N = 10000$, $\omega_r = 2 \times 50\,\mathrm{Hz}$, $\omega_z = 2 \times 130\,\mathrm{Hz}$. Adapted under CC-BY-4.0 license from [54]. Copyright 2023, The Author(s)

as in non-dipolar systems (see Fig. 12.9a). Tilting the dipoles breaks the in-plane symmetry, favoring the formation of linear, stripe-like vortex patterns that maximize head-to-tail alignment [42, 55, 56].

The first direct observation of vortices in a dipolar Bose-Einstein condensate was reported only recently by Klaus and Bland et al. [57], employing a technique they termed *magnetostirring*. This method exploits the natural tendency of dipolar atoms to align along the magnetic field; by dynamically rotating the polarization axis, angular momentum is transferred to the system, eventually generating a turbulent atomic cloud. Over time, this cloud relaxes into a stripe-ordered vortex lattice, as illustrated in Fig. 12.10.

12.9.2 Solitons

A series of theoretical investigations have shown that dipolar interactions in Bose-Einstein condensates (BECs) significantly enrich the physics of solitons, leading to behavior not present in systems governed solely by contact interactions. Much like vortices, the properties of solitons in dipolar gases can be understood by modeling them as effective quasi-particles with dipolar character. In this picture, solitons experience long-range, anisotropic interactions in addition to the conventional short-range interactions familiar from non-dipolar systems [58–61]. The interplay between these two contributions gives rise to a variety of novel solitonic phenomena. For instance, under suitable conditions, dipolar interactions can stabilize bound states of multiple solitons, both in the form of bright soliton molecules [62,63] and, surprisingly, dark soliton complexes [59–61], which are not accessible to single component non-dipolar systems, see Fig. 12.11. In harmonically trapped condensates, the presence of dipolar interactions introduces an additional degree of control over soliton dynamics. One clear manifestation is the modification of the oscillation frequency of dark solitons in elongated traps, which becomes sensitive to the strength and orientation of the dipolar interaction [64].

Moreover, dipolar interactions can stabilize higher-dimensional solitonic structures that would otherwise be unstable in non-dipolar gases. In particular, dipolar BECs are predicted to support self-trapped two-dimensional bright solitons in quasi-2D geometries [65,66], circumventing the collapse typically associated with attractive contact interactions in higher dimensions, see Fig. 12.11. Experimental evidence consistent with this prediction has recently been reported [67], where localized matter-wave structures were observed in a dipolar gas loaded into a one-dimensional lattice. Dipolar interactions also modify the stability of dark solitons in three-dimensional geometries. Specifically, they are predicted to suppress the well-known transverse modulational (or "snaking") instability that typically leads to soliton decay into vortex rings or other excitations [68]. This suppression opens the door to longer-lived, coherent dark soliton dynamics in dipolar systems, potentially enabling the study of soliton turbulence and solitonic excitations in regimes previously inaccessible.

Problems

12.1 *Manipulating the DDI:*

Starting with the unpolarised form of the DDI

$$U_{\mathrm{dd}}(\boldsymbol{r}) = \frac{C_{\mathrm{dd}}}{4\pi} \left(\frac{(\hat{\boldsymbol{e}}_1 \cdot \hat{\boldsymbol{e}}_2)r^2 - 3(\hat{\boldsymbol{e}}_1 \cdot \boldsymbol{r})(\hat{\boldsymbol{e}}_2 \cdot \boldsymbol{r})}{r^5} \right),$$

prove the following properties:

(a) Show that the z-polarized form ($\hat{\boldsymbol{e}}_1 = \hat{\boldsymbol{e}}_2 = (0, 0, 1)$, $\hat{\boldsymbol{r}} = (\sin\theta, 0, \cos\theta)$), satisfies Eq. (12.2).
(b) Taking time-dependent polarization $\hat{\boldsymbol{e}}_1 = \hat{\boldsymbol{e}}_2 = (\sin\varphi\cos\Omega t, \sin\varphi\sin\Omega t, \cos\varphi)$, show that the anti-dipolar interaction satisfies Eq. (12.24) after time-averaging over one period of oscillation $T = 2\pi/\Omega$.

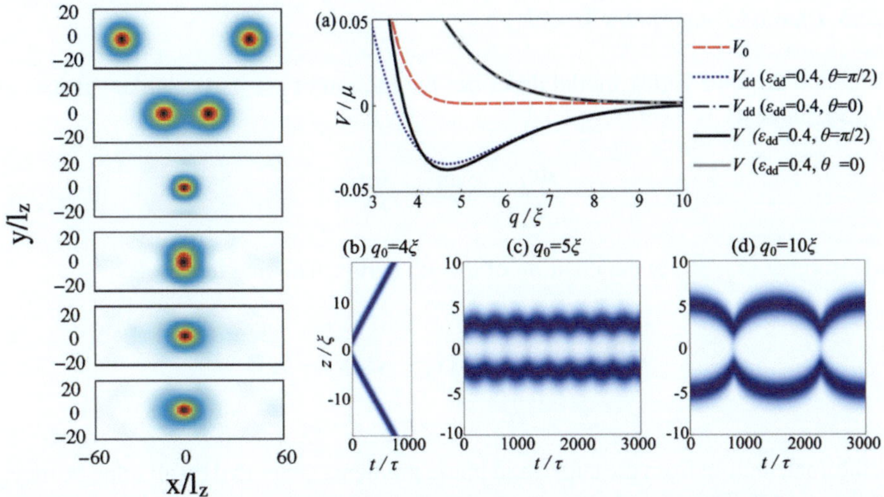

Fig. 12.11 Dipolar solitons. Left column: Fusion dynamics of two-dimensional bright solitons, stabilized by dipole-dipole interactions. Reprinted figure with permission from [65]. Copyright (2005) by the American Physical Society. Right column: **a** Negative of the effective interaction potential, $V = V_0 + V_{dd}$, between two dark solitons as a function of their separation for various values of ε_{dd} and polarization angles θ. The emergence of a minimum in the potential indicates the possibility of forming bound states. **b** If solitons are initially placed too close together, they experience strong repulsion and escape the bound state. **c–d** Varying the initial separation alters the trajectory within the potential well, allowing different dynamical behaviors to emerge. Reprinted figure with permission from [60]. Copyright (2015) by the American Physical Society

12.2 *Fourier Transform of the DDI*:

Derive the polarized dipole-dipole interaction in momentum space, Eq. (12.9). The following relations will be required:

- The plane wave expansion

$$e^{i\mathbf{k}\cdot\mathbf{r}} = 4\pi \sum_{\ell=0}^{\infty} i^{\ell} j_{\ell}(kr) \sum_{m=-\ell}^{\ell} Y_{\ell}^{m}(\hat{\mathbf{k}})^{*} Y_{\ell}^{m}(\hat{\mathbf{r}}),$$

where $j_{\ell}(kr)$ are spherical Bessel functions, and Y_{ℓ}^{m} are orthonormal spherical harmonics.
- The identities

$$3\cos^{2}\theta - 1 = \sqrt{\frac{16\pi}{5}}\, Y_{2}^{0}(\hat{\mathbf{r}})$$

$$\int_{0}^{\infty} \mathrm{d}r\, \frac{j_{2}(kr)}{r} = \frac{1}{3}.$$

12.3 *Dipolar Point Vortex Model*:

The dipolar point vortex model describes the motion of N_v particle-like vortices in the xy-plane as

$$\frac{d\mathbf{r}_j}{dt} = \mathbf{v}_j^{(P)} + \mathbf{v}_j^{(DD)},$$

where $\mathbf{r}_j \equiv (x_j, y_j)$ is the position of the jth vortex. The first term,

$$\mathbf{v}_j^{(P)} = \frac{\hbar}{m} \sum_{\substack{k=1 \\ k \neq j}}^{N_v} \nabla S(x_j - x_k, y_j - y_k),$$

is the contribution from the superfluid phase gradients generated by all other vortices in the system. The second term,

$$\mathbf{v}_j^{(DD)} = q_j \sum_{\substack{k=1 \\ k \neq j}}^{N_v} \hat{\mathbf{z}} \times \nabla U_{DD}(x_j - x_k, y_j - y_k),$$

represents the additional velocity induced by long-range dipolar interactions. This form of the interaction arises from the magnetic or electric dipoles associated with the vortices, oriented along the z-axis, which are *spinning* objects, and can attract or repel other vortices depending on their relative configuration.

Taking Eq. (12.38) for $S(x, y)$ and an approximation for the 2D DDI with dipoles along x polarized at an angle α from the z axis

$$U_{DD}(x, y) = \frac{\hbar^2}{m} \frac{\xi_v \varepsilon_{dd}}{(x^2 + y^2)^{3/2}} \left(1 - 3 \sin^2 \alpha \frac{x^2}{x^2 + y^2} \right), \tag{12.40}$$

with vortex core size ξ_v, calculate the velocity terms for the dipolar point vortex model.

References

1. A. Griesmaier, J. Werner, S. Hensler, J. Stuhler, T. Pfau, Phys. Rev. Lett. **94**, 160401 (2005)
2. M. Lu, N.Q. Burdick, S.H. Youn, B.L. Lev, Phys. Rev. Lett. **107**, 190401 (2011)
3. K. Aikawa, A. Frisch, M. Mark, S. Baier, A. Rietzler, R. Grimm, F. Ferlaino, Phys. Rev. Lett. **108**, 210401 (2012)
4. T. Lahaye, C. Menotti, L. Santos, M. Lewenstein, T. Pfau, Rep. Prog. Phys. **72**, 126401 (2009)
5. L. Chomaz, I. Ferrier-Barbut, F. Ferlaino, B. Laburthe-Tolra, B.L. Lev, T. Pfau, Rep. Prog. Phys. **86**, 026401 (2022)

6. Y. Shilo, K. Cohen, B. Laikhtman, K. West, L. Pfeiffer, R. Rapaport, Nat. Commun. **4**, 2335 (2013)
7. K. Cohen, Y. Shilo, K. West, L. Pfeiffer, R. Rapaport, Nano Lett. **16**, 3726–3731 (2016)
8. M. Stern, V. Umansky, I. Bar-Joseph, Science **343**, 55–57 (2014)
9. M. Combescot, O. Betbeder-Matibet, R. Combescot, Phys. Rev. Lett. **99**, 176403 (2007)
10. S. Misra, M. Stern, A. Joshua, V. Umansky, I. Bar-Joseph, Phys. Rev. Lett. **120**, 047402 (2018)
11. Y. Mazuz-Harpaz, K. Cohen, M. Leveson, K. West, L. Pfeiffer, M. Khodas, R. Rapaport, Proc. Natl. Acad. Sci. U.S.A. **116**, 18328–18333 (2019)
12. G.W. Burg, N. Prasad, K. Kim, T. Taniguchi, K. Watanabe, A.H. MacDonald, L.F. Register, E. Tutuc, Phys. Rev. Lett. **120**, 177702 (2018)
13. L. Ma, P.X. Nguyen, Z. Wang, Y. Zeng, K. Watanabe, T. Taniguchi, A.H. MacDonald, K.F. Mak, J. Shan, Nature **598**, 585–589 (2021)
14. N. Bigagli, W. Yuan, S. Zhang, B. Bulatovic, T. Karman, I. Stevenson, S. Will, Nature **631**, 289–293 (2024)
15. T. Karman, J.M. Hutson, Phys. Rev. Lett. **121**, 163401 (2018)
16. W. Li, M.W. Noel, M.P. Robinson, P.J. Tanner, T.F. Gallagher, D. Comparat, B. Laburthe Tolra, N. Vanhaecke, T. Vogt, N. Zahzam, et al., Phys. Rev. A **70**, 042713 (2004)
17. W. Li, T. Pohl, J.M. Rost, S.T. Rittenhouse, H.R. Sadeghpour, J. Nipper, B. Butscher, J.B. Balewski, V. Bendkowsky, R. Löw, T. Pfau, Science **334**, 1110–1114 (2011)
18. M. Lu, N.Q. Burdick, B.L. Lev, Phys. Rev. Lett. **108**, 215301 (2012)
19. A. Trautmann, P. Ilzhöfer, G. Durastante, C. Politi, M. Sohmen, M.J. Mark, F. Ferlaino, Phys. Rev. Lett. **121**, 213601 (2018). arXiv:1807.07555
20. P. Ilzhöfer, G. Durastante, A. Patscheider, A. Trautmann, M.J. Mark, F. Ferlaino, Phys. Rev. A **97**, 023633 (2018)
21. E.T. Davletov, V.V. Tsyganok, V.A. Khlebnikov, D.A. Pershin, D.V. Shaykin, A.V. Akimov, Phys. Rev. A **102**, 011302 (2020)
22. Y. Miyazawa, R. Inoue, H. Matsui, G. Nomura, M. Kozuma, Phys. Rev. Lett. **129**, 223401 (2022)
23. J. Miao, J. Hostetter, G. Stratis, M. Saffman, Phys. Rev. A **89**, 041401 (2014)
24. K. Góral, K. Rzażewski, T. Pfau, Phys. Rev. A **61**, 051601 (2000)
25. P.B. Blakie, D. Baillie, S. Pal, Commun. Theor. Phys. **72**, 085501 (2020)
26. S. Ronen, D.C.E. Bortolotti, J.L. Bohn, Phys. Rev. A **74**, 013623 (2006)
27. S. Giovanazzi, A. Görlitz, T. Pfau, Phys. Rev. Lett. **89**, 130401 (2002)
28. F. Claude, L. Lafforgue, J.J.A. Houwman, M.J. Mark, F. Ferlaino, Phys. Rev. Res. **6**, L042016 (2024)
29. Y. Tang, K. Wil, K.-Y. Li, B.L. Lev, Phys. Rev. Lett. **120**, 230401 (2018)
30. S.B. Prasad, T. Bland, B.C. Mulkerin, N.G. Parker, A.M. Martin, Phys. Rev. Lett. **122**, 050401 (2019)
31. D. Baillie, P.B. Blakie, Phys. Rev. A **101**, 043606 (2020)
32. N.N. Bogoliubov, J. Phys. (USSR) **11**, 23 (1947)
33. P.G. de Gennes, Benjamin, New York (1966)
34. G. Bismut, B. Laburthe-Tolra, E. Maréchal, P. Pedri, O. Gorceix, L. Vernac, Phys. Rev. Lett. **109**, 155302 (2012)
35. M. Wenzel, F. Böttcher, J.-N. Schmidt, M. Eisenmann, T. Langen, T. Pfau, I. Ferrier-Barbut, Phys. Rev. Lett. **121**, 030401 (2018)
36. L. Santos, G.V. Shlyapnikov, M. Lewenstein, Phys. Rev. Lett. **90**, 250403 (2003)
37. S. Giovanazzi, D.H.J. O'Dell, Eur. Phys. J. D **31**, 439–445 (2004)
38. L. Chomaz, R.M.W. van Bijnen, D. Petter, G. Faraoni, S. Baier, J.H. Becher, M.J. Mark, F. Wächtler, L. Santos, F. Ferlaino, Nat. Phys. **14**, 442 (2018)
39. D. Petter, G. Natale, R.M.W. van Bijnen, A. Patscheider, M.J. Mark, L. Chomaz, F. Ferlaino, Phys. Rev. Lett. **122**, 183401 (2019)
40. J.-N. Schmidt, J. Hertkorn, M. Guo, F. Böttcher, M. Schmidt, K.S.H. Ng, S.D. Graham, T. Langen, M. Zwierlein, T. Pfau, Phys. Rev. Lett. **126**, 193002 (2021)
41. A.M. Martin, N.G. Marchant, D.H.J. O'Dell, N.G. Parker, J. Phys.: Condens. Matter **29**, 103004 (2017)

42. S. Yi, H. Pu, Phys. Rev. A **73**, 061602 (2006)
43. M. Abad, M. Guilleumas, R. Mayol, M. Pi, D.M. Jezek, Phys. Rev. A **79**, 063622 (2009)
44. C. Ticknor, R.M. Wilson, J.L. Bohn, Phys. Rev. Lett. **106**, 065301 (2011)
45. B.C. Mulkerin, R.M.W. van Bijnen, D.H.J. O'Dell, A.M. Martin, N.G. Parker, Phys. Rev. Lett. **111**, 170402 (2013)
46. B.C. Mulkerin, D.H.J. O'Dell, A.M. Martin, N.G. Parker, J. Phys: Conf. Ser. **497**, 012025 (2014)
47. R.M. Wilson, S. Ronen, J.L. Bohn, H. Pu, Phys. Rev. Lett. **100**, 245302 (2008)
48. M. Jona-Lasinio, K. Łakomy, L. Santos, Phys. Rev. A **88**, 013619 (2013)
49. L. Fuchs, D. Kochan, J. Schmidt, N. Hüttner, C. Baumgartner, S. Reinhardt, S. Gronin, G.C. Gardner, T. Lindemann, M.J. Manfra, C. Strunk, N. Paradiso, Phys. Rev. X **12**, 041020 (2022)
50. R. Doran, T. Bland (2025). arXiv:2507.02779
51. S. Serafini, L. Galantucci, E. Iseni, T. Bienaimé, R.N. Bisset, C.F. Barenghi, F. Dalfovo, G. Lamporesi, G. Ferrari, Phys. Rev. X **7**, 021031 (2017)
52. S. Gautam, J. Phys. B: At. Mol. Opt. Phys. **47**, 165301 (2014)
53. S.B. Prasad, N.G. Parker, A.W. Baggaley, Phys. Rev. A **109**, 063323 (2024)
54. T. Bland, G. Lamporesi, M.J. Mark, F. Ferlaino, C. R. Physique **24**, 133–152 (2023)
55. Y. Cai, Y. Yuan, M. Rosenkranz, H. Pu, W. Bao, Phys. Rev. A **98**, 023610 (2018)
56. S.B. Prasad, T. Bland, B.C. Mulkerin, N.G. Parker, A.M. Martin, Phys. Rev. A **100**, 023625 (2019)
57. L. Klaus, T. Bland, E. Poli, C. Politi, G. Lamporesi, E. Casotti, R.N. Bisset, M.J. Mark, F. Ferlaino, Nat. Phys. **18**, 1453–1458 (2022)
58. J. Cuevas, B.A. Malomed, P.G. Kevrekidis, D.J. Frantzeskakis, Phys. Rev. A **79**, 053608 (2009)
59. K. Pawlowski, K. Rzazewski, New J. Phys. **17**, 105006 (2015)
60. T. Bland, M.J. Edmonds, N.P. Proukakis, A.M. Martin, D.H.J. O'Dell, N.G. Parker, Phys. Rev. A **92**, 063601 (2015)
61. M.J. Edmonds, T. Bland, D.H.J. O'Dell, N.G. Parker, Phys. Rev. A **93**, 063617 (2016)
62. B.B. Baizakov, S.M. Al-Marzoug, H. Bahlouli, Phys. Rev. A **92**, 033605 (2015)
63. M.J. Edmonds, T. Bland, R. Doran, N.G. Parker, New J. Phys. **19**, 023019 (2017)
64. T. Bland, K. Pawlowski, M.J. Edmonds, K. Rzazewski, N.G. Parker, Phys. Rev. A **95**, 063622 (2017)
65. P. Pedri, L. Santos, Phys. Rev. Lett. **95**, 200404 (2005)
66. I. Tikhonenkov, B.A. Malomed, A. Vardi, Phys. Rev. Lett. **100**, 090406 (2008)
67. G. Natale, T. Bland, S. Gschwendtner, L. Lafforgue, D.S. Grün, A. Patscheider, M.J. Mark, F. Ferlaino, Commun. Phys. **5**, 227 (2022)
68. R. Nath, P. Pedri, L. Santos, Phys. Rev. Lett. **101**, 210402 (2008)

Droplets and Supersolids

13

Abstract

Up to this point, our discussion has centered on mean-field descriptions of BECs, where the macroscopic behavior of the condensate is governed by the balance between kinetic energy and interparticle interactions. However, when systems are pushed beyond this dilute limit, quantum fluctuations and higher-order correlation effects begin to play a crucial role. These beyond-mean-field corrections fundamentally alter the stability and structure of the condensate, leading to the emergence of novel states of matter. In this chapter, we explore how such quantum corrections stabilise self-bound droplets–quantum liquids that exist without external confinement–and how, in dipolar systems, the competition between long-range anisotropic forces and quantum fluctuations gives rise to supersolidity. These phases provide a unique window into the interplay between superfluidity and crystalline order, offering new opportunities to probe collective excitations, phase coherence, and the role of quantum correlations in strongly interacting Bose gases.

13.1 Beyond-Mean-Field Effects

In the derivation of the Gross-Pitaevskii equation, we make the approximation that the many-body wavefunction can be separated into a mean-field contribution representing the condensate, $\Psi(\mathbf{r}, t) = \langle \hat{\Psi}(\mathbf{r}, t)\rangle$, and a non-condensate part, $\hat{\Psi}_{\mathrm{nc}}$, which contains information about thermal and quantum fluctuations. At zero temperature, quantum fluctuations still play a role, and for a homogeneous, non-dipolar gas, they lead to a correction to the ground-state energy per unit volume $\mathcal{V}$. The total condensate energy is then given by

$$\frac{E}{\mathcal{V}} = \frac{E_0 + \Delta E}{\mathcal{V}} = \frac{1}{2} g n_0^2 \left[1 + \frac{128}{15\sqrt{\pi}} \left(a_s^3 n_0\right)^{1/2} \right], \tag{13.1}$$

© The Author(s), under exclusive license to Springer Nature Switzerland AG 2026   259
C. F. Barenghi et al., *Quantum Fluids, Solitons, and Vortices*, Lecture Notes in Physics 1050, https://doi.org/10.1007/978-3-032-20171-3_13

valid for $a_s > 0$. The first term, E_0, corresponds to the mean-field energy of a homogeneous condensate with density n_0 (see Sect. 3.6), while the second term, ΔE, is the Lee–Huang–Yang (LHY) correction arising from "zero-point" quantum fluctuations. Zero-point energy refers to the residual energy present in a quantum system even at absolute zero temperature, due to the Heisenberg uncertainty principle. For instance, a particle confined in a trap can never be completely at rest and always retains some motion–and thus energy–in its ground state. This irreducible contribution is the zero-point energy.

Typically, the LHY correction is negligible compared to the mean-field energy, as the GPE is valid in the dilute regime where the gas parameter $n_0 a_s^3 \ll 1$. For single-component gases with contact interactions, this condition remains well satisfied, and mean-field theory provides an accurate description. However, as we will see in this chapter, the situation changes in two-component and dipolar systems. In these cases, by carefully tuning the interaction parameters, it is possible to cancel the mean-field energy while retaining a non-zero LHY term. This results in a collapse of the system that is arrested by the repulsive energy from quantum fluctuations, giving rise to new self-stabilised states.

The derivation of the quantum fluctuation energy shift ΔE is beyond an entry book to BEC, but we state here the general formula for any interaction potential $\tilde{U}_{\text{int}}(\mathbf{k})$ as

$$\frac{\Delta E}{V} = \frac{1}{2} \int \frac{d^3 \mathbf{k}}{(2\pi)^3} \left[\epsilon(\mathbf{k}) - \frac{\hbar^2 k^2}{2m} - n_0 \tilde{U}_{\text{int}}(\mathbf{k}) + \frac{m \left(n_0 \tilde{U}_{\text{int}}(\mathbf{k}) \right)^2}{\hbar^2 k^2} \right], \qquad (13.2)$$

with low-energy excitation spectrum $\epsilon(\mathbf{k})$. From this energy contribution, a so-called *extended Gross-Pitaevskii equation* (eGPE) can be derived for the system of interest, where an extra beyond-mean-field term is added to the Hamiltonian through a correction to the chemical potential

$$\Delta \mu = \frac{\partial \Delta E}{\partial n_0}. \qquad (13.3)$$

For example, the LHY term for a single component contact interacting gas is given by the derivative of Eq. (13.1),

$$\Delta \mu = \frac{32}{3} g \, (a_s n_0)^{3/2}, \qquad (13.4)$$

which, following the local-density approximation $\{n_0 \to n(\mathbf{r})\}$ gives the eGPE for a single component contact interacting gas

$$i\hbar \frac{\partial \Psi(\mathbf{r}, t)}{\partial t} = \left[-\frac{\hbar^2}{2m} \nabla^2 + V(\mathbf{r}, t) + g|\Psi(\mathbf{r}, t)|^2 + \frac{32}{3} g a_s^{3/2} |\Psi(\mathbf{r}, t)|^3 \right] \Psi(\mathbf{r}, t). \qquad (13.5)$$

13.2   Quantum Droplets

13.2.1   Bose-Bose Systems

We have discussed self-bound solutions to the GPE in the form of solitons. Solitons arise from a balance between dispersive effects due to kinetic energy and localization due to attractive interactions. As we have seen, interaction-based effects can lead to collapse in systems with dimensionality higher than one, and thus solitons are, at best, metastable in two or more dimensions.

Here, we introduce a new type of self-bound solution known as a *quantum droplet*. Such a state can exist due to a balance between attractive mean-field interactions and repulsive quantum fluctuations, independent of dimensionality. The one-component system discussed above does not permit such a setup due to the requirement that $a_s > 0$. We therefore turn to the two-component Bose–Bose gas introduced in Chap. 11, and follow the work of Petrov in 2015 [1]. The full LHY correction for the Bose–Bose system is quite involved, but it can be simplified under certain assumptions. The *same-shape approximation* assumes that the wavefunction for each component can be written as a shared spatial mode with different amplitudes, i.e., $\Psi_j(\mathbf{r}, t) = \sqrt{n_j}\,\phi(\mathbf{r}, t)$. Further assuming equal particle numbers in both components, a simple effective equation for ϕ can be derived.

We are particularly interested in the regime where $\delta g = g_{12} - \sqrt{g_{11}g_{22}}$ is small and negative, with $g_{11} > 0$ and $g_{22} > 0$. In this case, the mean-field interaction is overall attractive and the ground state is miscible, while the LHY term is repulsive. Under carefully chosen units, the eGPE under radial symmetry for $\phi(\tilde{r}, \tilde{t})$ takes the remarkably simple form

$$i\frac{\partial \phi}{\partial \tilde{t}} = \left[-\frac{\nabla_{\tilde{r}}^2}{2} - 3|\phi|^2 + \frac{5}{2}|\phi|^3 - \tilde{\mu} \right] \phi, \tag{13.6}$$

where $\tilde{\mu}$ is the dimensionless chemical potential and

$$\nabla_r^2 f = \frac{1}{r^2}\frac{\partial}{\partial r}\left(r^2 \frac{\partial f}{\partial r} \right), \tag{13.7}$$

is the Laplacian for a spherically symmetric function. Figure 13.1 shows ground state solutions ϕ_0 to this equation for increasing atom number, $\tilde{N} = 4\pi \int \tilde{r}^2 |\phi|^2 d\tilde{r}$. Unlike single component three-dimensional Bose gases in free space which permit no self-bound solution, this figure shows that for atom numbers exceeding a critical one, a spherically symmetric ball-like solution becomes stable, through the balance of attractive mean-field and repulsive LHY contributions. The critical atom number arises due to a need for the attractive binding energy to exceed the dispersive kinetic energy. For larger atom numbers, the peak density saturates, and adding more atoms simply increases the radial extent of the cloud. Indeed, this is where the term *droplet* comes from. If you were to merge two raindrops, the density of the water stays the same, but the raindrop becomes twice as large. It should be noted that the peak

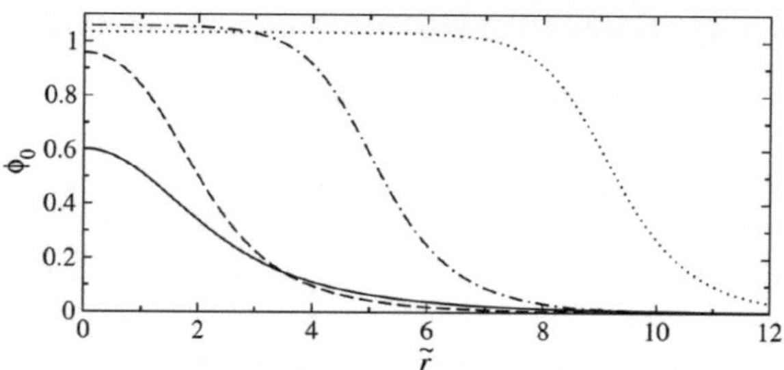

Fig. 13.1 Droplet solutions for $\tilde{N} = \tilde{N}_c \approx 18.65$ (solid), $\tilde{N} = 30$ (dashed), $\tilde{N} = 500$ (dash dotted), and $\tilde{N} = 3000$ (dotted). Reprinted figure with permission from [1]. Copyright (2015) by the American Physical Society

density here is 10 times that of normal superfluids, and hence three-body recombination losses as discussed in Sect. 3.9.2 do dominate. However, quantum droplets in delicately balanced mixtures of Bose-Bose gases have now been seen in many experiments [2–7].

If a cigar-shaped trap is added, it should be noted that for smaller atom numbers than the previously reported critical one a soliton solution is possible. Previously, we saw that increasing the atom number leads to a mean-field collapse of the state; however, in this case, quantum fluctuations stabilise the system, resulting in the formation of a stable droplet instead. In essence, the soliton exists due to a balance between dispersive kinetic energy and attractive mean-field interactions, whereas the droplet—exhibiting a flat-top density profile and thus minimal kinetic energy contribution—exists as a balance between attractive mean-field interactions and repulsive quantum fluctuations. The crossover between these two states was experimentally observed by Cheiney et al. [8].

13.2.2 Dipolar Gases

The inclusion of dipolar interactions offers a route to cancel the mean-field contributions even within a single-component gas. While it was Petrov's 2015 work [1] that ignited the race to observe quantum droplets, equivalent equations for dipolar systems were derived earlier in a series of papers by Lima and Pelster [9, 10]. Using

$$\tilde{U}_{\text{int}}(\mathbf{k}) = g + g_{\text{dd}}(3\cos^2\theta_k - 1), \tag{13.8}$$

from the previous chapter, we can evaluate Eq. (13.2) to obtain

$$\frac{\Delta E}{V} = \frac{64}{15} g n_0^2 Q_5(\varepsilon_{\text{dd}}) \sqrt{\frac{n_0 a_s^3}{\pi}}, \tag{13.9}$$

where the auxiliary function $Q_l(x) = \int_0^1 du\,(1 - x + 3xu^2)^{l/2}$ is monotonic, growing from $Q_5(0) = 1$ to $Q_5(1) = 3\sqrt{3}/2 \approx 2.6$. A commonly used approximation is $Q_5(x) \approx 1 + \frac{3}{2}x^2$, which holds well for $0 < \varepsilon_{\text{dd}} \lesssim 2$, provided the imaginary part of $Q_5(\varepsilon_{\text{dd}})$ is discarded for $\varepsilon_{\text{dd}} > 1$.

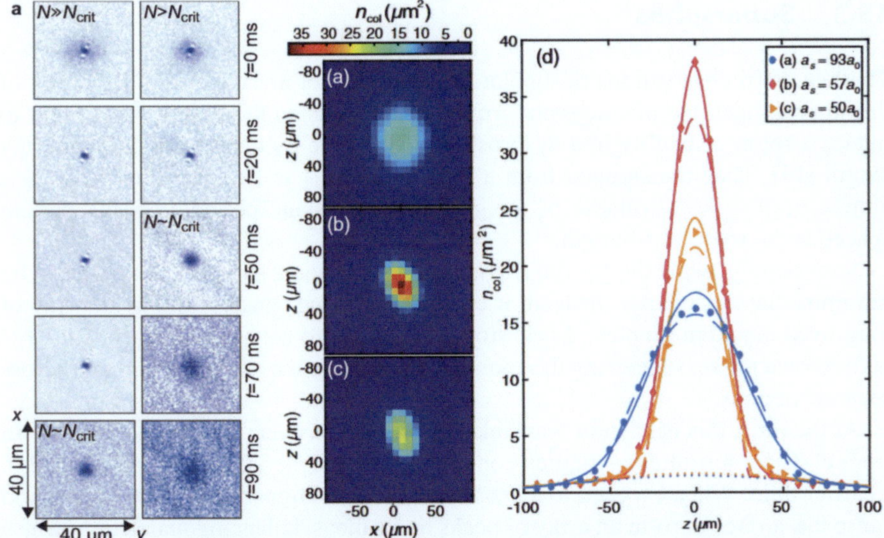

Fig. 13.2 Experimental observations of single dipolar droplets. Left two columns: Self-bound dysprosium droplet state remaining static for increasing hold times. Reprinted with permission from [11]. Copyright 2016, Macmillan Publishers Limited, part of Springer Nature. All rights reserved. Right: Density profile of an erbium droplet for decreasing scattering lengths. Reprinted under CC-BY-3.0 license from [12]. Copyright 2016, The Author(s)

From this, one can derive the eGPE suitable for dipolar atoms

$$i\hbar \frac{\partial \Psi}{\partial t} = \left(-\frac{\hbar^2 \nabla^2}{2m} + V_{\text{ext}} + g|\Psi|^2 + \Phi_{\text{dd}} + \gamma_{\text{QF}}|\Psi|^3 \right)\Psi, \tag{13.10}$$

where

$$\gamma_{\text{QF}} = \frac{32}{3}g a_s^{3/2} \mathcal{Q}_5(\varepsilon_{\text{dd}}). \tag{13.11}$$

Similar to the Bose–Bose system, this equation also supports droplet solutions in free space ($V_{\text{ext}} = 0$) for $\varepsilon_{\text{dd}} > 1$, i.e., when the mean-field interaction becomes attractive. However, several key differences arise. Firstly, due to the anisotropic nature of dipolar interactions, the resulting droplets are not spherically symmetric, but instead exhibit magnetostriction along the polarization (attractive) axis. Secondly, as this is a single-component system, there is no need to finely tune interaction parameters or particle number balance, as is required in Bose–Bose mixtures. Droplets will form provided the particle number is large enough that the attractive dipolar binding energy exceeds the dispersive kinetic energy. In 2016, the Pfau group [11] and Ferlaino group [12] observed these single macrodroplets in dysprosium and erbium, respectively, and results from these experiments are shown in Fig. 13.2.

13.3   Supersolids

The late 2016 observations of dipolar droplets did not mark the true beginning of the story. In fact, the first key hint arrived in 2015, when the Pfau group sought to induce a roton instability in a dysprosium condensate by quenching the scattering length [13]. They transitioned from a regime with $\epsilon_{dd} < 1$ to one where $\epsilon_{dd} \gg 1$, aiming to provoke a collapse of the gas into a transient, periodic density pattern spaced at the roton wavelength.

Surprisingly, while the gas did fragment into a periodic structure, this pattern did not immediately collapse. Instead, it persisted for long times, forming an array of long-lived quantum droplets. Each droplet appeared to act as an independent BEC with its own phase, suggesting the spontaneous emergence of local coherence across separate regions.

At the time, this behaviour was linked to the Rosensweig instability [14] known from classical ferrofluids—liquids containing tiny magnetic particles suspended in a carrier fluid. When exposed to a magnetic field, the aligned dipoles in a ferrofluid cause the surface to form an array of peaks and valleys, balancing magnetic, surface tension, and gravitational forces. This striking resemblance led to the description of dipolar quantum gases as *quantum ferrofluids*. A comparison between the experimental image of dipolar droplets from Kadau et al. [13] and a classical ferrofluid above a magnet is shown in Fig. 13.3.

Over the following year, a series of experimental and theoretical works established that the ground state of this system was actually a single quantum droplet, as discussed in the previous section [15]. However, this initial observation sparked a deeper idea: if one could stabilise the periodic droplet structure while maintaining global phase coherence between the individual density peaks, the system would behave as a single macroscopic wavefunction exhibiting both crystalline order and superfluidity. This would constitute a new phase of matter: the supersolid.

A *supersolid* is a phase of matter that simultaneously exhibits the properties of a solid and a superfluid. This requires the spontaneous breaking of two fundamental symmetries:

- **U(1) phase symmetry**: A global phase is selected across the system, as in a Bose–Einstein condensate.
- **Translational symmetry**: A periodic density modulation forms spontaneously, characterized by an order parameter such as $\cos(kx + \phi)$, where the phase ϕ varies randomly with each experiment.

The coexistence of these broken symmetries leads to a quantum state with both long-range coherence and crystalline structure.

Theoretical proposals for supersolidity date back to the 1950s, when two complementary approaches were developed [16–20].

The first approach considers a conventional crystalline solid, such as a lattice of atoms, and attempts to endow it with superfluid properties. This was the guiding motivation behind the extensive search for supersolidity in solid helium. In particu-

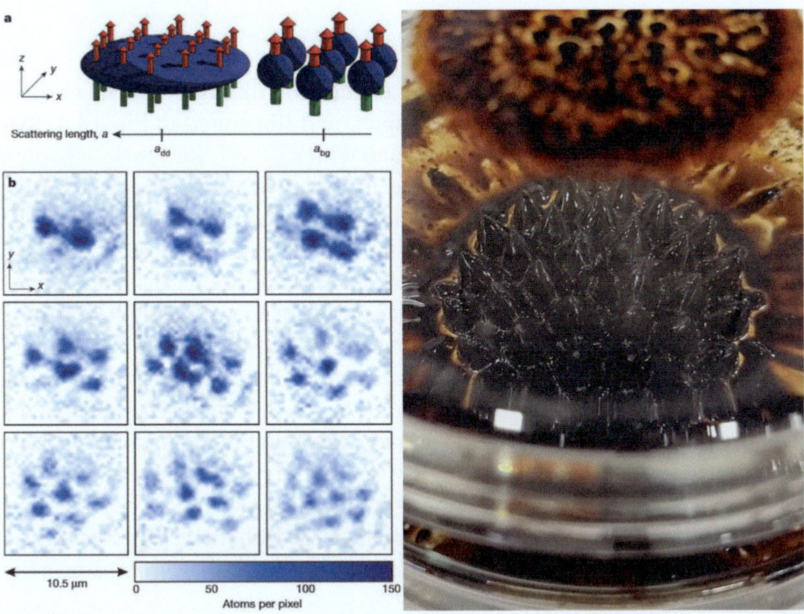

Fig. 13.3 Quantum and classical Rosensweig instabilities. Left: **a** Schematic representation of the state following an interaction quench. **b** Density snapshots of arrays of quantum droplets. Each image (left to right and top to bottom) shows an increasing number of atoms before the quench. Reprinted with permission from [13]. Copyright 2016, Springer Nature Limited. All rights reserved. Right: Image of a classical ferrofluid sitting above a magnet displaying the iconic Rosensweig instability

lar, torsional oscillator experiments sought to detect non-classical rotational inertia as evidence of superfluidity. These experiments observed a sudden increase in the shear modulus of solid helium at low temperatures [21], consistent with a potential transition to a supersolid state, where a fraction of the system behaves as an irrotational superfluid. However, subsequent studies revealed that this effect was instead due to a structural rearrangement of atoms within the solid [22].

In contrast, the second approach—more accessible in ultracold atom systems— starts with a superfluid and seeks to introduce crystalline (solid-like) order. Over the past decade, this paradigm has been realized with remarkable success. Supersolids have been observed in various platforms, including dipolar quantum gases [23–25], cavity-mediated systems [26], spin-orbit coupled condensates [27], and most recently in exciton-polariton systems [28].

In this chapter, we focus on supersolids formed via dipolar interactions, which provide a natural route to induce both long-range order and global phase coherence in a single-component system.

13.4   Dipolar Supersolids

The origin of dipolar supersolidity can be understood through the lens of the roton instability, introduced in the previous Chapter, and illustrated in Fig. 13.4. Starting with an ultracold dipolar gas in free space and with $\varepsilon_{dd} > 1$, the atoms self-organize into a single, elongated quantum droplet due to attractive head-to-tail dipolar interactions (Fig. 13.4a). When confinement is added along the polarization axis—such as a harmonic or box potential—the head-to-tail attraction is suppressed, encouraging the atoms to arrange side-by-side. For sufficiently strong dipolar interactions, the ground state may consist of multiple smaller droplets (Fig. 13.4b), each behaving as an independent condensate with its own phase. This configuration is often referred to as the *isolated droplet* or *independent droplet* regime. The spacing between the droplets is related to the roton wavelength of the corresponding homogeneous gas, which is typically $\sqrt{2}l_z$, where l_z is the harmonic oscillator length along the polarization direction (see Sect. 12.7.2). Finally, by tuning the contact interactions—such as by increasing the scattering length—the droplets can broaden and begin to overlap. If the overlap is sufficient to establish global phase coherence, the resulting state breaks both the U(1) phase symmetry and translational symmetry, realising a dipolar supersolid.

Comprehensive reviews on the state and future of supersolidity can be found in Refs. [29,30]. Here, we wish to give a flavour of what interesting new physics the solid component of a supersolid brings to the system.

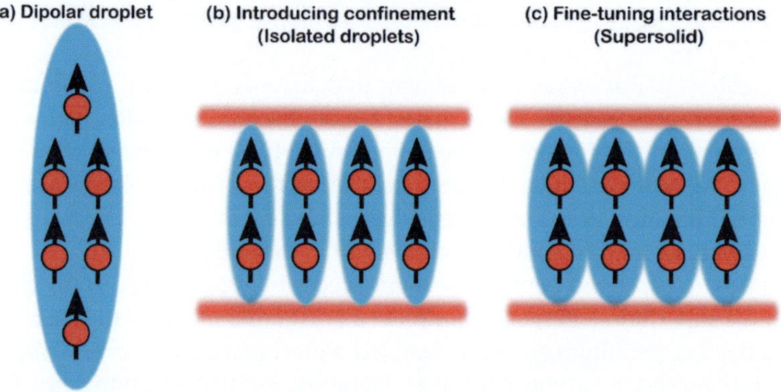

Fig. 13.4 Schematic illustration of the emergence of dipolar supersolidity. **a** In free space, strong dipolar interactions lead to the formation of a single, self-bound quantum droplet aligned along the polarization axis. **b** Adding axial confinement (red lines) suppresses the head-to-tail alignment and results in multiple smaller, side-by-side droplets, each with an independent phase. **c** Tuning the interactions to increase the droplet width can lead to wavefunction overlap between droplets, establishing global phase coherence and giving rise to a supersolid state

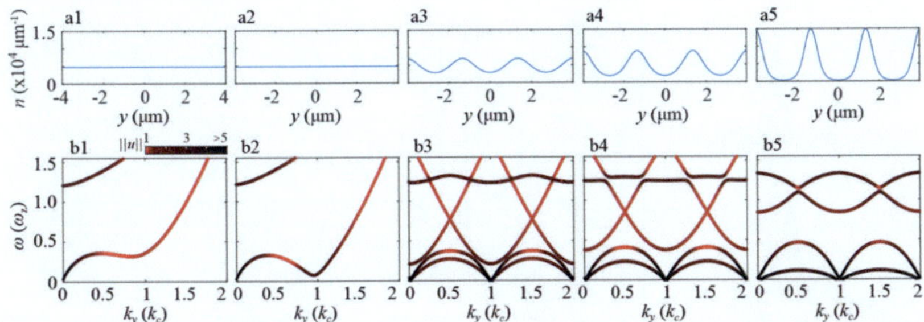

Fig. 13.5 Excitation spectra of dipolar gases. Top row: Density solutions to the eGPE, with increasing ε_{dd} from left to right. Bottom row: Excitation spectrum of the corresponding state from the top row. When the crystallization occurs the spectrum becomes a Brillouin zone structure. The number of branches emanating from $k = 0$ corresponds to the number of broken symmetries. Reprinted with permission from [31]. Copyright 2021, American Physical Society. All rights reserved

13.4.1 One-Dimensional Supersolids

The first experiments to show dipolar supersolidity did so in a cigar-shaped trap [23–25]. In this section, we will refer to these as one-dimensional supersolids, but wish to highlight that they are three-dimensional objects, with broken translational symmetry only along a single axis. In the scenario of an infinite tube potential, the ground state can take on three possible solutions: a homogeneous density, a supersolid, and an array of independent droplets. These states are shown as ground state solutions to the eGPE in the top row of Fig. 13.5.

To discuss excitations of the supersolid, we need to introduce the concept of Goldstone modes. A hallmark of supersolidity is the simultaneous breaking of two continuous symmetries: the $U(1)$ phase symmetry (as in superfluids) and translational symmetry (as in crystals). According to Goldstone's theorem, each broken continuous symmetry gives rise to a gapless excitation branch. As a result, supersolids exhibit *two* Goldstone modes, manifesting as two distinct low-energy phonon branches emanating from $k = 0$, as shown in the second row of Fig. 13.5b3–b5. Note, that due to the crystalline structure the excitation spectrum now takes on a repeating Brillouin Zone structure [32].

One branch corresponds to *superfluid phonons*. These excitations preserve the overall density modulation, but involve coherent flow of atoms through the supersolid. That is, the droplets (or density peaks) remain stationary, while atoms move through them in a collective oscillation, much like sound waves in a regular superfluid.

The second branch corresponds to *crystalline phonons*. Here, the density peaks themselves oscillate in space, shifting side-to-side, while the atoms within each peak largely remain localized. This is akin to the lattice vibrations of a conventional solid, where the periodic structure itself moves.

To illustrate the difference: a density peak moving to the right can occur in two conceptually distinct ways. In a superfluid phonon, atoms within the peak flow to the right, transporting the peak itself. In a crystal phonon, atoms from the neighboring

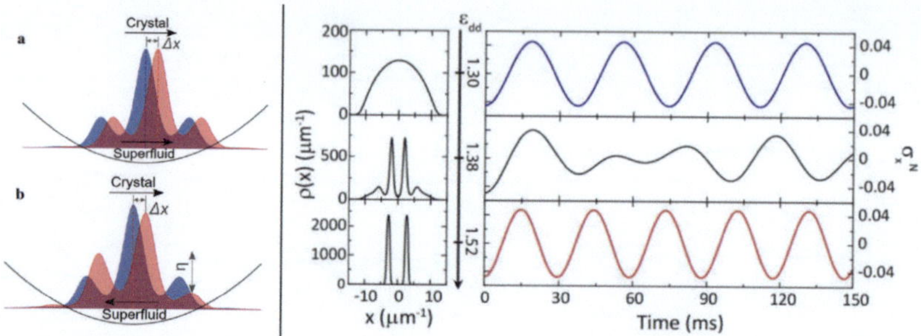

Fig. 13.6 Excitation modes of linear supersolids. Left: **a** a superfluid phonon in a finite sized system. **b** a crystal phonon in a finite sized system. The density lumps move right, whilst the superfluid moves left. Note, in the infinite case, there are an infinite number of atoms to the right to "fill in" the droplets moving to the right, hence the density amplitude will not change. Reprinted with permission from [33]. Copyright 2019, The Author(s), under exclusive licence to Springer Nature Limited. Right: Breathing modes in a superfluid (top), supersolid (middle), and independent droplet phases (bottom). Note the multimode reponse for the supersolid. Reprinted with permission from [34]. Copyright 2019, The Author(s), under exclusive licence to Springer Nature Limited

left peak move rightward and accumulate, effectively pushing the density maximum to the right without requiring atoms within the original peak to move significantly. In finite-sized systems this is referred to as the out-of-phase Goldstone mode, and has been observed in experiments from the Pfau group [33], and schematically shown in Fig. 13.6.

A consequence of these multiple branches is multiple modes. For example, attempting to excite the breathing mode of a supersolid by suddenly increasing or decreasing the axial trap frequency excites a breathing oscillation of the superfluid and another of the crystal, giving a multimode response. This was observed in an experiment from the Modugno group [34], and is shown in Fig. 13.6.

As ε_{dd} is increased from Fig. 13.5b3–b5, the energy of the superfluid phonon branch decreases, and the corresponding superfluid speed of sound tends towards zero. This reflects the transition from a supersolid to an array of independent droplets. Naturally, this raises the question: how can we quantify the degree of supersolidity in such a system?

A useful measure is the *superfluid fraction* f_s, which characterizes the portion of the system that exhibits non-classical (i.e., superfluid) behaviour. For a one-dimensional supersolid with broken translational symmetry along z, the superfluid fraction can be defined as

$$f_s = 1 - \lim_{v_z \to 0} \frac{\langle \hat{P}_z \rangle}{m v_z}, \tag{13.12}$$

where $\langle \hat{P}_z \rangle$ is the expectation value of the linear momentum operator, given by

$$\langle \hat{P}_z \rangle = \frac{i\hbar}{N} \int \psi^*(z) \frac{\partial \psi(z)}{\partial z} \, \mathrm{d}z. \tag{13.13}$$

This definition can be understood by comparing the quantum and classical responses to motion. The denominator of Eq. (13.12) corresponds to the classical momentum per particle, $p = mv$, while the numerator gives the quantum mechanical momentum response of the system. A perfect superfluid does not respond to a slow imposed velocity, so $\langle \hat{P}_z \rangle = 0$, yielding $f_s = 1$. On the other hand, a classical solid responds fully with $\langle \hat{P}_z \rangle = mv_z$, giving $f_s = 0$.

A key identifying feature of a supersolid is therefore that $0 < f_s < 1$. In other words, when a supersolid is pushed through a pipe, its response lies between that of a pure superfluid and a classical solid. More generally, Eq. (13.12) is known as the fraction of non-classical translational inertia $f_s \equiv f_{\mathrm{NCTI}}$, and there are other formulations of the superfluid fraction suitable for other geometries, which we will discuss later. Obtaining the superfluid fraction has been a key observable sought by experimentalists, with probes into the superfluid nature of the scissors mode response [35,36] and Josephson oscillations [37]. To test Eq. (13.12) directly experiments are required in a ring-trap geometry, where the periodic ring can emulate infinite space.

A simpler, yet less rigourous, measure of the superfluidity is the *density contrast* $\mathcal{C}$. This measure looks only at the density, and therefore states nothing on the global phase coherence, only on the degree of modulation. For a one-dimensional chain, it is defined as

$$\mathcal{C} = \frac{n_{\max} - n_{\min}}{n_{\max} + n_{\min}}, \tag{13.14}$$

where $n_{\max}$ is the density evaluated at the density peak, and $n_{\min}$ in the density trough. For a homogeneous density, $n_{\max} = n_{\min}$ and $\mathcal{C} = 0$, and in the independent droplet regime the density connection is zero, hence $\mathcal{C} = 1$. In practice, a conservative estimate is placed on this quantity, defining supersolidity only between $0 < \mathcal{C} \le 0.95$.

13.4.2 Two-Dimensional Supersolids

The transition from cigar-shaped to pancake-shaped traps brings with it a wealth of new and interesting density patterns. In experiments by the Ferlaino group [38,39], a structural phase transition was observed—from a linear chain of droplets to a triangular lattice—by tuning the trap aspect ratio $\alpha_t = \omega_x/\omega_y$ from $\alpha_t \ll 1$ up to $\omega_x = \omega_y$, as shown in Fig. 13.7, in agreement with eGPE simulations.

To definitively demonstrate two-dimensional supersolidity, time-of-flight (TOF) images must exhibit two key features: a central peak corresponding to macroscopic occupation of the $k = 0$ quantum state, and satellite peaks spaced at $60°$ intervals, reflecting both the triangular lattice symmetry and global phase coherence. A crucial ingredient for observing this transition was conducting the experiment at constant trap density. That is, loosening one trap frequency (to round out the trap) must be accompanied by an increase in atom number to maintain a fixed density in the new volume. This was described in Ref. [40] as maintaining supersolidity as long as the

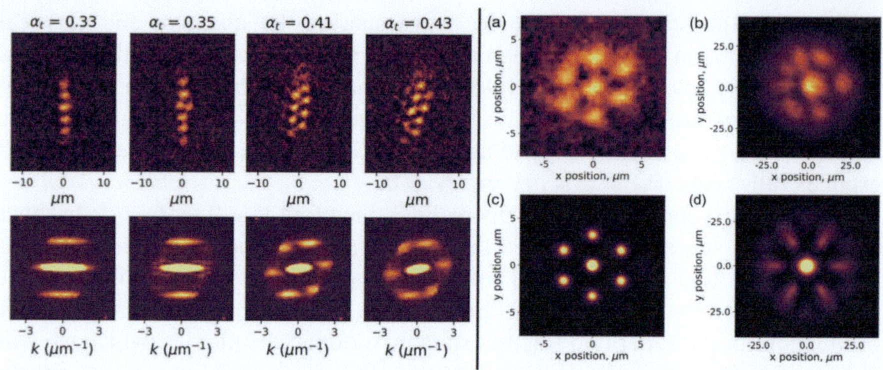

Fig. 13.7 Two-dimensional supersolidity. Left: In situ densities (top row) and time-of-flight (TOF) images (bottom row), showing the evolution from a cigar-shaped to a pancake-shaped trap from left to right. Each droplet contains approximately 5–10 thousand dysprosium atoms. Reprinted with permission from [38]. Copyright 2021, The Author(s), under exclusive licence to Springer Nature Limited. Right: **a** Experimental image of a 2D supersolid in a pancake-shaped trap. **b** Corresponding TOF image. (c,d) Numerical results from solving the eGPE under experimental conditions. Reprinted figure with permission from [39]. Copyright (2022) by the American Physical Society

trap density $\rho = N/(l_x l_y)$ is held fixed, where l_x and l_y are the harmonic oscillator lengths in the xy-plane.

Theory predicts a plethora of other intriguing new density structures that appear at higher densities than current experiments [42, 43], see Fig. 13.8. These structural phase transitions occur due to the finite space restricting the number of atoms per droplet. The droplet-to-stripe transition fills in the density minima between pairs of droplets along one axis, and the stripe-to-honeycomb transition fills in more gaps, regaining the triangular symmetry. Note that here the dipolar angle is fixed perpendicularly to the plane, and so the rotational orientation of these states is arbitrary. So far, no experiments have observed these higher density states. This would be a phenomenal observation, due to their remarkable similarity to nuclear pasta states expected in the heart of neutron stars [41].

In finite systems, the interplay between the trap and these high density states breaks many of the clean symmetries, giving rise to many degeneracies in the orientation of the state. Primarily, this affects the stripe state, which becomes more labyrinthine. Such solutions to the eGPE are shown in Fig. 13.9.

13.4.3 Vortices in Supersolids

Experiments on dipolar supersolids over the past six years have focused on extracting and confirming the superfluid nature of these states. One key approach is the observation of quantized vortices, which can only exist due to the single-valued nature of the underlying wavefunction. In this section, we discuss several properties of supersolid vortices and present evidence for their experimental observation.

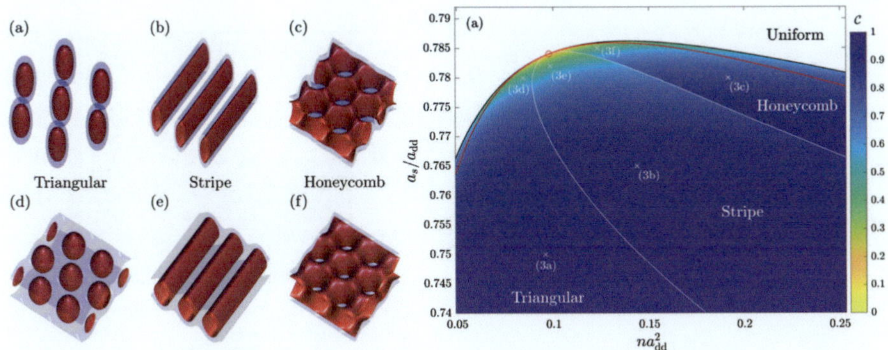

Fig. 13.8 High density phases of infinite supersolids. Left: Triangular, stripe, and honeycomb states as ground states to the eGPE with increasing density from left to right. **d**–**f** here show equivalent states with higher superfluid fraction f_s. Right: phase diagram of the density contrast for $1/\varepsilon_{dd}$ vs density. The highlighted states "(3x)" correspond to the left-hand-side of this figure. Reprinted figure with permission from [43]. Copyright (2023) by the American Physical Society

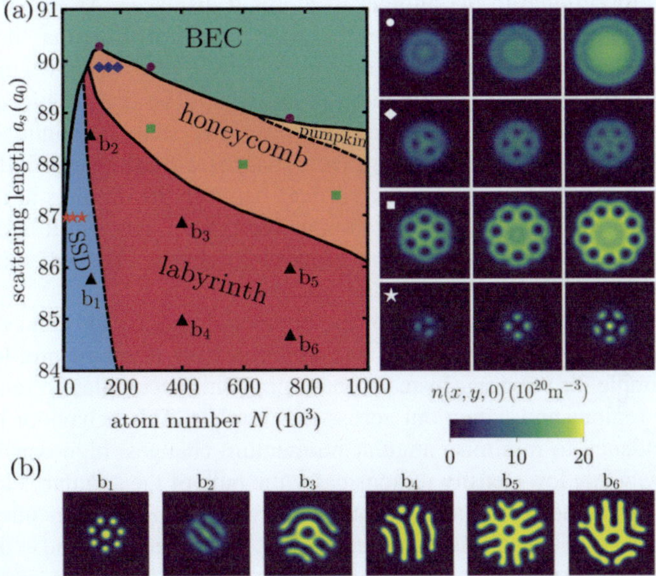

Fig. 13.9 High density phases of finite supersolids. Phase diagram from the eGPE, with highlighted states marked by symbols on the diagram. Supersolid droplets (SSD), transition to labyrinthine (b3–b6), and honeycomb (squares) states for increasing densities. Reprinted under CC-BY-4.0 license from [44]. Copyright 2021, The Author(s)

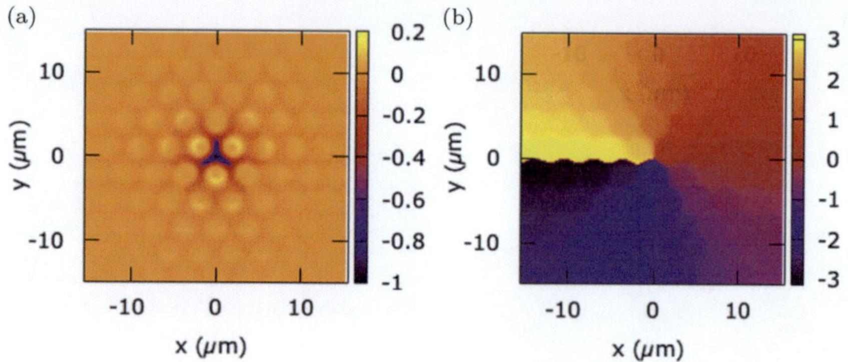

Fig. 13.10 Vortices in an infinite two-dimensional triangular supersolid. **a** Density difference between a vortex state and a vortex-free supersolid. **b** Phase profile of the vortex state. Reprinted figure with permission from [45]. Copyright (2021) by the American Physical Society

First, we consider the position of vortices. Naturally, the energy cost of hosting a vortex is proportional to the number of atoms it displaces. As a result, vortices tend to reside in the low-density regions between the density maxima. In the infinite system, this behaviour was clearly demonstrated in Ref. [45], as shown in Fig. 13.10 (see Ref. [46] for extensions to finite-sized systems). The density plot reveals the difference between the vortex state and the vortex-free state, highlighting the local depletion caused by vortex imprinting. Note that the overall change in the density is extremely subtle—can you spot the vortices between the droplets in the top row of Fig. 13.7? The average TOF images tell us there aren't any, but it is likely some experimental images did contain them by random chance. We will return to this challenge when discussing experimental detection.

The vortex core adopts the shape of the interstitial regions and can even slightly displace droplets up to five lattice sites away. Interestingly, the phase profile no longer exhibits a simple 2π winding. Instead, phase gradients accumulate primarily in the low-density regions and flatten out across the droplets. This behaviour reflects the system's tendency to minimise angular momentum changes: high-density regions move slowly, while low-density regions carry the bulk of the angular velocity.

The angular momentum of a singly-quantized vortex in a homogeneous superfluid is known to be $L_z = N\hbar$, as discussed in Chap. 7. In a supersolid, due to the reduced superfluidity, this instead becomes $L_z = f_s N\hbar$, where the superfluid fraction can be approximated via the non-classical rotational inertia (NCRI):

$$f_s \equiv f_{\text{NCRI}} = 1 - \lim_{\Omega \to 0} \frac{\langle \hat{L}_z \rangle}{\Omega I_{\text{rig}}}, \tag{13.15}$$

for rotation about the z axis with angular frequency Ω. Similar to Eq. (13.12), a non-zero f_{NCRI} signals non-classical behavior. Here, $I_{\text{rig}} = \int r^2 n(\mathbf{r})\, \mathrm{d}V$ is the classical (rigid-body) moment of inertia, and

$$\langle \hat{L}_z \rangle = i\hbar \int \Psi^* \left(y\frac{\partial}{\partial x} - x\frac{\partial}{\partial y} \right) \Psi \, dV \tag{13.16}$$

is the expectation value of the angular momentum operator. Note that the equivalence $f_s = f_{\text{NCRI}}$ strictly holds only in infinite systems or ring geometries.

In finite-sized systems, the critical rotation frequency for vortex nucleation and the resulting angular momentum have been studied numerically in Ref. [47]. Key results from that work are shown in Fig. 13.11. Several insightful features are evident. Firstly, the energy cost for inducing a vortex decreases with increasing ε_{dd}. This can be intuitively understood: as ε_{dd} increases, the connecting superfluid density between droplets diminishes, lowering the energy barrier for vortex formation. Note that in the ID regime, where there is no phase coherence, vortices cannot exist.

Figure 13.11b shows how the reduced superfluid fraction leads to a smaller angular momentum per vortex. In finite systems, the angular momentum jump Δ is approximately equal to f_{NCRI}, offering a possible experimental method for determining the superfluid fraction via angular momentum measurements.

Perhaps the most revealing insight comes from the final row of Fig. 13.11. In the superfluid phase (Fig. 13.11c), there is no angular momentum response to rotation until a critical frequency is reached, beyond which a quantized jump of $\hbar$ is observed. In contrast, the ID regime (Fig. 13.11e) shows a purely classical response: a linear increase in angular momentum with slope equal to I_{rig}. The supersolid response (Fig. 13.11d) lies between these extremes. At low rotation frequencies, the solid component rotates classically with a reduced effective moment of inertia, giving an initial slope related to Eq. (13.15). At a critical frequency—reduced from the superfluid case—a vortex nucleates with an angular momentum jump smaller than $\hbar$, reflecting the diminished superfluid fraction.

Before turning to experimental detection, let us briefly explore some additional features of supersolid vortices. Vortex–antivortex annihilation occurs similarly to conventional superfluids; however, in supersolids, the vortices are constrained to follow paths dictated by the underlying density profile [45]. By analyzing correlations in vortex trajectories during such annihilation events, it has been proposed that one can extract an estimate of the superfluid fraction [48]. Vortex lattices in supersolids tend to form in the density minima between droplets. Depending on the rotation frequency, the resulting vortex lattice can vary from triangular to honeycomb structures, or even more densely packed configurations [47,49].

Interestingly, supersolid phases are also expected to occur naturally in the dense interiors of neutron stars. So-called neutron star "glitches"—sudden transfers of angular momentum from the superfluid core to the rotating solid crust—require the existence of vortex pinning. These pinning sites are thought to arise from the periodic density modulations of a supersolid-like phase in the inner crust. A recent numerical study demonstrated that dipolar supersolids can exhibit glitch-like behavior [50], where vortices pinned between density maxima are suddenly released during a deceleration of the external rotation. This result suggests that laboratory realisations of

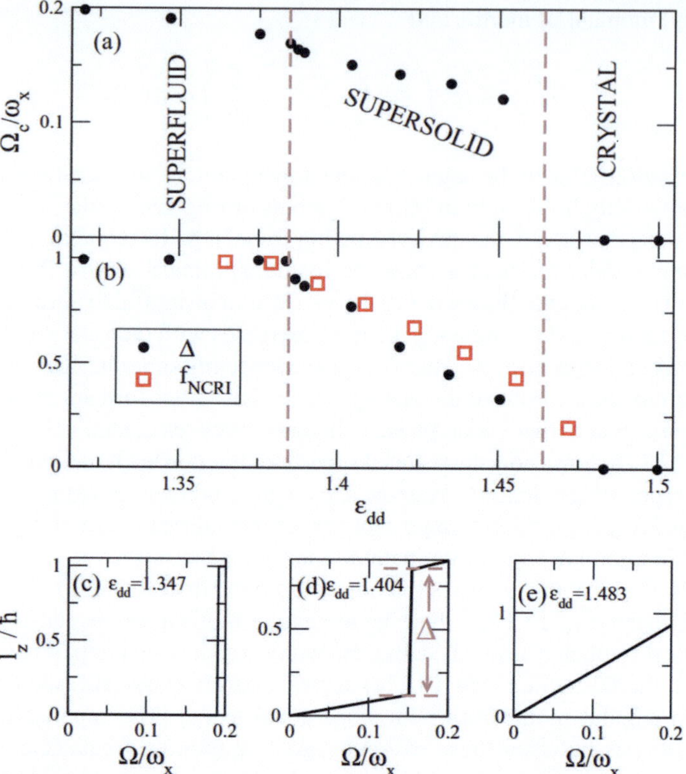

Fig. 13.11 Critical rotation frequency and angular momentum of supersolids. **a** Critical rotation frequency for vortex nucleation across the three phases. Note that this simulation used parameters corresponding to a three-droplet SS and ID state. **b** Angular momentum per particle ($L_z/N\hbar$) of an induced vortex and the non-classical rotational inertia fraction f_{NCRI}, which differ in finite systems. Total angular momentum as a function of rotation frequency for **c** a superfluid, **d** a supersolid, and **e** an independent droplet state. Reprinted figure with permission from [47]. Copyright (2020) by the American Physical Society

dipolar supersolids may provide a powerful platform for modeling and understanding the dynamics of such astrophysical phenomena.

Let us return to the central question: how can one observe a vortex-induced density hole in a system where density minima already exist due to supersolidity? The key insight comes from experiments on the BEC–BCS crossover in Fermi gases, where superfluidity in the BCS regime was confirmed by rotating the gas, quenching the interactions to the BEC side, and then immediately imaging the system. Because vortices are topologically protected, such a quench—as long as it connects two phase-coherent states—preserves the vortex configuration. Thus, the observation of vortices in the post-quench BEC state served as indirect evidence of superfluidity in the initial BCS state [51].

The same principle can be applied to dipolar supersolids. If one can nucleate a vortex in a supersolid, and then quench the scattering length such that the system

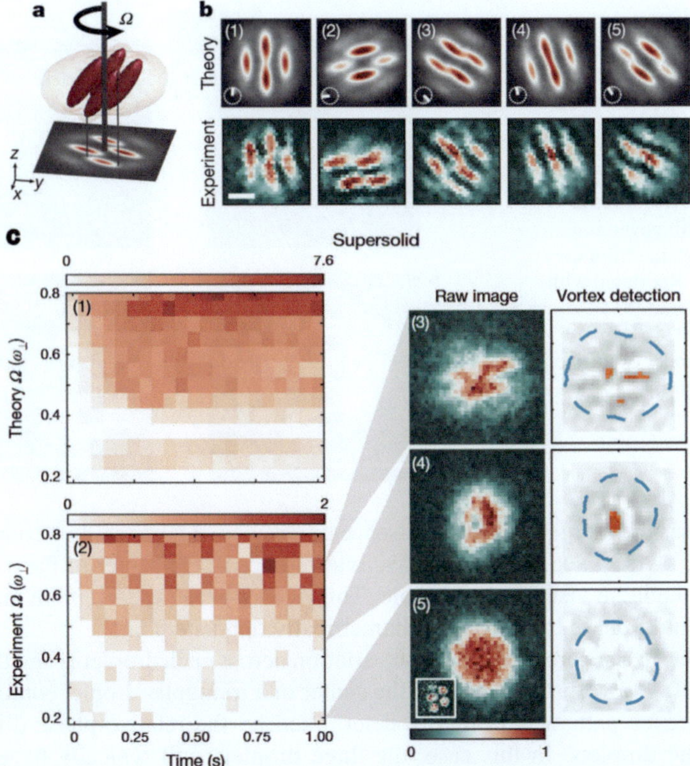

Fig. 13.12 Observation of vortices in a dipolar supersolid. **a** eGPE simulation showing the 3D density distribution (high density in red, low density transparent) and its column density projection. **b** Theoretical and experimental images of the column density during magnetostirring. **c** After a variable stirring time at fixed frequency, a projection to the BEC state is performed, followed by destructive imaging. Vortices appear as density holes in the resulting BEC images (see insets). Color in the phase diagram shows the average vortex number after several runs. Reprinted with permission from [53]. Copyright 2024, The Author(s), under exclusive licence to Springer Nature Limited

transitions to a superfluid phase, the density modulation will vanish but the vortex will persist. In this case, the homogeneous background density of the superfluid enhances the contrast, making the vortex core clearly visible [52].

This precise idea was realized experimentally by the Ferlaino group [53]. First, vortices were nucleated via magnetostirring (as introduced in the previous chapter), which involves tilting the polarization axis to $\theta = 30°$ from the tight z-axis of the pancake trap and rotating around the z axis at a variable frequency. After a variable stirring duration, the scattering length was rapidly quenched from a regime with $\varepsilon_{dd} \gtrsim 1.3$ to one with $\varepsilon_{dd} < 1$, thereby melting the density modulation and revealing the surviving vortex. This technique provided the first indirect observation of a vortex in a dipolar supersolid, as shown in Fig. 13.12.

A second experimental approach was used to observe vortex signatures via TOF measurements. Rather than destroying the supersolid, this method allowed interfer-

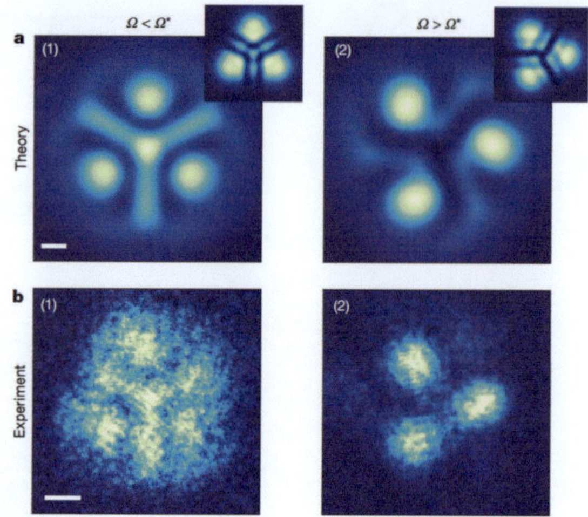

Fig. 13.13 Direct observation of supersolid vortices in time-of-flight. **a** Images after 36ms TOF from the eGPE without (left) and with (right) a vortex. The inset shows the same pattern from an analytic Ansatz, which we will cover as a Problem. **b** Same from the experiment. Reprinted with permission from [53]. Copyright 2024, The Author(s), under exclusive licence to Springer Nature Limited

ence patterns to emerge due to the presence or absence of a vortex. In a conventional superfluid, a vortex leads to destructive interference at the center of the expanding cloud, producing a visible density hole. This occurs because each point around the vortex is out of phase by π from its diametrically opposite point.

However, in a supersolid, the phase variation across each droplet is small, as seen in Fig. 13.10. If a vortex is located at the center of a triangular droplet configuration, the interference pattern is primarily determined by the relative phase differences between the droplets. In this case, the three droplets will typically have relative phases separated by $60°$ around the circle. Upon expansion, these phase differences give rise to an interference pattern that reveals the vortex indirectly. A comparison of the density after TOF with and without a vortex, from both eGPE theory and experiment, is shown in Fig. 13.13.

Though we have seen the effects that a vortex has on a supersolid, an important question remains: how does a vortex nucleate in a supersolid? This was addressed in the most recent experimental observation by the Ferlaino group [54]. In their experiment, a static supersolid with a polarization tilt of $30°$ was subjected to a sudden jump in rotation frequency from zero to a finite value. Initially, only the solid part of the supersolid—i.e., the density modulation—responds by rotating. The superfluid component, being irrotational, does not immediately react. As a result, the droplets begin to spin around their own axes while their centers of mass remain stationary. Over time, the asymmetry of the superfluid component causes an increasing transfer of angular momentum, eventually setting the droplets' center of mass into motion. After further stirring, a vortex nucleates, allowing both the superfluid and solid parts of the supersolid to rotate synchronously with the driving frequency. This effect was observed experimentally by following the centre-of-mass motion of the droplets in destructive measurements, and an exemplar theoretical simulation is shown in Fig. 13.14.

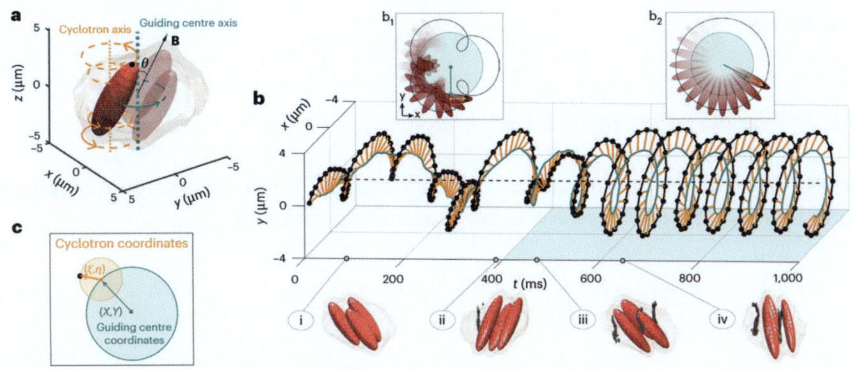

Fig. 13.14 Synchronisation of the solid and superfluid components of a supersolid to external driving. **a** Each droplet's motion can be decomposed into it's rotation around itself (cyclotron motion) and it's centre-of-mass (guiding centre motion), see also (**c**). **b** Black points follow the rotation of the tip of a droplet in time. For times $< 400\,\text{ms}$ the droplet goes through a double rotation motion (b1), for later times the guiding centre and cyclotron coordinates rotate together (b2). This transition is caused by a vortex nucleation (black lines in (i) to (iv)). Reprinted under CC-BY-4.0 license from [54]. Copyright 2025, The Author(s)

A useful analogy for this process is a glass of water containing ice cubes. If a straw is inserted and used to stir the liquid, the ice cubes quickly bounce off one another and begin rotating with the straw. However, the water initially remains at rest. With continued stirring, a vortex eventually forms in the water at the center of the glass, and the entire system reaches an equilibrium with the stirring motion [55]. This analogy closely mirrors the supersolid system —except here, the "solid" (ice) and "superfluid" (water) components are indistinguishable and composed of the same atomic gas.

We conclude this section with a brief note on solitons in supersolids. Bright solitons can be considered the low-atom-number counterparts of quantum droplets. Therefore, in supersolid systems where the atom number far exceeds that of a single droplet, bright solitons are unlikely to play a significant role.

Dark solitons, like vortices, arise from phase coherence across the system, and their observation would be a further proof of the superfluidity of supersolids. However, their behavior within supersolids remains largely unexplored. Imprinting a π phase jump across a linear supersolid confined in a harmonic trap generates a dark soliton. Unlike in a conventional superfluid, where such solitons undergo oscillatory motion, the soliton in a supersolid exits the system, transferring its momentum to the crystal structure and setting the droplets into motion [56].

Problems

13.1 *Self-bound solutions in 3D free-space:*

The per-particle energy functional for the radially symmetric single-component approximation of a balanced Bose–Bose mixture is given by

$$E[\phi] = 4\pi \int_0^\infty dr\, r^2 \left[\frac{1}{2}\phi^* \nabla_r^2 \phi - \frac{3}{2}|\phi|^4 + |\phi|^5 \right]. \tag{13.17}$$

(a) For a Gaussian ball solution with radial width l, given by

$$\phi(r) = \sqrt{\frac{N}{l^3 \pi^{3/2}}}\, e^{-\frac{r^2}{2l^2}}, \tag{13.18}$$

calculate the energy as a function of N and l.

(b) Setting $N = 100$, plot this function with and without the LHY correction. What do you notice? With LHY, numerically find the width of the stable solution.

13.2 *Interference pattern of dipolar droplets in TOF:*

(a) Given the free-particle propagator in three dimensions,

$$K(\mathbf{r}, t; \mathbf{r}', 0) = \left(\frac{m}{2\pi i \hbar t}\right)^{3/2} \exp\left[\frac{im(\mathbf{r} - \mathbf{r}')^2}{2\hbar t}\right], \tag{13.19}$$

find the time-dependent solution $\psi(\mathbf{r}, t)$ of the Schrödinger equation for a non-interacting Gaussian wavepacket in free space, evolved under the equation

$$\psi(\mathbf{r}, t) = \int_{\mathbb{R}^3} d^3\mathbf{r}'\, \psi(\mathbf{r}', 0)\, K(\mathbf{r}, t; \mathbf{r}', 0). \tag{13.20}$$

Assume the initial wavefunction at $t = 0$ is a spherically symmetric Gaussian ball with N atoms and radial width σ,

$$\psi(\mathbf{r}, 0) = \sqrt{\frac{N}{(2\pi\sigma^2)^{3/2}}} \exp\left(-\frac{|\mathbf{r}|^2}{4\sigma^2} - i\phi_0\right). \tag{13.21}$$

Compute $\psi(\mathbf{r}, t)$ explicitly, and show that it remains a Gaussian under free expansion, with time-dependent width $\sigma(t)$. What is the explicit form of $\sigma(t)$?

(b) A minimal model of a three-droplet supersolid can be generated by taking the sum of three Gaussians at positions $\mathbf{r}_j = (x_j, y_j, z_j)$, and solutions from (a) with $\psi(\mathbf{r} - \mathbf{r}_j, t)$. Assuming the droplets have the same atom number, are the same width, and are initially placed at the corners of an equilateral triangle, show that the total density $|\Psi|^2$ formed from

$$\Psi(\mathbf{r}, t) = \sum_{j=1}^{3} \psi(\mathbf{r} - \mathbf{r}_j, t),$$

is nonzero in time-of-flight when $\phi_{0,j} = 0$ and zero when $\phi_{0,j} = \arctan(y_j/x_j)$. This can be done numerically, or by considering the behaviour of the central point. What does this tell you about the TOF of supersolids with and without vortices?

References

1. D.S. Petrov, Phys. Rev. Lett. **115**, 155302 (2015)
2. C.R. Cabrera, L. Tanzi, J. Sanz, B. Naylor, P. Thomas, P. Cheiney, L. Tarruell, Science **359**, 301 (2018)
3. G. Semeghini, G. Ferioli, L. Masi, C. Mazzinghi, L. Wolswijk, F. Minardi, M. Modugno, G. Modugno, M. Inguscio, M. Fattori, Phys. Rev. Lett. **120**, 235301 (2018)
4. G. Ferioli, G. Semeghini, L. Masi, G. Giusti, G. Modugno, M. Inguscio, A. Gallemí, A. Recati, M. Fattori, Phys. Rev. Lett. **122**, 090401 (2019)
5. C. D'Errico, A. Burchianti, M. Prevedelli, L. Salasnich, F. Ancilotto, M. Modugno, F. Minardi, C. Fort, Phys. Rev. Res. **1**, 033155 (2019)
6. Z. Guo, F. Jia, L. Li, Y. Ma, J.M. Hutson, X. Cui, D. Wang, Phys. Rev. Res. **3**, 033247 (2021)
7. L. Cavicchioli, C. Fort, F. Ancilotto, M. Modugno, F. Minardi, A. Burchianti, Phys. Rev. Lett. **134**, 093401 (2025)
8. P. Cheiney, C.R. Cabrera, J. Sanz, B. Naylor, L. Tanzi, L. Tarruell, Phys. Rev. Lett. **120**, 135301 (2018)
9. A.R.P. Lima, A. Pelster, Phys. Rev. A **84**, 041604(R) (2011)
10. A.R.P. Lima, A. Pelster, Phys. Rev. A **86**, 063609 (2012)
11. M. Schmitt, M. Wenzel, F. Böttcher, I. Ferrier-Barbut, T. Pfau, Nature **539**, 259 (2016)
12. L. Chomaz, S. Baier, D. Petter, M.J. Mark, F. Wächtler, L. Santos, F. Ferlaino, Phys. Rev. X **6**, 041039 (2016)
13. H. Kadau, M. Schmitt, M. Wenzel, C. Wink, T. Maier, I. Ferrier-Barbut, T. Pfau, Nature **530**, 194 (2016)
14. R. E. Rosensweig, (Courier Corporation, 2013)
15. R.N. Bisset, R.M. Wilson, D. Baillie, P.B. Blakie, Phys. Rev. A **94**, 033619 (2016)
16. E.P. Gross, Phys. Rev. **106**, 161 (1957)
17. A.F. Andreev, I.M. Lifshitz, J. Exp. Theo. Phys. **56**, 2057 (1969)
18. D.J. Thouless, Ann. Phys. **52**, 403 (1969)
19. G.V. Chester, Phys. Rev. A **2**, 256 (1970)
20. A.J. Leggett, Phys. Rev. Lett. **25**, 1543 (1970)
21. E. Kim, M.H.W. Chan, Nature **427**, 225 (2004)
22. D.Y. Kim, M.H.W. Chan, Phys. Rev. Lett. **109**, 155301 (2012)
23. L. Tanzi, E. Lucioni, F. Famà, J. Catani, A. Fioretti, C. Gabbanini, R.N. Bisset, L. Santos, G. Modugno, Phys. Rev. Lett. **122**, 130405 (2019)

24. L. Chomaz, D. Petter, P. Ilzhöfer, G. Natale, A. Trautmann, C. Politi, G. Durastante, R.M.W. van Bijnen, A. Patscheider, M. Sohmen, M.J. Mark, F. Ferlaino, Phys. Rev. X **9**, 021012 (2019)
25. F. Böttcher, J.-N. Schmidt, M. Wenzel, J. Hertkorn, M. Guo, T. Langen, T. Pfau, Phys. Rev. X **9**, 011051 (2019)
26. J. Léonard, A. Morales, P. Zupancic, T. Esslinger, T. Donner, Nature **543**, 87 (2017)
27. J.-R. Li, J. Lee, W. Huang, S. Burchesky, B. Shteynas, F.Ç. Top, A.O. Jamison, W. Ketterle, Nature **543**, 91 (2017)
28. D. Trypogeorgos, A. Gianfrate, M. Landini, D. Nigro, D. Gerace, I. Carusotto, F. Riminucci, K.W. Baldwin, L.N. Pfeiffer, G.I. Martone, M. De Giorgi, D. Ballarini, D. Sanvitto, Nature **639**, 337–341 (2025)
29. L. Chomaz, I. Ferrier-Barbut, F. Ferlaino, B. Laburthe-Tolra, B.L. Lev, T. Pfau, Rep. Prog. Phys. **86**, 026401 (2022)
30. A. Recati, S. Stringari, Nat. Rev. Phys. **5**, 735 (2023)
31. D. Petter, A. Patscheider, G. Natale, M.J. Mark, M.A. Baranov, R. van Bijnen, S.M. Roccuzzo, A. Recati, B. Blakie, D. Baillie et al., Phys. Rev. A **104**, L011302 (2021)
32. C. Kittel, *Introduction to Solid State Physics*, 8th edn. (Wiley, Hoboken, NJ, 2004)
33. M. Guo, F. Böttcher, J. Hertkorn, J.-N. Schmidt, M. Wenzel, H.P. Büchler, T. Langen, T. Pfau, Nature **574**, 386 (2019)
34. L. Tanzi, S.M. Roccuzzo, E. Lucioni, F. Famà, A. Fioretti, C. Gabbanini, G. Modugno, A. Recati, S. Stringari, Nature **574**, 382 (2019)
35. L. Tanzi, J.G. Maloberti, G. Biagioni, A. Fioretti, C. Gabbanini, G. Modugno, Science **371**, 1162 (2021)
36. M.A. Norcia, E. Poli, C. Politi, L. Klaus, T. Bland, M.J. Mark, L. Santos, R.N. Bisset, F. Ferlaino, Phys. Rev. Lett. **129**, 040403 (2022)
37. G. Biagioni, N. Antolini, B. Donelli, L. Pezzè, A. Smerzi, M. Fattori, A. Fioretti, C. Gabbanini, M. Inguscio, L. Tanzi, G. Modugno, Nature **629**, 773 (2024)
38. M.A. Norcia, C. Politi, L. Klaus, E. Poli, M. Sohmen, M.J. Mark, R.N. Bisset, L. Santos, F. Ferlaino, Nature **596**, 357 (2021)
39. T. Bland, E. Poli, C. Politi, L. Klaus, M.A. Norcia, F. Ferlaino, L. Santos, R.N. Bisset, Phys. Rev. Lett. **128**, 195302 (2022)
40. E. Poli, T. Bland, C. Politi, L. Klaus, M.A. Norcia, F. Ferlaino, R.N. Bisset, L. Santos, Phys. Rev. A **104**, 063307 (2021)
41. D.G. Ravenhall, C.J. Pethick, J.R. Wilson, Phys. Rev. Lett. **50**, 2066 (1983)
42. Y.-C. Zhang, F. Maucher, T. Pohl, Phys. Rev. Lett. **123**, 015301 (2019)
43. B.T.E. Ripley, D. Baillie, P.B. Blakie, Phys. Rev. A **108**, 053321 (2023)
44. J. Hertkorn, J.-N. Schmidt, M. Guo, F. Böttcher, K.S.H. Ng, S.D. Graham, P. Uerlings, T. Langen, M. Zwierlein, T. Pfau, Phys. Rev. Res. **3**, 033125 (2021)
45. F. Ancilotto, M. Barranco, M. Pi, L. Reatto, Phys. Rev. A **103**, 033314 (2021)
46. S.M. Roccuzzo, A. Gallemí, A. Recati, S. Stringari, Phys. Rev. Lett. **124**, 045702 (2020)
47. A. Gallemí, S.M. Roccuzzo, S. Stringari, A. Recati, Phys. Rev. A **102**, 023322 (2020)
48. S. Das, V.W. Scarola, Phys. Rev. Lett. **134**, 163401 (2025)
49. T. Bland, G. Lamporesi, M.J. Mark, F. Ferlaino, C. R. Physique **24**(S3), 133–152 (2023)
50. E. Poli, T. Bland, S.J.M. White, M.J. Mark, F. Ferlaino, S. Trabucco, M. Mannarelli, Phys. Rev. Lett. **131**, 223401 (2023)
51. M.W. Zwierlein, J.R. Abo-Shaeer, A. Schirotzek, C.H. Schunck, W. Ketterle, Nature **435**, 1047 (2005)
52. M. Šindik, A. Recati, S.M. Roccuzzo, L. Santos, S. Stringari, Phys. Rev. A **106**, L061303 (2022)
53. E. Casotti, E. Poli, L. Klaus, A. Litvinov, C. Ulm, C. Politi, M.J. Mark, T. Bland, F. Ferlaino, Nature **635**, 327–331 (2024)
54. E. Poli, A. Litvinov, E. Casotti, C. Ulm, L. Klaus, M.J. Mark, G. Lamporesi, T. Bland, F. Ferlaino, Nat. Phys. **21**, 1820 (2025). https://doi.org/10.1038/s41567-025-03065-7
55. Y.-K. Lu, Nat. Phys. **21**, 1692 (2025). https://doi.org/10.1038/s41567-025-03066-6
56. G.A. Bougas, T. Bland, H.R. Sadeghpour, S.I. Mistakidis, Phys. Rev. A **113**, L041305 (2026). arXiv:2506.22290

Appendix
Simulating the 1D GPE

The GPE is a nonlinear partial differential equation, and it solution must, in general, be obtained numerically. A variety of numerical methods exist to solve the GPE, including those based on Runge-Kutta methods, the Crank-Nicolson method and the split-step Fourier method.[1] The latter (also known as the time-splitting spectral method) is particularly compact and efficient, and here we apply it to the 1D GPE. Furthermore, we introduce the imaginary time method for obtaining ground state solutions. Basic Matlab code is provided.

A.1   Split-Step Fourier Method

The split-step fourier method is well-established for numerically solving the time-dependent Schrodinger equation, written here in one-dimension,

$$i\hbar \frac{\partial \psi(x,t)}{\partial t} = \hat{H}\psi(x,t). \tag{A.1}$$

The Hamiltonian $\hat{H}$ can be expressed as $\hat{H} = \hat{T} + \hat{V}$, where $\hat{T} \equiv -\frac{\hbar^2}{2m}\frac{\partial^2}{\partial x^2}$ and $\hat{V} \equiv V(x)$ are the kinetic and potential energy operators. Integrating from t to $t + \Delta t$ (and noting the time-independence of the Hamiltonian) leads to the time-evolution equation,

$$\psi(x, t + \Delta t) = e^{-i\Delta t \hat{H}/\hbar}\psi(x,t). \tag{A.2}$$

The operators T and V do not commute, hence $e^{-i\Delta t \hat{H}/\hbar} \neq e^{-i\Delta t \hat{T}/\hbar}e^{-i\Delta t \hat{V}/\hbar}$. Nonetheless, the following approximation,

$$e^{-i\Delta t \hat{H}/\hbar}\psi \approx e^{-i\Delta t \hat{V}/2\hbar}e^{-i\Delta t \hat{T}/\hbar}e^{-i\Delta t \hat{V}/2\hbar}\psi, \tag{A.3}$$

---

[1] A. Minguzzi, S. Succi, F. Toschi, M. P. Tosi, P. Vignolo, Phys. Rep. **395**, 223 (2004).

C. F. Barenghi et al., *Quantum Fluids, Solitons, and Vortices*, Lecture Notes in Physics 1050, https://doi.org/10.1007/978-3-032-20171-3

holds with error $\mathcal{O}(\Delta t^3)$. In position space $\hat{V}$ is diagonal, and so the operation $e^{-i\Delta t \hat{V}/2\hbar}\psi$ simply corresponds to multiplication of $\psi(x, t)$ by $e^{-i\Delta t V(x)/2\hbar}$. Although $\hat{T}$ is not diagonal in position space, it becomes diagonal in reciprocal space. Conversion to reciprocal space is achieved by taking the Fourier transform $\mathcal{F}$ of the wavefunction $\tilde{\psi}(k, t) = \mathcal{F}[\psi(x, t)]$, where k denotes the 1D wavevector. Then the kinetic energy operation corresponds to multiplication of $\tilde{\psi}(k, t)$ by $e^{-i\hbar\Delta t k^2/2m}$. Thus Eq. (A.3) can be written as,

$$\psi(x, t + \Delta t) \approx e^{-\frac{i}{2\hbar}V(x)\Delta t} \cdot \mathcal{F}^{-1}\left[e^{-\frac{i\hbar k^2}{2m}\Delta t} \cdot \mathcal{F}[e^{-\frac{i}{2\hbar}V(x)\Delta t} \cdot \psi(x, t)] \right]. \quad \text{(A.4)}$$

In practice, the computational expense of performing forward and backward Fourier transforms to evaluate Eq. (A.3) is small (particularly when using numerical fast Fourier transform techniques) compared to the significant expense of evaluating the kinetic energy term directly in position space. Note that the split-step method naturally incorporates periodic boundary conditions.

The above method was developed for the linear Schrodinger equation with time-independent Hamiltonian. Remarkably, it holds for the GPE (despite its nonlinearity and time-dependent Hamiltonian) under the replacement $V(x) \mapsto V(x) + g|\psi|^2$. Errors of $\mathcal{O}(\Delta t^3)$ are maintained, providing the most up-to-date ψ is always employed during the sequential operations in Eq. (A.4).[2]

A.2   1D GPE Solver

We now outline the approach to solve the 1D GPE using the split-step method, with reference to the Matlab code included below. To make the numbers more convenient, the GPE is divided through by $\hbar$ (equivalent to considering energy in units of $\hbar$). We consider a 1D box, discretized into grid points with spacing Δx (dx), and extending over the spatial range $x = [-M\Delta x, M\Delta x]$, where M (M) is a positive integer. Position is described by a vector x_i (x), defined as $x_i = -M\Delta x + (i - 1)\Delta x$, with $i = 1, ..., 2M + 1$. The potential $V(x)$ is defined as the vector $V_i = V(x_i)$. Starting from the initial time, the wavefunction $\psi(x)$, represented by the vector $\psi_i = \psi(x_i)$ (psi), is evolved over the time interval Δt (dt) by evaluating Equation (A.4) numerically by replacing the Fourier transform $\mathcal{F}$ (and its inverse $\mathcal{F}^{-1}$) by the discrete fast Fourier transform. Here, wavenumber is discretized into a vector k_i (k), defined as $k_i = -M\Delta k + (i - 1)\Delta k$, with $\Delta k = \pi/M\Delta x$ (dk). This time iteration step is repeated N_t (Nt) times to find the solution at the desired final time.

The Matlab code below simulates a BEC of 5000 ^{87}Rb atoms with $a_s = 5.8$ nm and trapping frequencies $\omega_\perp = 2\pi \times 100$ Hz and $\omega_x = 2\pi \times 40$ Hz. Starting from the narrow non-interacting ground state (Gaussian) profile, the condensate undergoes oscillating expansions and contractions, due to the competition between repulsive

---

[2] J. Javanainen, J. Ruostekoski, J. Phys. A **39**, L179 (2006).

interactions and confining potential. Note—under different scenarios, reduced time and grid spacings may be required to ensure numerical convergence.

```matlab
% SOLVES THE 1D GPE VIA THE SPLIT-STEP FOURIER METHOD
clear all;clf; %Clear workspace and figure

hbar=1.054e-34;amu=1.660538921e-27; %Physical constants
m=87*amu;as=5.8e-9; %Atomic mass; scattering length
N=1000;wr=100*2*pi;wx=40*2*pi; %Atom number; trap frequencies

M=200; Nx=2*M+1;
dx=double(2e-7); x=(-M:1:M)*dx; %Define spatial grid
dk=pi/(M*dx); k=(-M:1:M)*dk; %Define k-space grid
dt=double(10e-8); Nt=200000; %Define time step and number

lr=sqrt(hbar/(m*wr)); lx=sqrt(hbar/(m*wx)); %HO lengths
g1d=2*hbar*hbar*as/(m*lr^2); %1D interaction coefficient

V=0.5*m*wx^2*x.^2/hbar; %Define potential
psi_0=sqrt(N/lx)*(1/pi)^(1/4)*exp(-x.^2/(2*lx^2)); %Initial wavefunction

%[psi_0,mu] = get_ground_state(psi_0,dt,g1d,x,k,m,V); %Imaginary time

Nframe=100; %Data saved every Nframe steps
t=0; i=1; psi=psi_0; spacetime=[]; %Initialization

for itime=1:Nt %Time-stepping with split-step Fourier method
    psi=psi.*exp(-0.5*1i*dt*(V+(g1d/hbar)*abs(psi).^2));
    psi_k=fftshift(fft(psi)/Nx);
    psi_k=psi_k.*exp(-0.5*dt*1i*(hbar/m)*k.^2);
    psi=ifft(ifftshift(psi_k))*Nx;
    psi=psi.*exp(-0.5*1i*dt*(V+(g1d/hbar)*abs(psi).^2));
    if mod(itime,Nt/Nframe) == 0  %Save wavefunction every Nframe steps
      spacetime=vertcat(spacetime,abs(psi.^2)); t
    end
    t=t+dt;
end

subplot(1,3,1); %Plot potential
plot(x,V,'k'); xlabel('x (m)'); ylabel('V (J/hbar)');

subplot(1,3,2); %Plot initial and final density
plot(x,abs(psi_0).^2,'k',x,abs(psi).^2,'b');
legend('\psi(x,0)','\psi(x,T)');xlabel('x (m)');ylabel('|\psi|^2 (m^{-1})');

subplot(1,3,3); % Plot spacetime evolution as pcolor plot
dt_large=dt*double(Nt/Nframe);
pcolor(x,dt_large*(1:1:Nframe),spacetime); shading interp;
xlabel('x (m)'); ylabel('t (s)');
```

A.3   Imaginary Time Method

A convenient numerical method for obtaining ground state solutions of the Schrodinger equation/GPE is through imaginary time propagation. The wavefunction $\psi(x, t)$ can be expressed as a superposition of eigenstates $\phi_m(x)$ with time-dependent amplitudes $a_m(t)$ and energies $E_m(t)$, i.e. $\psi(x, t) = \sum_m a_m(t)\phi_m(x)$, for which, after the substitution $t \rightarrow -i\,\Delta t$, the evolution equation (A.3) becomes,

$$\psi(t + \Delta t) = e^{-\Delta t \hat{H}/\hbar}\psi(x, t) = \sum_m a_m(t)\phi_m(x)e^{-\Delta t E_m/\hbar}. \qquad (A.5)$$

The amplitude of each eigenstate contribution decays over time, with the ground state (with lowest E_m) decaying the slowest. Thus, by renormalizing ψ after each iteration (to ensure the conservation of the desired norm/number of particles), ψ will evolve towards the ground state.

Convergence may be assessed by monitoring the chemical potential. This is conveniently evaluated using the relation $\mu = (\hbar/\Delta t) \ln |\psi(x, t)/\psi(x, t + \Delta t)|$ at some coordinate within the condensate; this relation is obtained by introducing the eigenvalue μ and imaginary time into Eq. (A.3).

The Matlab function `get_ground_state` below obtains the GPE ground state via imaginary time propagation. Uncommenting line 19 in the above GPE solver calls this function prior to real time propagation; as one expects, the profile remains static in time.

```
% SOLVES THE 1D GPE IN IMAGINARY TIME USING THE SPLIT-STEP METHOD
function [psi,mu] = get_ground_state(psi,dt,g1d,x,k,m,V)

hbar=1.054e-34; dx=x(2)-x(1); dk=2*pi/(x(end)-x(1));
N=dx*norm(psi).^2; Nx=length(x);
psi_mid_old=psi((Nx-1)/2); mu_old=1; j=1; mu_error=1;
 while mu_error > 1e-8
    psi=psi.*exp(-0.5*dt*(V+(g1d/hbar)*abs(psi).^2));
    psi_k=fftshift(fft(psi))/Nx;
    psi_k=psi_k.*exp(-0.5*dt*(hbar/m)*k.^2);
    psi=ifft(ifftshift(psi_k))*Nx;
    psi=psi.*exp(-0.5*dt*(V+(g1d/hbar)*abs(psi).^2));

    psi_mid=psi((Nx-1)/2);
    mu=log(psi_mid_old/psi_mid)/dt; mu_error=abs(mu-mu_old)/mu;

    psi=psi*sqrt(N)/sqrt((dx*norm(psi).^2));
 if mod(j,5000) == 0
        mu_error
    end
    if j > 1e8
```

```
            'no solution found'
            break
        end
        psi_mid_old=psi((Nx-1)/2); mu_old=mu; j=j+1;
    end
end
```

Solutions

Problems of Chap. 1

1.1 *Temperature of the beam of the LHC:*
Recall that 1 eV $= 1.602 \times 10^{-19}$ J. Thus $E = 6.5$ TeV $= 10.4 \times 10^{-7}$ J. Set $E = k_B T$ where $k_B = 1.38 \times 10^{-23}$ J/K is Boltzmann's constant and T is the absolute temperature, obtaining $T = 7.5 \times 10^{16}$ K.

Problems of Chap. 2

2.1 *Boltzmann distribution I:*

Cell energy E	Macrostates											$\bar{N}(E)$
	1	2	3	4	5	6	7	8	9	10	11	
6ϵ	1	0	0	0	0	0	0	0	0	0	0	6/411 = 0.015
5ϵ	0	1	0	0	0	0	0	0	0	0	0	30/411 = 0.07
4ϵ	0	0	1	1	0	0	0	0	0	0	0	90/411 = 0.22
3ϵ	0	0	0	0	2	1	1	0	0	0	0	210/411 = 0.51
2ϵ	0	0	1	0	0	1	0	3	2	1	0	420/411 = 1.02
1ϵ	0	1	0	2	0	1	3	0	2	4	6	756/411 = 1.84
0ϵ	5	4	4	3	4	3	2	3	2	1	0	1260/411 = 3.07
Statistical weighting W	6	30	30	60	15	120	60	20	90	30	1	Total: 462

© The Editor(s) (if applicable) and The Author(s), under exclusive license to Springer Nature Switzerland AG 2026
C. F. Barenghi et al., *Quantum Fluids, Solitons, and Vortices*, Lecture Notes in Physics 1050, https://doi.org/10.1007/978-3-032-20171-3

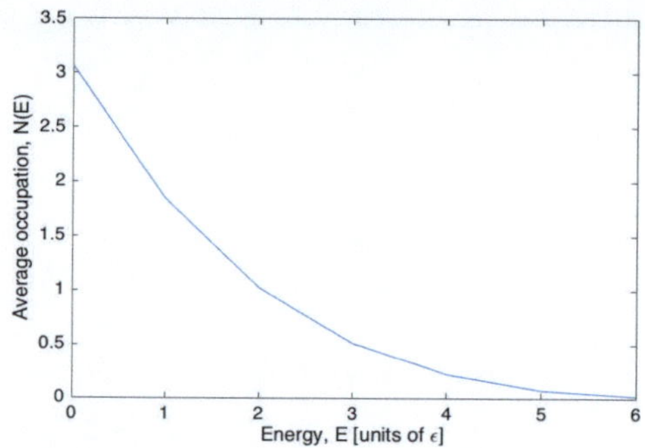

2.2 Boltzmann distribution II:

(a) Each macrostate must satisfy the number and energy constraints:

$$N_1 + N_2 + N_3 = N$$

$$N_2\epsilon + 2N_3\epsilon = 0.5\,N\epsilon.$$

We can express N_1 and N_2 in terms of the other parameters:

$$N_2 = 0.5\,N - 2N_3$$

$$N_1 = N - 0.5N + N_3 = 0.5N + N_3$$

The number of microstates in a given macrostate is,

$$W = \frac{N!}{N_1!N_2!N_3!},$$

which, in terms of N and N_3 becomes,

$$W(N, N_3) = \frac{N!}{(0.5N + N_3)!(0.5N - 2N_3)!N_3!}$$

Providing we specify some energy constraint we can solve this numerically. For $E = 0.5N\,\Delta E$, this becomes:

$$W = \frac{N!}{(0.5N + N_3)!(0.5N - 2N_3)!N_3!}$$

(b) The distribution gets rapidly narrower and more pronounced with N—see figure below. Note the number of microstates!

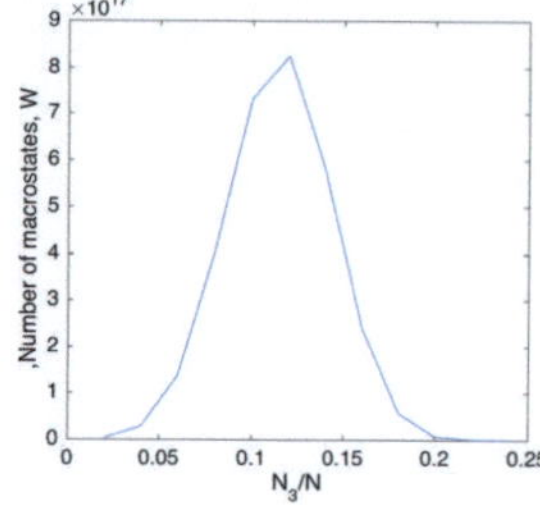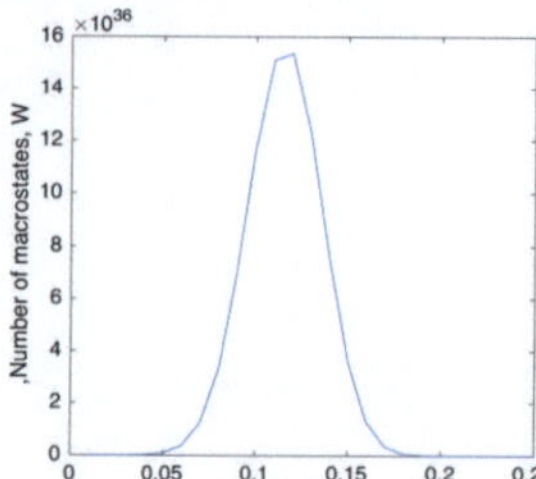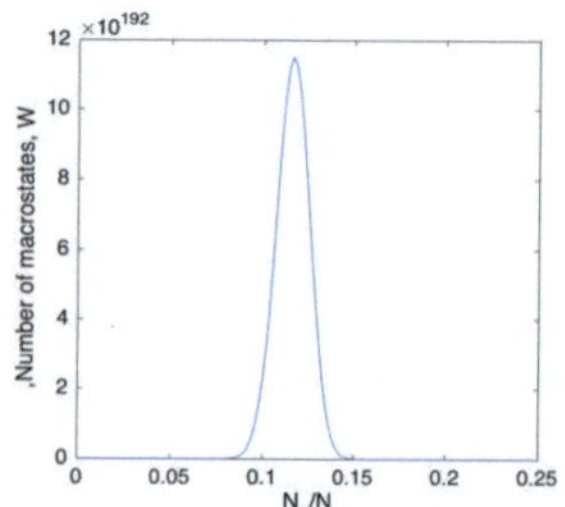

2.3 *Bose-Einstein condensation in 2D:*

(a) We are considering a 2D system, i.e. phase space has 4 coordinates: x, y, p_x and p_y. The volume of momentum space between p and $p + dp$ is a circular shell of inner radius p and thickness dp, with volume $2\pi\, p\, dp$. The volume in phase space V_{ph} is thus,

$$V_{ph} = \mathcal{V}_{2D} 2\pi\, p\, dp$$

In 2D, each state in phase space occupies a region of volume h^2, such that the number of states between p and $p + dp$ is,

$$g(p)dp = \frac{2\pi \mathcal{V}_{2D}\, p\, dp}{h^2}.$$

We relate this expression from momentum to energy using the relation $E = p^2/2m$, giving,

$$g(p)dp = g(E)dE = \frac{2\pi m \mathcal{V}_{2D}}{h^2}dE,$$

Thus the density of states for a 2D gas of bosons in a box is,

$$g(E) = \frac{2\pi m \mathcal{V}_{2D}}{h^2}.$$

(b) The number of particles in the system is given by integrating the Bose-Einstein distribution over all energies,

$$N_{ex} = \int_0^\infty \frac{g(E)}{e^{(E-\mu)/k_B T} - 1}dE = \frac{2\pi m \mathcal{V}_{2D}}{h^2} \int_0^\infty \frac{dE}{e^{(E-\mu)/k_B T} - 1},$$

where we have introduced $g(E)$ from above. Substituting for z and x gives,

$$N_{ex} = \frac{2\pi m \mathcal{V}_{2D} k_B T}{h^2} \int_0^\infty \frac{ze^{-x}}{1 - ze^{-x}}dx.$$

Letting $y = ze^{-x}$ gives,

$$dy = -ze^{-x}dx = -ydx.$$

Inserting into the above integral gives,

$$N_{ex} = \frac{2\pi m \mathcal{V}_{2D} k_B T}{h^2} \int_z^0 \frac{-dy}{1 - y} = -\frac{2\pi m \mathcal{V}_{2D} k_B T}{h^2} \ln(1 - z).$$

(c) z is confined to the range $0 < z < 1$. Inserting $z = e^{\mu/k_B T}$ and rearranging for μ we get,

$$\mu = k_B T \ln\left(1 - \exp\left[-\frac{h^2 N_{ex}}{2\pi m \mathcal{V}_{2D} k_B T}\right]\right)$$

Bose-Einstein condensation occurs when $\mu = 0$. The ln-term cannot equal zero (this would require N_{ex} to be negative and infinite, which is clearly unphysical!) and so μ can only become zero (and thus the 2D gas of bosons can only undergo Bose-Einstein condensation) if $T = 0$.

2.4 Internal energy and heat capacity:

Assuming $T < T_c$ we can take $z = 1$. Then, inserting expression (2.16) for N_{ex} into the total energy given by Eq. (2.24) we obtain,

$$U = \frac{2}{\pi^{1/2}\zeta(3/2)} N k_B T \left(\frac{T}{T_c}\right)^{3/2} \int_0^\infty \frac{x^{3/2}}{e^x - 1} dx.$$

Using the general result (2.15) the definite integral in the above equation is given by,

$$\int_0^\infty \frac{x^{3/2}}{e^x - 1} dx = \frac{3\sqrt{\pi}}{4} \zeta(5/2),$$

such that we obtain,

$$U = \frac{3}{2} \frac{\zeta(5/2)}{\zeta(3/2)} N k_B T \left(\frac{T}{T_c}\right)^{3/2} \approx 0.770 \, N k_B T \left(\frac{T}{T_c}\right)^{3/2}.$$

At high temperature $(T \gg T_c)$, $z \ll 1$, and we can approximate $e^y/z - 1 \approx e^y/z$, such that the energy is,

$$U = \frac{2}{\pi^{1/2}\zeta(3/2)} N k_B T \left(\frac{T}{T_c}\right)^{3/2} z \int_0^\infty \frac{x^{3/2}}{e^x} dx.$$

Using the relation $\int_0^\infty x^{3/2} e^{-x} dx = \Gamma(5/2)$,

$$U = \frac{2\Gamma(5/2)}{\pi^{1/2}\zeta(3/2)} N k_B T \left(\frac{T}{T_c}\right)^{3/2} z.$$

Now for small z, $g_{3/2}(z) \approx z$. Meanwhile, using Eqs. (2.16) and (2.18) we can obtain the relation,

$$g_{3/2}(z) \left(\frac{T}{T_c}\right)^{3/2} = \zeta(3/2).$$

For $z \ll 1$ this becomes,

$$z \left(\frac{T}{T_c}\right)^{3/2} = \zeta(3/2).$$

Inserting this into U gives,

$$U = \frac{3}{\pi^{1/2}} \Gamma(5/2) N k_B T \approx \frac{3}{2} N k_B T.$$

It is then straightforward to differentiate to give the corresponding expressions for the heat capacity $C_V = \partial U/\partial t$.

2.5 *Bose gas in a harmonic trap:*

(a) Starting from the expression $N_{ex} = \int N(E)\,dE = \int f_{BE}(E)g(E)\,dE$, we introduce the Bose-Einstein distribution function and the density of states in a harmonic trap $g(E) = E^2/(2\hbar^3 \omega_x \omega_y \omega_z)$, giving,

$$N_{ex} = 4\frac{\pi^3}{h^3\bar{\omega}^3} \int_0^\infty \frac{E^2}{e^{(E-\mu)/k_B T} - 1}\,dE,$$

where $\bar{\omega} = (\omega_x \omega_y \omega_z)^{1/3}$. Using the general integral result (2.15) gives,

$$N_{ex} = 4\left(\frac{\pi}{h\bar{\omega}}\right)^3 (k_B T)^3 \Gamma(3) g_3(z).$$

At the point at which BEC occurs, N_{ex} saturates to its maximum value. This occurs when $g_3(z) = \zeta(3)$. Substituting in this relation, and setting $N_{ex} = N_c$, we obtain the critical number of particles to be,

$$N_c = 4\left(\frac{\pi}{h\bar{\omega}}\right)^3 (k_B T)^3 \Gamma(3)\zeta(3). \tag{A.6}$$

(b) At the critical temperature T_c, the number of particles in excited states N_{ex} *just* equals the total number of particles. Replacing N_{ex} with N above and rearranging for T gives,

$$T_c = \frac{\hbar}{k_B}\bar{\omega}[N/\zeta(3)]^{1/3}. \tag{A.7}$$

(c) We wish to determine an expression for N_0/N, where N_0 is the number of particles in the zero-energy level. We can express the total number of particles as $N = N_0 + N_{ex}$. Rearranging for N_0/N gives,

$$\frac{N_0}{N} = 1 - \frac{N_{ex}}{N} \tag{A.8}$$

We know that above the critical temperature for BEC, $N_0 \approx 0$ but what about below the critical temperature? Below the transition, N_{ex} is equivalent to Eq. (2.16) with $z = 1$, i.e.,

$$N_{ex} = 4\left(\frac{\pi}{h\bar{\omega}}\right)^3 (k_B T)^3 \Gamma(3)\zeta(3). \tag{A.9}$$

At the critical temperature T_c, the number of particles in excited states N_{ex} *just* equals the total number of particles, and so we can write N_{ex} at the critical temperature as,

$$N = 4\left(\frac{\pi}{h\bar{\omega}}\right)^3 (k_B T_c)^3 \Gamma(3)\zeta(3). \tag{A.10}$$

Inserting Eqs. (B.5) and (B.4) into Eq. (B.3) gives,

$$\frac{N_0}{N} = 1 - \frac{N_{ex}}{N} = 1 - \left(\frac{T}{T_c}\right)^3.$$

2.6 Compressibility of the Bose gas:

From Eq. (2.25) we have an expression for the energy of a BEC below the critical temperature,

$$U = \frac{3}{2}\frac{\zeta(5/2)}{\zeta(3/2)}NkT(T/T_c)^{3/2}$$

We need to re-introduce the full form of T_c. Recall that $T_c^{3/2} = (h/2\pi mk)^{3/2}N/(\zeta(3/2)V)$. Inserting this into the expression for U gives,

$$U = \frac{3}{2}k\zeta(5/2)(2\pi mk/h)^{3/2}VT^{5/2}$$

Evaluating the pressure $P = 2U/3V$ gives,

$$P = \frac{2U}{3V} = k\zeta(5/2)(2\pi mk/h)^{3/2}T^{5/2}$$

To determine the compressibility it is easier to first derive its reciprocal $1/\beta$. Inverting the compressibility definition and evaluating,

$$\frac{1}{\beta} = -V\frac{\partial P}{\partial V} = -V \times 0 = 0$$

This implies the compressibility of the BEC β is infinite! This is not surprising since we are assuming an ideal gas with no particle-particle interactions, and so there is nothing to stop the gas from compressing!

Problems of Chap. 3

3.1 Dimensions:

We use S.I. units of metre, kilogram and second (m, kg and s), indicating the dimensions of a quantity x as $[x]$. We recall that the S.I. unit of energy is the Joule: $J = kg\ m^2/s^2$.

(a) From the normalization condition $\int_v |\psi|^2 dV = N$, since $[V] = m^3$, we conclude $[\psi] = m^{-3/2}$.
(b) Since $[\hbar] = J\ s = kg\ m^2/s$, $[g] = [4\pi\hbar^2 a_s/m] = kg\ m^5/s^2$ and $[\mu] = J$ (this is easily noticed from Eq. (3.8): the argument of the exponential, $\mu t/\hbar$, must be dimensionless), each term of the GPE has dimensions of $kg\ m^{1/2}/s^2$.
(c) $[g|\psi|^2] = J$.

3.2 *Spherical BEC:*

(a) Using $n(r) = m\omega^2 (R_r^2 - r^2)/2g$ we normalise in spherical polar coordinates:

$$N = \int_0^{R_r} 4\pi r^2 \frac{m\omega^2}{2g}(R_r^2 - r^2)\, dr = \ldots = \frac{4\pi m\omega^2 R_r^5}{15g}.$$

Rearranging for R_r and inserting $g = 4\pi \hbar^2 a_s/m$ gives:

$$R_r = (15\, N l_r^4 a_s)^{1/5}.$$

(b) Recognising that the peak density is $n_0 = \mu/g$ and using $\mu = m\omega^2 R_r^2/2$ and the above expression for R_r we obtain:

$$n_0 = \frac{15\, N}{8\pi\, R_r^3}.$$

(The same result is obtained by normalizing $n(r) = n_0(1 - r^2/R_r^2)$.)

(c) Using the above expression for R_r we construct:

$$\left(\frac{R_r}{l_r}\right)^5 = \frac{15\, N a_s}{l_r} \rightarrow \frac{R_r}{l_r} = \left(\frac{15\, N a_s}{l_r}\right)^{1/5}.$$

For large N, $R_r \gg l_r$, i.e. the condensate becomes very broad relative to the non-interacting width.

(d) Direct integration of the Thomas-Fermi profile within the energy integral (3.14) gives:

$$E = \frac{\pi m^2 \omega^4 R_r^7}{10\, g}.$$

3.3 *Variational solutions:*

Plugging the 3D Ansatz (3.42) into the energy integral (3.14) and evaluating the integral leads directly to Eq. (3.43).

In the 2D case, we adopt an Ansatz $\psi(x, y, z) = A \exp(-(x^2 + y^2)/2\sigma^2 l_r^2)$. In the z direction the condensate is uniform, and we consider a length L. Assuming N atoms in this length, normalization in polar coordinates gives the coefficient A to be:

$$A = \frac{N}{\pi \sigma^2 l_r^2 L}.$$

Plugging the Ansatz into the energy integral and evaluating in polar coordinates leads to:

$$E(\sigma) = \hbar \omega_r N \left[\frac{1}{2\sigma^2} + \frac{\sigma^2}{2} + \frac{1}{L}\left(\frac{N a_s}{\ell_r}\right)\frac{l_r}{\sigma^2}\right]. \tag{A.11}$$

Now for the 1D case, we adopt an Ansatz $\psi(x, y, z) = A \exp(-x^2/2\sigma^2 l_r^2)$. In the y, z plane the condensate is uniform, and we consider an $\mathcal{A}$. Assuming N atoms in this area, normalization gives the coefficient A to be:

$$A = \frac{1}{\pi^{1/4}} \frac{N}{\sigma l_r \mathcal{A}}.$$

Plugging the Ansatz into the energy integral and evaluating leads to:

$$E(\sigma) = \hbar \omega_r N \left[\frac{1}{4\sigma^2} + \frac{\sigma^2}{4} + \frac{1}{L^2} \left(\frac{N a_s}{\ell_r} \right) \frac{\sqrt{2\pi} l_r^2}{\sigma} \right]. \tag{A.12}$$

The crucial difference between the 3D result and 1D result is that, for $\sigma \to 0$, the interaction term always wins in 3D (since it goes like $1/\sigma^3$, and hence beats the other terms in the energy expression, notably the $1/\sigma^2$ kinetic term). This means that for $a_s < 0$ there is a tendency to collapse. However, in 1D the kinetic $1/\sigma^2$ term always wins in this limit; this term is always positive, and so prevents any collapse instability. 2D is a borderline case where the kinetic term and interaction term have the same scaling $(1/\sigma^2)$, and so the system parameters $(N, a_s, ...)$ will then determine which term wins, and hence whether a collapse can occur.

3.4 *Numerical ground states:*

One can use the code to solve the 1D GPE given in the Appendix, including calling the `get_ground_state` function. Using the parameters in the code provided ($N = 1000, a_s = 5.8\text{nm}$) the ground state is a poor approximation to the Thomas-Fermi solution. The approximation becomes valid for larger atom number, e.g. $N = 50,000$. For the non-interacting case (setting $a_s = 0$ in the code), the ground state is much narrower than for $a_s > 0$ and corresponds to the Gaussian harmonic oscillator solution. For attractive interactions (e.g. setting $a_s = 1\text{nm}$) the solution is even narrower than the Gaussian state (due to the binding effect of the attractive interactions).

3.5 *Absorption image:*
The density is,

$$n(x, y, z) = |\psi|^2 = n_0 \, e^{-x^2/l_x^2} e^{-y^2/l_y^2} e^{-z^2/l_z^2}$$

Performing the column-integration over z,

$$n_{\text{CI}}(x, y) = \int_\infty^\infty n(x, y, z) \, dz = n_0 \, e^{-x^2/l_x^2} e^{-y^2/l_y^2} \times 2 \int_0^\infty e^{-z^2/l_z^2} \, dz$$

$$= n_0 \, e^{-x^2/l_x^2} e^{-y^2/l_y^2} \times 2 \left(\frac{1}{2} \sqrt{\pi l_z^2} \right) = \sqrt{\pi l_z^2} \, n_0 \, e^{-x^2/l_x^2} e^{-y^2/l_y^2}$$

Here column-integrated density maintains the same functional dependence in x and y, i.e. a Gaussian profile. This is not always the case,

3.6 *Static and moving uniform BEC:*
The density is uniform, i.e. $n(x) = n_0$. In the energy integral (3.14) the kinetic energy is therefore zero. Since $V(x) = 0$ the potential energy is also zero. Only the interaction energy term remains:

$$E = \frac{g}{2} \int_0^L n(x)^2 \, dx = \frac{g}{2} L n_0^2.$$

Using Eq. (3.51) we construct the flowing solution as:

$$\psi(x) = \sqrt{n_0} \exp\left[\frac{i m v_0 x}{\hbar} - \frac{m v_0^2 t}{2\hbar}\right] \exp\left[-\frac{i \mu t}{\hbar}\right].$$

Plugging into the time-dependent GPE, and noting that the LHS = RHS, we confirm that this solution indeed satisfies the GPE.

Using the fluid velocity equation, Eq. (3.19), in 1D we confirm that:

$$v(x, t) = \frac{\hbar}{m} \frac{\partial S(x, t)}{\partial x} = v_0.$$

Plugging the solution into the energy integral (3.14) we obtain:

$$E = \frac{g}{2} L n_0^2 + \frac{1}{2} N m v_0^2,$$

and using the momentum integral (3.15) we obtain:

$$P = N m v_0.$$

These are consistent with the classical results for a fluid parcel with mass Nm flowing with speed v_0,

3.7 *Dimensionless GPE:*
Consider the GPE

$$i\hbar \frac{\partial \psi}{\partial t} = -\frac{\hbar^2}{m} \nabla^2 \psi + g|\psi|^2 \psi - \mu\psi,$$

The steady homogeneous solution is $\psi_0 = \sqrt{\mu/g}$. Let $\psi = \psi_0 \psi'$, $x = x'\xi$, $y = \xi y'$, $z = \xi z'$ and $t = \tau t'$ where $\xi = \alpha\hbar/\sqrt{\mu m}$ and $\tau = \hbar/\mu$ are the units of length and time, α being a dimensionless parameter. All prime quantities are thus dimensionless. Substituting into the GPE we find

$$i \frac{\partial \psi'}{\partial t'} = -\frac{1}{2\alpha^2} \nabla'^2 \psi' + |\psi'|^2 \psi' - \psi',$$

where ∇'^2 is the Laplacian operator with respect to x', y' and z'. We conclude that if we choose $\alpha = 1/\sqrt{2}$, that is $\xi = \hbar/\sqrt{2\mu m}$, the dimensionless form of the GPE is

$$i\frac{\partial \psi'}{\partial t'} = -\nabla'^2 \psi' + |\psi'|^2 \psi' - \psi',$$

as required.

3.8 *Moving frame:*
Note that $\psi(x, t) = \psi(x(x', t'), t(x', t'))$ therefore

$$\frac{\partial \psi}{\partial x} = \frac{\partial \psi}{\partial x'}\frac{\partial x'}{\partial x} + \frac{\partial \psi}{\partial t'}\frac{\partial t'}{\partial x} = \frac{\partial \psi}{\partial x'},$$

hence $\partial^2 \psi / \partial x^2 = \partial^2 \psi / \partial x'^2$. Similarly

$$\frac{\partial \psi}{\partial t} = \frac{\partial \psi}{\partial x'}\frac{\partial x'}{\partial t} + \frac{\partial \psi}{\partial t'}\frac{\partial t'}{\partial t} = -v_x \frac{\partial \psi}{\partial x'} + \frac{\partial \psi}{\partial t'}.$$

Substitute into the GPE

$$i\hbar\frac{\partial \psi}{\partial t} = -\frac{\hbar^2}{2m}\frac{\partial^2 \psi}{\partial x^2} + g|\psi|^2\psi + V\psi,$$

and find the GPE in the moving frame of reference:

$$i\hbar\frac{\partial \psi}{\partial t'} = -\frac{\hbar^2}{2m}\frac{\partial^2 \psi}{\partial x'^2} + g|\psi|^2\psi + V\psi + i\hbar v_x \frac{\partial \psi}{\partial x'}.$$

3.9 *Three-body loss:*
We seek $\dfrac{\partial N}{\partial t}$. Introducing the density $n(\mathbf{r}, t) = |\psi|^2$ we write this as,

$$\frac{\partial N}{\partial t} = \frac{\partial}{\partial t}\int n\, dr = \int \frac{\partial n}{\partial t}\, dr = \int \left(\psi^*\frac{\partial \psi}{\partial t} + \psi\frac{\partial \psi^*}{\partial t}\right) dr, \qquad \text{(A.13)}$$

where the latter step is by applying the product rule for differentiation. Next we write the GPE with three-body loss as, and its complex conjugate as,

$$i\hbar\frac{\partial \psi}{\partial t} = H_{gp}\psi - \frac{i\hbar K_3}{2}|\psi|^2\psi, \qquad -i\hbar\frac{\partial \psi^*}{\partial t} = H_{gp}\psi^* + \frac{i\hbar K_3}{2}|\psi|^2\psi^*, \quad \text{(A.14)}$$

where we have used that the GP Hamiltonian is Hermitian such that $(H_{gp}\psi^*) = H_{gp}\psi^*$. We can write these equivalently as,

$$\frac{\partial \psi}{\partial t} = \frac{1}{i\hbar}H_{gp}\psi - \frac{K_3}{2}|\psi|^2\psi, \qquad \frac{\partial \psi^*}{\partial t} = \frac{-1}{i\hbar}H_{gp}\psi^* - \frac{K_3}{2}|\psi|^2\psi^*. \qquad \text{(A.15)}$$

Inserting into the RHS integral of the top equation,

$$\frac{\partial N}{\partial t} = \frac{1}{i\hbar} \int \left[\psi^*(H_{gp}\psi) - \psi(H_{gp}\psi^*) \right] dr - \frac{K_3}{2} \int \left[\psi^*|\psi|^4\psi + \psi|\psi|^4\psi^* \right] dr. \tag{A.16}$$

Since H_{gp} is Hermitian, $\int \psi^*(H_{gp}\psi)dr = \int \psi(H_{gp}\psi^*)$, and so the first integral vanishes. Rearranging the second term gives,

$$\frac{\partial N}{\partial t} = -\frac{K_3}{2} \int \left[|\psi|^6 + |\psi|^6 \right] dr = -K_3 \int n^3 dr. \tag{A.17}$$

Problems of Chap. 4

4.1 *Condensate oscillations:*

1. In steady state ($\partial/\partial t = 0$), if $\mathbf{v} = \mathbf{0}$, then the equation for the density is satisfied, and the equation for the velocity reduces to

$$\nabla(gn_{eq} + V) = 0,$$

where $n_{eq}(r)$ is the equilibrium density which we seek. Since n_{eq} depends only on r and $V = m\omega_0^2 r^2/2$ we have $gdn_{eq}/dr = -m\omega_0^2 r$. Integrate: $n_{eq}(r) = -\omega_0^2 r^2/(2g) + C$ where C is a constant. Clearly $C = \hat{n}_0$ (the density at the centre). Since the density cannot be negative, we set $n_{eq}(r)$ equal to zero for $-\omega_0^2 r^2/(2g) + \hat{n}_0 < 0$, concluding that

$$n_{eq}(r) = \begin{cases} \hat{n}_0 \left(1 - \frac{r^2}{R_0^2} \right) & \text{if } r \leq R_0 \\ 0 \ \text{else} \end{cases}$$

where

$$R_0^2 = \frac{2\,g\hat{n}_0}{m\omega_0^2}, \qquad \hat{n}_0 = n_{eq}(0).$$

To find $\hat{n}_0$, impose the normalization that $N = \int n_{eq}(r)d^3\mathbf{x}$. In spherical coordinates we have

$$N = 4\pi \int_0^{R_0} \hat{n}_0 \left(1 - \frac{r^2}{R_0^2} \right) r^2\, dr = 4\pi\hat{n}_0 \left(\frac{R_0^3}{3} - \frac{R_0^5}{5R_0^2} \right) = \frac{8\pi}{15}\hat{n}_0 R_0^3.$$

We conclude that $\hat{n}_0$ is

$$\hat{n}_0 = \frac{15\,N}{8\pi\,R_0^3}.$$

Combining this result with $\hat{n}_0 = m\omega_0^2 R_0^2/(2g)$ found previously, we have

$$R_0 = \left(\frac{15N}{4\pi m\omega_0^2}\right)^{1/5}.$$

Using the last two equations, we conclude that, given the properties of the cloud of atoms and of the trap (N, m, g, ω_0), we find firstly the radius of the condensate, R_0, and from R_0 then value of the central density $\hat{n}_0$.

2. Let $n = n_{eq} + n' + \cdots$ and $\mathbf{v} = \mathbf{v}' + \cdots$ where the perturbations n' and $\mathbf{v}'$ (indicated by primed symbols) are small quantities. The first equation of motion becomes

$$\frac{\partial n'}{\partial t} + \nabla \cdot [(n_{eq} + n')\mathbf{v}'] \approx \frac{\partial n'}{\partial t} + \nabla \cdot (n_{eq}\mathbf{v}') = 0,$$

because terms which are quadratic in the perturbations can be neglected. The second equation of motion gives

$$m\frac{\partial \mathbf{v}'}{\partial t} = -\nabla\left(gn_{eq} + gn' + V\right),$$

But $gn_{eq} + V = 0$ in the equilibrium state, so the equation reduces to

$$m\frac{\partial \mathbf{v}'}{\partial t} = -g\nabla n'.$$

In conclusions the equations for the perturbations are

$$\frac{\partial n'}{\partial t} = -\nabla \cdot \left(n_{eq}\mathbf{v}'\right), \qquad \frac{\partial \mathbf{v}'}{\partial t} = -\frac{g}{m}\nabla n'.$$

3. Now take the time derivative of the equation for $\partial n'/\partial t$ and substitute the equation for $\partial \mathbf{v}'/\partial t$:

$$\frac{\partial^2 n'}{\partial t^2} = \frac{g}{m}\nabla \cdot (n_{eq}\nabla n').$$

Assuming that $n' \sim e^{i\omega t}$ we find

$$-\omega^2 n' = \frac{g}{m}\left(n_{eq}\nabla \cdot \nabla n' + \nabla n_{eq} \cdot \nabla n'\right)$$

Now use $n_{eq} = \hat{n}_0(1 - r^2/R_0^2)$, $\nabla n_{eq} = -2\hat{n}_0 r/R_0^2\hat{\mathbf{r}}$ (where $\hat{\mathbf{r}}$ is the unit vector in the r-direction) and $\nabla \cdot \nabla n' = \nabla^2 n'$ and $R_0^2 = 2g\hat{n}_0/(m\omega_0^2)$ to conclude that

$$\omega^2 n' = -\frac{\omega_0^2}{2}(R_0^2 - r^2)\nabla^2 n' + \omega_0^2 r\frac{dn'}{dr}.$$

4. Let $n' = Cr^\ell e^{im\phi} P(\theta)$ where m is integer, $P(\theta)$ is a function of θ to be determined, ℓ is an exponent to be determined and C is a constant. We have $rdn'/dr = rC\ell r^{\ell-1} e^{im\phi} P(\theta) = \ell n'$, thus

$$\omega^2 n' = -\frac{\omega_0^2}{2}(R_0^2 - r^2)\nabla^2 n' + \omega_0^2 \ell n'.$$

Since $d(r^\ell)/dr = \ell r^{\ell-1}$ and $d^2(r^\ell)/dr^2 = \ell(\ell-1)r^{\ell-2}$ we have

$$\nabla^2 n' = Cr^{\ell-2} e^{im\phi} \left[\ell(\ell-1)P + 2\ell P + \frac{1}{\sin\theta}\frac{d}{d\theta}\left(\sin\theta\frac{dP}{d\theta}\right) - \frac{m^2 P}{\sin^2\theta} \right] =$$

$$= Cr^{\ell-2} e^{im\phi} \left[\ell(\ell+1)P + \frac{1}{\sin\theta}\frac{d}{d\theta}\left(\sin\theta\frac{dP}{d\theta}\right) - \frac{m^2 P}{\sin^2\theta} \right].$$

Let $\zeta = \cos\theta$. Then using the chain rule we have

$$\frac{d}{d\theta} = \frac{d}{d\zeta}\frac{d\zeta}{d\theta} = -\sin\theta\frac{d}{d\zeta}.$$

Therefore, since $\sin^2\theta = 1 - \cos^2\theta = 1 - \zeta^2$, we have

$$\frac{1}{\sin\theta}\frac{d}{d\theta}\left(\sin\theta\frac{dP}{d\theta}\right) == \frac{1}{\sin\theta}(-\sin\theta)\frac{d}{d\zeta}\sin\theta(-\sin\theta)\frac{dP}{d\zeta} =$$

$$= \frac{d}{d\zeta}(1-\zeta^2)\frac{dP}{d\zeta} = (1-\zeta^2)\frac{d^2 P}{d\zeta^2} - 2\zeta\frac{dP}{d\zeta}$$

and

$$\nabla^2 n' = Cr^{\ell-2} e^{im\phi} \left[(1-\zeta^2)\frac{d^2 P}{d\zeta^2} - 2\zeta\frac{dP}{d\zeta} + \ell(\ell+1)P - \frac{m^2 P}{1-\zeta^2} \right]$$

If we choose $P = P_\ell^m(\cos\theta)$ where P_ℓ^m are the associated Legendre polynomials and ℓ is an integer, then $\nabla^2 n' = 0$ and we are left with

$$\omega^2 n' = \omega_0^2 \ell n'.$$

We conclude that

$$\omega = \sqrt{\ell}\omega_0.$$

4.2 *Simulating collective modes:*

You can use the 1D GPE code provided in the Appendix with the parameters provided and calling the `get_ground_state` function to numerically obtain the ground state of the system. To trigger centre-of-mass oscillations, you can apply an instantaneous shift of the potential along x. For example, you can implement this in the code by adding the line SPSVERBc4immediately after calling the `get_ground_state` function; this shifts the potential by 5 microns along x at $t = 0$. The wavepacket should undergo clear centre-of-mass oscillations about the new trap centre at twice the trap frequency.

Similarly, you can excite the monopole mode by suddenly changing the trap frequency at $t = 0$. For example, introduce the command `V=0.5*m*(1.1*wx)^2*x.^2/hbar` immediately after calling the `get_ground_state` function; this redefine the trap frequency to be 10% larger. The condensate should visibly undergo symmetric oscillations in its width and peak density.

4.3 *Generation and propagation of sound waves:*

One can use the 1D GPE solver provided in the Appendix. To broaden the condensate, increase the number of atoms to say $N = 20,000$ and decrease the axial trap frequency to say $\omega_x = 2\pi \times 5$Hz. Immediately after obtaining the ground state, add a narrow Gaussian potential centred at the origin, e.g. $V(x) = V_0 \exp(-x^2/\sigma^2)$ with $V_0 = 1 \times 10^{-31}$J and $\sigma = 2$ microns. This can be performed by adding the following lines immediately after calling the `get_ground_state` function to :

```
sigma=2e-6; V0=1e-31/hbar; V=V+V0*exp(-x.^2/sigma^2);
```

This suddenly introduction of the potential at $t = 0$ causes atoms to be expelled from a small region in the centre of the system; these atoms propagate outwards as a sound wave. In the code you can calculate an estimate of the speed of sound based on the peak density of the numerically-obtained ground state; this should be in reasonably agreement with the wave speed seen by eye.

4.4 *Quantum carpets:*

One can use the 1D GPE solver provided in the Appendix. For the non-interacting case, set $a_s = 0$ in the code, set the trap to zero (set $\omega_x = 0$) and comment out the `get_ground_state` function such that the initial condition is specified by the Gaussian initial state. The initial wavepacket will spread to fill the box, eventually interfering with reflections from the edge to create a complex set of patterns over time. You may need to run for long times to see the eventual perfect revival of the original wavepacket.

Repeat with repulsive interactions, say $a_s = 5.8$nm. Narrower features are supported by the repulsive interactions, with the narrower low density features taking the form of dark solitons. The nonlinearity destroys the perfect revival of the original wavepacket.

Problems of Chap. 5

5.1 *Integrals of motion:*
Direct integration leads to the results:

$$N_s = 2\xi n_0 \beta,$$

$$P_s = -\frac{2\hbar n_0 u \beta}{c} + 2\hbar n_0 \arctan\left(\frac{c\beta}{u}\right),$$

and,

$$E_s = \frac{4}{3} n_0 \hbar c \beta^3,$$

where $\beta = \sqrt{1 - u^2/c^2}$.

5.2 *Turning points:*
Recall Eq. (5.7) relating the soliton depth to its speed in a uniform system, $n_d = n_0(1 - u^2/c^2)$. Now we have a background density (and hence speed of sound) which varies, while the soliton depth stays constant:

$$n_d = n(x)\left(1 - \frac{u(x)^2}{c(x)^2}\right),$$

with $n(x) = n_0(1 - x^2/R_x^2)$ and $c(x) = \sqrt{n(x)g/m}$. Simplifying this down leads to:

$$u(x) = \pm c_0 \left(1 - \frac{x^2}{R_x^2} - \frac{n_d}{n_0}\right)^{1/2}.$$

At the turning points, $u(x) = 0$, which occurs when:

$$1 - \frac{x^2}{R_x^2} - \frac{n_d}{n_0} = 0,$$

or,

$$x = \pm R_x \left(1 - \frac{n_d}{n_0}\right)^{1/2}.$$

It is clear from this that a black soliton ($n_d = n_0$) is fixed at the origin, while in the shallow limit $n_d \to 0$, the soliton extends right up to the condensate boundary.

5.3 *Simulating a dark soliton:*
Modify the parameters in the GPE solver code as specified in the Problem. After calling the `get_ground_state` function, the following lines of code can be used to multiply in the $u = 0.5c$ dark soliton solution at the centre of the system:

```
n0=max(max(psi_0.^2));        %extract peak density (at trap centre)
xi=hbar/sqrt((m*n0*g1d));  % calculate healing length at peak density
c=sqrt(n0*g1d/m);            % calculate speed of sound at peak density
u=0.5;                    % define soliton speed (relative to c)
B=sqrt(1-u^2);                % calculate B parameter of soliton solution
psi_sol=B*tanh(B*(x)/xi)+1i*u;  %define soliton solution
                                % (on uniform n=1 background)
psi=psi_0.*psi_sol;        %multiply onto numerically-obtained ground state
```

The simulated soliton oscillates in the system with a period $0.0352\,\text{s}$. The soliton is predicted to oscillate with frequency $\omega_s = \omega_x/\sqrt{2} = 2\pi \times 28.28$ (2 d.p.), which corresponds to a period of $0.0354\,\text{s}$, in close agreement with the simulated result.

5.4 *Soliton and barrier:*

- (a) In the GPE solver code, modify the definition of the potential to include the Gaussian barrier:

```
V0=2e-31/hbar;          % define amplitude of barrier
x0=5e-6;                  % define position of barrier
sigma=2e-6;              % define width of barrier
V=0.5*m*wx^2*x.^2/hbar + V0*exp(-((x-x0)/sigma).^2)/hbar;
                                %define new potential
```

The soliton traverses the barrier with emission of sound waves during the interaction.
- (b) Double V_0 in the code. The soliton is reflected at the barrier with emission of sound waves.
- (c) We estimate the density at the centre of the barrier through the Thomas-Fermi approach. We write the density in the TF approximation of the 1D system (dropping 1D notation for simplicity) as,

$$n(x) = n_0 - \frac{V(x)}{g},$$

where $V(x) = \frac{1}{2}m\omega_x^2 x^2 + V_0\exp(-x^2/\sigma^2)$. The reduced density at the centre of the barrier is $n(x_0)$. Now recall that the density depth of a dark soliton with speed u is $n_d = n_0(1 - u^2/c^2)$. The dark soliton can only traverse the barrier if there is sufficient local density to support the dark soliton density dip, i.e. the critical point is when $n_d = n(x_0)$. Working this through,

$$n_0(1 - u^2/c^2) = n_0 - \frac{m\omega^2 x_0^2}{2g} - \frac{V_0}{g}.$$

Rearranging gives,

$$V_0 = \frac{n_0 g u^2}{c^2} - \frac{1}{2}m\omega_x^2 x_0^2.$$

For the parameters used in the simulation we get the critical V_0 for the transition between soliton reflection/transmission at the barrier to be 2.8×10^{-31}J. This is consistent with the results of the simulations which showed transmission for $V_0 = 2 \times 10^{-31}$J and reflection for $V_0 = 4 \times 10^{-31}$J.

5.5 *Imprinting a soliton in imaginary time:*

- (a) In the `get_ground_state` function add the phase-imprinting line:

```
psi=abs(psi).*exp(0.5*1i*pi*sign(x));
```

  towards the end of the while loop, e.g. after or before the renormalisation step. This forms a stationary black soliton at the trap centre. which remains there over time. This is to be expected since a black soliton at the trap centre is a stationary solution of the system (in fact, it is the first-excited state).
- (b) Repeat as in (a) but replace `sign(x)` with `sign(x-10e-6)` to create the spatial offset of the phase step. The black soliton is now formed off-centre. When propagated in time, the soliton accelerates towards the trap centre and undergoes oscillations at the expected frequency $\omega_s \approx \omega_x/\sqrt{2}$.
- (c) Modify the phase-imprinting line to create impose two phase steps:

```
psi=abs(psi).*exp(0.5*1i*pi*sign(x-10e-6)).*exp(-0.5*1i*pi*sign(x+10e-6));
```

  The two solitons, initially off-centre and black, undergo out-of-phase oscillations, passing through each other unscathed, as expected for soliton collisions.
  When the initial offset is reduced to 1×10^{-6}m the two dark solitons remain approximately stationary, forming a "bound state". This is because each soliton is now close enough to feel the short-range repulsion from the other soliton; this, combined with the opposing force from the external potential, means that each soliton lies at an approximate equilibrium position.

5.6 *Chemical potential of soliton:*
Plugging the soliton solutions into the GPE, using the derivatives $\frac{d}{dx}(\text{sech}(x)) = -\text{sech}(x)\tanh(x)$ and $\frac{d}{dx}(\tanh(x)) = 1 - \tanh^2(x)$, and the identity $\text{sech}^2(x) + \tanh(x) = 1$, we obtain:

$$\mu = -\frac{\hbar^2}{2\,m\xi_s^2}.$$

5.7 *Soliton as a classical particle:*
We obtain:

$$P_s = Nmv,$$

$$E_s = N\left(\frac{\hbar^2}{6\,m\xi_s^2} + \frac{gN}{6\xi_s} + \frac{mv^2}{2}\right).$$

Hence the soliton behaves like a classical particle with mass Nm and rest mass energy given by the first two terms in E_s.

5.8 *Simulating a bright soliton:*
(a) In the 1D GPE solver code, modify the physical parameters to the given values, comment out the `get_ground_state` function and instead define the initial solution as per the 1D bright soliton solution,

```
xi_s=2*hbar^2/(m*abs(g1d)*N);                          %define soliton lengthscale
psi_0=sqrt(0.5*N/xi_s)*sech(x/xi_s);       %define soliton solution
```

The bright soliton remains stationary over time, confirming that it is indeed the ground state of the system.
(b) Give the initial soliton state a kick with the modified line:

```
v=1e-3;
psi_0=sqrt(0.5*N/xi_s)*sech(x/xi_s).*exp(1i*m*v*x/hbar);
```

The bright soliton translates in the positive x direction with speed v and with unchanging profile, as expected for a soliton.
(c) Add the harmonic potential to the code. The soliton undergoes centre-of-mass oscillations at the trap frequency, as expected. The soliton profile undergoes breathing oscillations. This is because the trap modifies the ground state solution of the bright soliton, and so the imposed initial profile (the untrapped bright soliton solution) is not the lowest energy profile.

5.9 *Bright solitons under axial trapping:*
(a) Using the Ansatz (5.34), evaluate the energy contribution from the axial harmonic potential. Combining with the existing terms in Eq. (5.35) gives the energy-per-particle,

$$\frac{E}{N} = \hbar\omega_\perp \left(\frac{1}{2\sigma_\perp^2} + \frac{\sigma_r^2}{2} + \frac{1}{6\sigma_x^2} + \frac{\pi^2}{24}\lambda^2\sigma_x^2 - \frac{k}{3\sigma_x\sigma_\perp^2} \right).$$

(b) The code below generates a heat map of the energy-per-particle (divided by $\hbar\omega_\perp$) vs σ_x and $\sigma_\perp$.

```
clear all;
Lx=(0.01:0.01:20);
Lperp=(0.01:0.01:2);

[lx,lperp]=meshgrid(Lx,Lperp);
lambda=0.1;
k=0.4;
E=0.5*1./lperp.^2+0.5*lperp.^2+(1/6)./lx.^2+...
    (pi*pi/24)*lambda*lambda*lx.^2-(k/3)*1./(lx.*lperp.^2);

TF = islocalmin2(E);
```

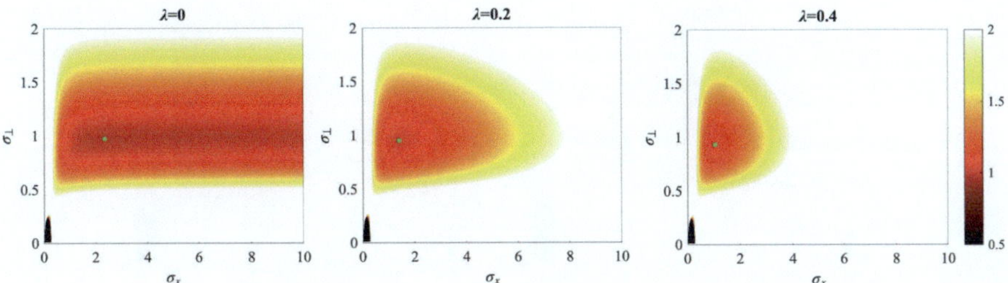

Fig. A.1 Variational energy landscape in the $(\sigma_\perp, \sigma_z)$ plane for $k = 0.4$ and various values of trap aspect ratio λ. The local minimum corresponding to the variational solution is shown by the green dot

```
pcolor(lx,lperp,E);
colormap(hot)
set(gca,'PlotBoxAspectRatio',[1.2 1 1]);
colorbar;

caxis([0.5 2])
axis([0 10 0 2])
shading interp;
hold on;
plot(lx(TF),lperp(TF),"green.",MarkerSize=12)
ylabel('$\sigma_\perp $','interpreter','latex')
xlabel('$\sigma_x$','interpreter','latex')
title('\lambda=0.1')
```

The figure shows how the energy landscape changes as axial trapping is introduced. The energy minimum becomes more tightly bound in σ_x and shifts to lower σ_x (Fig. A.1).

(c) The figure below shows how the energy landscape changes as expulsive axial trapping is introduced. The energy minimum becomes more weakly bound in σ_x and shifts to higher σ_x. Above a critical value of $|\lambda|$ the minimum vanishes—this reflects that the expulsive potential overcomes the attractive nonlinearity to tear the soliton apart (Fig. A.2).

Problems of Chap. 6

6.1 First and second sound:

1. Since ρ_0 is constant and $\boldsymbol{v}_{n0} = 0$, the density equation becomes

$$\frac{\partial \rho'}{\partial t} + \nabla \cdot \left((\rho_{n0} + \rho'_n) \boldsymbol{v}'_n + (\rho_{s0} + \rho'_s) \boldsymbol{v}'_s \right) = 0.$$

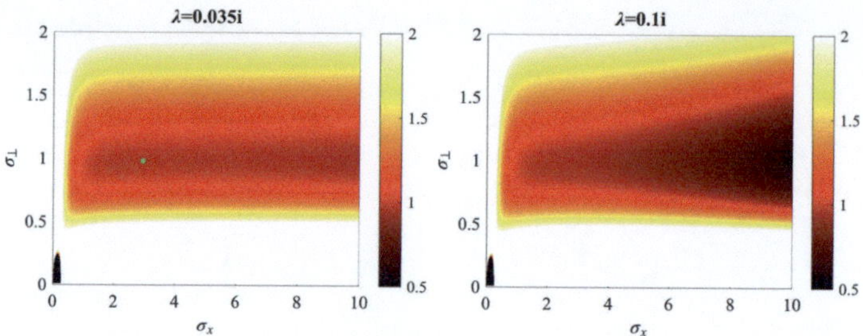

Fig. A.2 Variational energy landscape in the $(\sigma_\perp, \sigma_z)$ plane for $k = 0.4$ and for expulsive harmonic potentials $\lambda^2 < 0$. The local minimum corresponding to the variational solution is shown by the green dot

But ρ'_n and the components of $\boldsymbol{v}'_n$ are small quantities, therefore their products are negligible, and the equation reduces to

$$\frac{\partial \rho'}{\partial t} + \nabla \cdot (\rho_{n0} \boldsymbol{v}'_n + \rho_{s0} \boldsymbol{v}'_s) = 0,$$

as required. The other equations are obtained in a similar way:

$$\rho_0 \frac{\partial S'}{\partial t} + S_0 \frac{\partial \rho'}{\partial t} + \rho_0 S_0 \nabla \cdot \boldsymbol{v}'_n = 0,$$

$$\rho_{s0} \frac{\partial \boldsymbol{v}'_s}{\partial t} = -\frac{\rho_{s0}}{\rho_0} \nabla P' + \rho_s S_0 \nabla T',$$

$$\rho_{n0} \frac{\partial \boldsymbol{v}'_n}{\partial t} = -\frac{\rho_{n0}}{\rho_0} \nabla P' - \rho_s S_0 \nabla T'.$$

2. Take the time derivative of the equaution for the density perturbation:

$$\frac{\partial^2 \rho'}{\partial t^2} + \nabla \cdot (\rho_{n0} \frac{\partial \boldsymbol{v}'_n}{\partial t'} + \rho_{s0} \frac{\partial \boldsymbol{v}'_s}{\partial t'}) = 0,$$

then, since $\rho_0 = \rho_{s0} + \rho_{n0}$,

$$\frac{\partial^2 \rho'}{\partial t^2} = \nabla \cdot \nabla P' = \nabla^2 P',$$

as requested.

3. From the equations for the velocity perturbations we have:

$$\frac{\partial (\boldsymbol{v}'_n - \boldsymbol{v}'_s)}{\partial t'} = -\frac{\rho_0 S_0}{\rho_{n0}} \nabla T',$$

as requested.

4. The equation for the entropy perturbation becomes

$$\rho_0 \frac{\partial S'}{\partial t} - S_0 \nabla \cdot (\rho_{n0} \boldsymbol{v}'_n + \rho_{s0} \boldsymbol{v}'_s) + \rho_0 S_0 \nabla \cdot \boldsymbol{v}'_n = 0,$$

$$\rho_0 \frac{\partial S'}{\partial t} + \rho_{s0} S_0 \nabla \cdot \boldsymbol{v}'_n - \rho_{s0} S_0 \boldsymbol{v}'_s = 0,$$

$$\rho_0 \frac{\partial S'}{\partial t} + \rho_{s0} S_0 \nabla \cdot (\boldsymbol{v}'_n - \boldsymbol{v}'_s) = 0.$$

Take the time derivative

$$\rho_0 \frac{\partial^2 S'}{\partial t^2} + \rho_{s0} S_0 \nabla \cdot [\frac{\partial (\boldsymbol{v}'_n - \boldsymbol{v}'_s)}{\partial t}] = 0.$$

We find, as requested, that

$$\frac{\partial^2 S'}{\partial t^2} = \frac{\rho_{s0} S_0^2}{\rho_{n0}} \cdot \nabla^2 T',$$

5. If $\rho = \rho(P, T)$ and $S = S(P, T)$, then small density increments $\rho' = \Delta \rho$ can be expressed in terms of small increments $P' = \Delta P$ and $T' = \Delta T$ of pressure and temperature using partial derivatives:

$$\rho' = \left(\frac{\partial \rho}{\partial P} \right)_T P' + \left(\frac{\partial \rho}{\partial T} \right)_P T',$$

from which we have, taking the second time derivative, that

$$\frac{\partial^2 \rho'}{\partial t^2} = \left(\frac{\partial \rho}{\partial P} \right)_T \frac{\partial^2 P'}{\partial t^2} + \left(\frac{\partial \rho}{\partial T} \right)_P \frac{\partial^2 T'}{\partial t^2} = \nabla^2 P'.$$

Similarly

$$S' = \left(\frac{\partial S}{\partial P} \right)_T P' + \left(\frac{\partial S}{\partial T} \right)_P T',$$

hence

$$\frac{\partial^2 S'}{\partial t^2} = \left(\frac{\partial S}{\partial P} \right)_T \frac{\partial^2 P'}{\partial t^2} + \left(\frac{\partial S}{\partial T} \right)_P \frac{\partial^2 T'}{\partial t^2} = \frac{\rho_{s0} S_0^2}{\rho_{n0}} \nabla^2 T'.$$

6. Assume that

$$P' = \hat{P}' e^{i\omega t - ikx} = \hat{P}' e^{i\omega(t - x/u)},$$

$$T' = \hat{T}' e^{i\omega t - ikx} = \hat{T}' e^{i\omega(t - x/u)},$$

where $u = \omega/k$. We find the following system of two equations for two unknowns $\hat{P}'$ and $\hat{T}'$:

$$-\omega^2 \left(\frac{\partial \rho}{\partial P}\right)_T \hat{P}' - \omega^2 \left(\frac{\partial \rho}{\partial T}\right)_P \hat{T}' = -k^2 \hat{P}',$$

$$-\omega^2 \left(\frac{\partial S}{\partial P}\right)_T \hat{P}' - \omega^2 \left(\frac{\partial S}{\partial T}\right)_P \hat{T}' = -\frac{k^2 \rho_{s0} S_0^2}{\rho_{n0}} \hat{T}',$$

which is

$$\left(-\omega^2 \left(\frac{\partial \rho}{\partial P}\right)_T + k^2\right) \hat{P}' - \omega^2 \left(\frac{\partial \rho}{\partial T}\right)_P \hat{T}' = 0,$$

$$-\omega^2 \left(\frac{\partial S}{\partial P}\right)_T \hat{P}' + \left(-\omega^2 \left(\frac{\partial S}{\partial T}\right)_P + \frac{k^2 \rho_{s0} S_0^2}{\rho_{n0}}\right) \hat{T}' = 0.$$

The system has non trivial solutions $\hat{P}'$ and $\hat{T}'$ only if the determinant is zero, which is

$$\left(-\omega^2 \left(\frac{\partial \rho}{\partial P}\right)_T + k^2\right)\left(-\omega^2 \left(\frac{\partial S}{\partial T}\right)_P + \frac{k^2 \rho_{s0} S_0^2}{\rho_{n0}}\right) - \omega^4 \left(\frac{\partial \rho}{\partial T}\right)_P \left(\frac{\partial S}{\partial P}\right)_P = 0.$$

Simplify the notation using the Jacobian matrix,

$$\frac{\partial(S, \rho)}{\partial(T, P)} = \left(\frac{\partial \rho}{\partial P}\right)_T \left(\frac{\partial S}{\partial T}\right)_P - \left(\frac{\partial \rho}{\partial T}\right)_P \left(\frac{\partial S}{\partial P}\right)_T$$

then the condition of zero determinant is

$$\omega^4 \frac{\partial(S, \rho)}{\partial(T, P)} + \omega^2 \left(-\frac{k^2 \rho_{s0} S_0^2}{\rho_{n0}} \left(\frac{\partial \rho}{\partial P}\right)_T - k^2 \left(\frac{\partial S}{\partial T}\right)_P\right) + k^4 \frac{\rho_{s0} S_0^2}{\rho_{n0}} = 0.$$

Divide by k^4 using $u = \omega/k$

$$u^4 \frac{\partial(S, \rho)}{\partial(T, P)} - u^2 \left(\frac{\rho_{s0} S_0^2}{\rho_{n0}} \left(\frac{\partial \rho}{\partial P}\right)_T + \left(\frac{\partial S}{\partial T}\right)_P\right) + \frac{\rho_{s0} S_0^2}{\rho_{n0}} = 0.$$

Use the thermodynamics results that

$$\frac{\partial(S, \rho)}{\partial(T, P)} = \frac{C_v}{T}\left(\frac{\partial \rho}{\partial P}\right)_T,$$

$$\left(\frac{\partial S}{\partial T}\right)_P \frac{T}{C_v}\left(\frac{\partial P}{\partial \rho}\right)_T = \left(\frac{\partial P}{\partial \rho}\right)_S,$$

where C_v is the specific heat at constant volume. Then

$$u^4 - u^2\left(\left(\frac{\partial P}{\partial \rho}\right)_S + \frac{\rho_{s0}S_0^2}{\rho_{n0}}\frac{T}{C_v}\right) + \frac{\rho_{s0}S_0^2}{\rho_{n0}}\frac{T}{C_v}\left(\frac{\partial P}{\partial \rho}\right)_T = 0.$$

If we make the approximations $C_v \approx C_P = C$ and

$$\left(\frac{\partial P}{\partial \rho}\right)_S \approx \left(\frac{\partial P}{\partial \rho}\right)_T$$

then we find the equation

$$u^4 - u^2\left(\frac{\partial P}{\partial \rho} + \frac{\rho_{s0}S_0^2}{\rho_{n0}}\frac{T}{C}\right) + \frac{\rho_{s0}S_0^2 T}{C\rho_{n0}}\frac{\partial P}{\partial \rho} = 0,$$

which has the two roots

$$u_1 = \left(\frac{\partial P}{\partial \rho}\right)^{1/2}, \qquad u_2 = \left(\frac{\rho_{s0}T S_0^2}{\rho_{n0}}\right)^{1/2},$$

as requested.

6.2 *Thermal counterflow I:*
The mass flux along the channel is zero because the channel is blocked at the resistor's end. From Eq. (6.2) we have $\rho_n V_n + \rho_s V_s = 0$, hence $V_s = -(\rho_n/\rho_s)V_n$. The counterflow velocity is then

$$V_{ns} = V_n - V_s = V_n + \frac{\rho_n}{\rho_s}V_n = \frac{\rho}{\rho_s}V_n,$$

because $\rho_n + \rho_s = \rho$. From Eq. (6.3), $W = \rho S T V_n$, hence the average normal fluid velocity is proportional to the applied heat flux: $V_n = W/(\rho S T)$. from the above equation, so is the average counterflow velocity: $V_{ns} = W/(\rho_s S T)$.

6.3 *Thermal counterflow II:*

1. Since $\partial/\partial t = 0$ and the quadratic terms are negligible, the superfluid and normal fluid equations are

$$0 = -\frac{\rho_s}{\rho}\nabla P + \rho_s S \nabla T,$$

$$0 = -\frac{\rho_n}{\rho}\nabla P - \rho_s S \nabla T + \eta \nabla^2 \boldsymbol{v}_n,$$

which is (summing the equations)

$$\nabla P = \eta \nabla^2 \boldsymbol{v}_n, \qquad \nabla P = \rho S \nabla T.$$

2. Using cylindrical coordinates, the three components of the equation for $\boldsymbol{v}_n$ yield $\partial P/\partial r = 0$, $\partial P/\partial \theta = 0$ (meaning that P depends only on z) and

$$\frac{dP}{dz} = \eta \left(\frac{d^2 v_n}{dr^2} + \frac{1}{r}\frac{dv_n}{dr} \right).$$

The solution which satisfies the viscous no-slip boundary conditions $v_n(R) = 0$ and remains finite on $r = 0$ is

$$v_n(r) = \frac{\Delta P R^2}{4\eta \Delta z}\left(1 - \frac{r^2}{R^2} \right).$$

The average normal fluid velocity in the channel is

$$V_n = \frac{1}{\pi R^2}\int_0^{2\pi} d\theta \int_0^R dr\, r\, v_n(r) = \frac{\Delta P R^2}{8\eta \Delta z}.$$

3. Since $V_n = W/(\rho S T)$, then $\Delta P = 8\eta \Delta z V_n/R^2 = 8\eta \Delta z W/(R^2 \rho S T)$. Since $\Delta P = \rho S \Delta T$, then $\Delta T = 8\eta \Delta z W/(R^2 \rho^2 S^2 T)$. In conclusion, the viscosity of the normal fluid induces pressure and temperature drops ΔP and ΔT, along the counterflow channel, which are proportional to the applied heat flux W.

6.4 *Landau's critical velocity:*

1. The object creates an excitation only if its initial velocity v_i is larger than Landau's critical velocity $v_\mathrm{L} = (E/p)_{min}$. To find the minimum of E/p we set

$$\frac{d}{dp}\left(\frac{E}{p} \right) = -\frac{E}{p^2} + \frac{1}{p}\frac{dE}{dp} = 0,$$

which is

$$\frac{dE}{dp} = \frac{E}{p}.$$

2. In the phonon region we have $(E/p)_{min} = u_1$, however there is a lower minimum in the roton region. Use Eq. (6.17), setting $dE/dp = E/p$:

$$\frac{(p - p_0)}{\mu_0} = \frac{1}{p}\left(\Delta_0 + \frac{(p - p_0)^2}{2\mu_0}\right).$$

We expect that the root p_L of this equation is near $p = p_0$, hence

$$\frac{(p_L - p_0)}{\mu_0} \approx \frac{\Delta_0}{p_0},$$

which is

$$p_L \approx p_0 + \Delta\mu_0/p_0.$$

The point p_L is therefore just to the right of p_0 where a line from the origin is tangent to Landau's dispersion curve. The resulting critical velocity (using values of parameters at low pressure) is

$$v_L = \frac{E(p_L)}{p_L} \approx \frac{\Delta_0}{p_0} = 59 \text{ m/s}.$$

3. In an idea Bose gas $E = p^2/(2M)$ thus

$$v_L = \left(\frac{E}{p}\right)_{min} = \left(\frac{p}{2M}\right)_{min} = 0,$$

hence a moving impurity will always create excitations and lose energy.

6.5 Fountain effect:
At $T = 2$ K and SVP, the specific entropy of liquid helium is $S = 962.1$ J/(kg K) and the density is $\rho = 145.62$ kg/m^3. Using $\Delta P = \rho S \Delta T$ and $\Delta T = 10^{-3}$ K, we find $\Delta P = 140$ kg/(m s^2) $= 140$ Pa. Then, using $\Delta P = \rho g \Delta z$ with $g = 9.81$ m/s^2, we have $\Delta z = 0.098$ m $= 9.8$ cm.

Problems of Chap. 7

7.1 Critical rotation:

1. –
2. Evaluating the kinetic energy expression, Eq. (7.12), featuring an on-axis vortex gives:

$$E_{\text{kin}} = \frac{\pi n_0 H_0 \hbar^2}{m} \left[\ln\left(\frac{R_r}{a_0}\right) - \frac{1}{2} + \frac{a_0^2}{2R_r^2} \right].$$

Since $a_0 \ll R_r$ the last term can be neglected.

3. Evaluating the angular momentum in the presence of a vortex we obtain,

$$L_z = 2\pi n_0 \hbar H_0 \left(\frac{1}{4} R_r^2 - \frac{1}{2} a_0^2 + \frac{a_0^4}{4R_r^2} \right).$$

Since $a_0 \ll R_r$ the last two terms can be neglected.
A vortex is energetically favourable when:

$$\Delta F = E_{\text{kin}} - \Omega L_z < 0,$$

is satisfied, where Ω is the angular frequency of rotation. Subbing the above results in and manipulating gives that the vortex is energetically favourable for $\Omega > \Omega_c$, where the critical rotation frequency is,

$$\Omega_c = \frac{2\hbar}{m R_r^2} \left[\ln\left(\frac{R_r}{a_0}\right) - \frac{1}{2} \right]$$

7.2 Angular momentum:
The z-component of the angular momentum is

$$L_z = \int_0^{H_0} dz \int_0^{2\pi} d\theta \int_0^R dr \, r \, \rho_s r v_\theta(r).$$

Here $v_\theta(r)$, the azimuthal velocity at distance r, is found from computing the circulation at distance r and then applying Stokes's Theorem:

$$\Gamma(r) = \oint_{C(r)} \boldsymbol{v} \cdot d\boldsymbol{r} = 2\pi r v_\theta(r) = \int_{A(r)} \boldsymbol{\omega} \cdot d\boldsymbol{A} = 2\pi \int_0^r dr' \, r' \, \bar{\omega}_z(r'),$$

where $C(r)$ is the circle of radius r enclosing the area $A(r)$. We obtain

$$v_\theta(r) = \frac{1}{r} \int_0^r dr' \, r' \, \bar{\omega}_z(r').$$

The angular momentum becomes

$$L_z = 2\pi \rho_s H_0 \int_0^R dr\, r \int_0^r dr'\, r'\, \bar{\omega}_z(r').$$

Instead of integrating in r' from $r' = 0$ to $r' = r$ at given r in 0 to R, we integrate in r from $r = r'$ to $r = R$ at given r',

$$L_z = 2\pi \rho_s H_0 \int_0^R dr'\, r'\bar{\omega}_z(r') \int_{r'}^R dr\, r = 2\pi \rho_s H_0 \int_0^R dr'\, r'\bar{\omega}_z(r')(\frac{R^2}{2} - \frac{r'^2}{2}),$$

which is

$$L_z = \pi \rho_s H_0 R^2 \int_0^R dr'\, r'\, \bar{\omega}_z(r') W(r').$$

Renaming the dummy variable yields the required answer with

$$W(r) = 1 - \frac{r^2}{R^2}.$$

7.3 *Pseudo-vorticity:*

1. Since $\rho = m|\psi|^2$ and $\boldsymbol{u} = (\hbar/m)\nabla\phi$, we have

$$\boldsymbol{j} = \rho \boldsymbol{u} = m|\psi|^2 \frac{\hbar}{m}\nabla\phi = \hbar|\psi|^2\nabla\phi = \hbar n\nabla\phi.$$

But $\psi = \sqrt{n}e^{i\phi}$ and $\psi^* = \sqrt{n}e^{-i\phi}$, thus we have $\nabla\psi = i\nabla\phi\sqrt{n}e^{i\phi}$ and $\nabla\psi^* = -i\nabla\phi\sqrt{n}e^{-i\phi}$, hence

$$\boldsymbol{j} = \frac{\hbar}{2i}(\psi^*\nabla\psi - \psi\nabla\psi^*) = \frac{\hbar}{2i}n\left(i\nabla\phi e^{i\phi}e^{-i\phi} - (-i\nabla\phi)e^{-i\phi}e^{i\phi}\right) = \hbar n\nabla\phi,$$

which is the same.
2. We have

$$\boldsymbol{\omega}_p = \frac{1}{2}\nabla\times\boldsymbol{j} = \frac{\hbar}{4i}\nabla\times(\psi^*\nabla\psi - \psi\nabla\psi^*) = \frac{\hbar}{4i}\left(\psi^*\nabla\times\nabla\psi + \nabla\psi^*\times\nabla\psi - \psi\nabla\times\nabla\psi^* - \nabla\psi\times\nabla\psi^*\right)$$

Hence since $\nabla\times\nabla\psi = \nabla\times\nabla\psi^* = 0$ and $\nabla\psi\times\nabla\psi^* = -\nabla\psi^*\times\nabla\psi$, we conclude that

$$\boldsymbol{\omega}_p = \frac{\hbar}{2i}\nabla\psi^*\times\nabla\psi.$$

3. Using $\psi = \psi_r + i\psi_i$ and $\psi^* = \psi_r - i\psi_i$ we have

$$\boldsymbol{\omega}_p = \frac{\hbar}{2i}\left((\nabla\psi_r - i\nabla\psi_i) \times (\nabla\psi_r + i\nabla\psi_i)\right) = \hbar\nabla\psi_r \times \nabla\psi_i.$$

4. From $\psi = \sqrt{n}e^{i\theta}$ and $\psi^* = \sqrt{n}e^{-i\theta}$ we have

$$\nabla\psi = \left(\frac{\partial\psi}{\partial r}, \frac{1}{r}\frac{\partial\psi}{\partial\theta}, \frac{\partial\psi}{\partial z}\right) = \left(\frac{1}{2\sqrt{n}}\frac{\partial n}{\partial r}e^{i\theta}, \frac{i\sqrt{n}}{r}e^{i\theta}, 0\right),$$

$$\nabla\psi^* = \left(\frac{\partial\psi^*}{\partial r}, \frac{1}{r}\frac{\partial\psi^*}{\partial\theta}, \frac{\partial\psi^*}{\partial z}\right) = \left(\frac{1}{2\sqrt{n}}\frac{\partial n}{\partial r}e^{-i\theta}, \frac{-i\sqrt{n}}{r}e^{-i\theta}, 0\right),$$

hence

$$\boldsymbol{\omega}_p = \frac{\hbar}{2r}\frac{\partial n}{\partial r}\widehat{z}.$$

In our case since

$$\frac{1}{r}\frac{\partial n}{\partial r} \approx \frac{2n_0}{(1 + r^2/\xi^2)^2},$$

we conclude that in the core region ($r < \xi$) the pseudo-vorticity is constant, and it vanishes for large r.

Problems of Chap. 8

8.1 *Generalized Helmholtz Theorem for the GPE:*
Since $z - z_0 = (x - x_0) + i(y - y_0)$, then $\nabla(z - z_0) = (1, i)$. Therefore

$$\nabla[(z - z_0)\rho e^{i\phi}] = (1, i)\rho e^{i\phi} + (z - z_0)\nabla\rho e^{i\phi} + i\nabla\phi(z - z_0)\rho e^{i\phi}.$$

Use the identities $\nabla^2 f = \nabla \cdot \nabla f$ and $\nabla \cdot (f\boldsymbol{F}) = f\nabla \cdot \boldsymbol{F} + \nabla f \cdot \boldsymbol{F}$ where f and $\boldsymbol{F}$ are respectively a scalar field and a vector field. We find

$$\nabla^2\psi_0 = \nabla^2[(z - z_0)\rho e^{i\phi}] =$$
$$= \Big(2(1, i) \cdot \nabla\rho + 2i\nabla\phi \cdot \nabla\rho(z - z_0) + 2i\rho(1, i) \cdot \nabla\phi + (z - z_0)[\nabla^2\rho + i\rho\nabla^2\phi - \rho(\nabla\phi)^2]\Big)$$
$$e^{i\phi}.$$

Substitute into

$$\psi_1 = \left(1 - \frac{i\delta t\mathcal{H}}{\hbar}\right)\psi_0 = \psi_0 + \frac{i\delta t}{\hbar}2m\nabla^2\psi_0 - \frac{i\delta t\mathcal{U}}{\hbar}\psi_0,$$

where $\mathcal{H} = -\hbar^2/(2m)\nabla^2 + \mathcal{U}$, and get

$$\psi_1 = \rho(z-z_0)e^{i\phi} + \frac{i\delta t\hbar}{2m}\left(2(1,i)\cdot\nabla\rho + 2i\nabla\rho\cdot\nabla\phi(z-z_0) + 2i\rho(1,i)\cdot\nabla\phi + \right.$$
$$\left. +(z-z_0)[\nabla^2\rho + i\nabla^2\phi - \rho(\nabla\phi)^2]\right)e^{i\phi} - \frac{i\delta t\mathcal{U}}{\hbar}(z-z_0)\rho e^{i\phi}.$$

Evaluate ψ_1 at $z = z_0 + \delta z$, which is the new position of the vortex core. By definition, $\psi_1 = 0$ at this point. Divide by $e^{i\phi}$,

$$\rho\delta z + \frac{i\hbar\delta t}{2m}\left(2(1,i)\cdot\nabla\rho + 2i\nabla\rho\cdot\nabla\phi\delta z + 2i\rho(1,i)\cdot\nabla\phi + \right.$$
$$\left. +\delta z[\nabla^2\rho + i\nabla^2\phi - \rho(\nabla\phi)^2]\right) - \frac{i\mathcal{U}\rho\delta t\delta z}{\hbar} = 0.$$

Divide by δt and take the limit $\delta z \to 0$, $\delta t \to 0$. We are left with

$$-\rho\frac{\delta z}{\delta t} = \frac{i\hbar}{m}(1,i)\cdot\nabla\phi - \frac{\hbar}{m}\rho(1,i)\cdot\nabla\phi + \cdots,$$

where we have neglected smaller terms which vanish in this limit. Dividing by ρ and noticing that $(\nabla\rho)/\rho = \nabla\ln\rho$, we conclude that the velocity of the vortex core, $\delta z/\delta t = \dot{x}_v + i\dot{y}_v$, is

$$-\dot{x}_v - i\dot{y}_v = \frac{i\hbar}{m}(1,i)\cdot\nabla\ln\rho - \frac{\hbar}{m}(1,i)\cdot\nabla\phi,$$

where the dot represents the time derivative. Separating real and imaginary parts we have

$$\dot{x}_v = \frac{\hbar}{m}\partial_y\ln\rho + \frac{\hbar}{m}\partial_x\phi,$$
$$\dot{y}_v = -\frac{\hbar}{m}\partial_x\ln\rho + \frac{\hbar}{m}\partial_y\phi.$$

In vector notation, letting $\boldsymbol{r}_v = (x_v, y_v)$ be the position of the vortex core, we conclude

$$\dot{\boldsymbol{r}}_v = -\frac{\hbar}{m}\widehat{\boldsymbol{\kappa}} \times \nabla\ln\rho + \frac{\hbar}{m}\nabla\phi.$$

8.2 *Vortex-antivortex pair:*
The complex potential of a vortex of circulation k located at position $z_1 = x_1 + iy_1$ on the xy plane is

$$\Omega(z) = -\frac{ik}{2\pi}\log(z-z_1).$$

Using the superposition principle, the complex potential of a vortex (circulation k) and an anti-vortex (circulation $-k$) located respectively at $z_1 = D$ and $z_2 = -D$ is

$$\Omega(z) = -\frac{ik}{2\pi} \log(z - D) + \frac{ik}{2\pi} \log(z + D).$$

The resulting velocity field (at $z \neq \pm D$) has components (u, v) given by

$$u - iv = \frac{d\Omega}{dz} = -\frac{ik}{2\pi} \frac{1}{(z - D)} + \frac{ik}{2\pi} \frac{1}{(z + D)}.$$

The velocity (u_1, v_1) of the vortex arises only from the velocity field of the anti-vortex, which must be evaluated at $z = D$:

$$u_1 - iv_1 = \frac{ik}{4\pi D},$$

thus $u_1 = 0$, $v_1 = -k/(4\pi D)$.

The velocity (u_2, v_2) of the anti-vortex arises only from the velocity field of the vortex, which must be evaluated at $z = -D$:

$$u_2 - iv_2 = \frac{ik}{4\pi D}.$$

Thus $u_2 = 0$, $v_2 = -k/(4\pi D)$.

In conclusion, both vortex and anti-vortex move along the negative y-direction with speed $k/(4\pi D)$.

8.3 *Vortex-vortex pair.*
The complex potential of a vortex of circulation k located at position $z_1 = x_1 + iy_1$ on the xy plane is

$$\Omega(z) = -\frac{ik}{2\pi} \log(z - z_1).$$

Using the superposition principle, the complex potential of two vortices of the same circulation k placed at $z = z_1$ and $z = z_2$ is

$$\Omega(z) = -\frac{ik}{2\pi} \log(z - z_1) - \frac{ik}{2\pi} \log(z - z_2).$$

The velocity components (u, v) at position z ($z \neq z_1$ and $z \neq z_2$) are obtained from

$$u - iv = \frac{d\Omega}{dz} = -\frac{ik}{2\pi} \frac{1}{(z - z_1)} = -\frac{ik}{2\pi} \frac{1}{(z - z_2)}.$$

To find the velocity components (u_1, v_1) of the first vortex (located at $z = z_1$) we take only the contribution of the velocity field of the second vortex and evaluate it at $z = z_1$,

$$u_1 - iv_1 = -\frac{ik}{2\pi} \frac{1}{(z_1 - z_2)}.$$

Vice-versa, the find the velocity components (u_2, v_2) of the second vortex (located at $z = z_2$) we take only the contribution of the velocity field of the first vortex and evaluate it at $z = z_2$:

$$u_2 - iv_2 = -\frac{ik}{2\pi} \frac{1}{(z_2 - z_1)}.$$

Now assume that the two vortices are at positions $z_1 = D$ and $z_2 = -D$ at time $t = 0$ and rotate around their midpoint (the origin) with angular frequency ω,

$$z_1 = De^{i\omega t}, \qquad z_2 = De^{i(\omega t + \pi)}.$$

We obtain

$$u_1 - iv_1 = -\frac{ik}{2\pi} \frac{1}{(De^{i\omega t} - De^{i(\omega t + \pi)})} = -\frac{ik}{2\pi D} \frac{e^{-i\omega t}}{(1 - e^{i\pi})} = -\frac{ik}{4\pi D} (\cos(\omega t) - i \sin(\omega t)),$$

thus

$$u_1 = -\frac{k}{4\pi D} \sin(\omega t), \qquad v_1 = \frac{k}{4\pi D} \cos(\omega t).$$

The speed of the first vortex is $\sqrt{u_1^2 + v_1^2} = k/(4\pi D)$. This speed is equal to the distance $(2\pi D)$ travelled during one period $T = 2\pi/\omega$ divided by the time T, hence we have

$$\omega = \frac{k}{4\pi D^2}.$$

Similarly for the second vortex we find

$$u_2 = -\frac{k}{4\pi D} \sin(\omega t), \qquad v_2 = \frac{k}{4\pi D} \cos(\omega t).$$

that is $u_2 = u_1$ and $v_2 = v_1$. In conclusion, two vortices of the same circulation at distance $2D$ from each other rotate around each other with angular velocity $\omega = k/(4\pi D^2)$.

8.4 *Kelvin wave frequency:*
Consider the Kelvin wave of amplitude A, wavelength $\lambda = 2\pi/k$ and wavenumber k along a straight vortex aligned in the z-direction. The position along the vortex is

$$\mathbf{s} = (A \cos\phi, A \sin\phi, z),$$

where $\phi = kz - \omega t$. If the amplitude of the wave is small compared to the wavelength, the arclength is $\xi \approx z$, and the tangent vector $\mathbf{s}'$ at position $\mathbf{s}$ is

$$\mathbf{s}' = \frac{d\mathbf{s}}{d\xi} = (-Ak \sin\phi, Ak \cos\phi, 1).$$

Moreover

$$\mathbf{s}'' = \frac{d^2\mathbf{s}}{d\xi^2} = (-Ak^2 \sin\phi, -Ak^2 \cos\phi, 0).$$

Therefore, neglecting higher order terms in A,

$$\mathbf{s}' \times \mathbf{s}'' \approx (Ak^2 \sin\phi, -Ak^2 \cos\phi, 0).$$

We have also

$$\frac{d\mathbf{s}}{dt} = (A\omega \sin\phi, -A\omega \cos\phi, 0).$$

Substituting into $d\mathbf{s}/dt = \beta \mathbf{s}' \times \mathbf{s}''$, we conclude that the angular velocity of the wave is $\omega = \beta k^2$.

8.5 *Attenuation of second sound:*

1. The linearized equations for the perturbations are

$$\frac{\partial \rho'}{\partial t} + \rho_{n0}\nabla \cdot \mathbf{v}'_n + \rho_{s0}\nabla \cdot \mathbf{v}'_s = 0, \quad \rho_0 \frac{\partial S'}{\partial t} + S_0 \frac{\partial \rho'}{\partial t} + \rho_0 S_0 \nabla \cdot \mathbf{v}'_n = 0.$$

Substitute the first equation into the second, use $\rho_0 = \rho_{s0} + \rho_{n0}$ and find

$$\frac{\partial S'}{\partial t} = -\frac{\rho_{s0}S_0}{\rho_0}\nabla \cdot \mathbf{q}'.$$

From the definition of specific heat we have $S' = (C/T_0)T'$, hence

$$\frac{\partial T'}{\partial t} = -\frac{\rho_{s0}S_0 T_0}{C\rho_0}\nabla \cdot \mathbf{q}'.$$

2. The equations for the velocity perturbations are

$$\frac{\partial \mathbf{v}'_n}{\partial t} = -\frac{1}{\rho_0}\nabla p' - \frac{\rho_{s0}S_0}{\rho_{n0}}\nabla T' + \frac{\rho_{s0}}{\rho_0}\mathbf{F},$$

$$\frac{\partial \mathbf{v}'_s}{\partial t} = -\frac{1}{\rho_0}\nabla p' + S_0 \nabla T' - \frac{\rho_{n0}}{\rho_0}\mathbf{F}.$$

Subtract the two equations and use $\rho_0 = \rho_{s0} + \rho_{n0}$,

$$\frac{\partial \mathbf{q}'}{\partial t} = -\frac{\rho_0 S_0}{\rho_{n0}}\nabla T' + \mathbf{F}.$$

3. Add $-2\boldsymbol{\Omega} \times \boldsymbol{q}'$ to the right hand side of the previous equation; in the expression for $\boldsymbol{F}$, let $\boldsymbol{\omega}_s = 2\boldsymbol{\Omega}$ and $\widehat{\boldsymbol{\omega}}_s = \widehat{\boldsymbol{\Omega}}$,

$$\frac{\partial \boldsymbol{q}'}{\partial t} = -\frac{\rho_0 S_0}{\rho_{n0}} \nabla T' - 2\boldsymbol{\Omega} \times \boldsymbol{q}' + B\widehat{\boldsymbol{\Omega}} \times (\boldsymbol{\Omega} \times \boldsymbol{q}') + B'(\boldsymbol{\Omega} \times \boldsymbol{q}').$$

Take the time derivative,

$$\frac{\partial^2 \boldsymbol{q}'}{\partial t^2} = -\frac{\rho_0 S_0}{\rho_{n0}} \nabla \left(\frac{\partial T'}{\partial t}\right) - 2\boldsymbol{\Omega} \times \frac{\partial \boldsymbol{q}'}{\partial t} + B\widehat{\boldsymbol{\Omega}} \times (\boldsymbol{\Omega} \times \frac{\partial \boldsymbol{q}'}{\partial t}) + B'(\boldsymbol{\Omega} \times \frac{\partial \boldsymbol{q}'}{\partial t}).$$

Substitute the expression for $\partial T'/\partial t$ in terms of $\nabla \cdot \boldsymbol{q}'$ and conclude

$$\frac{\partial^2 \boldsymbol{q}'}{\partial t^2} = c_2^2 \nabla(\nabla \cdot \boldsymbol{q}') - (2 - B')(\boldsymbol{\Omega} \times \frac{\partial \boldsymbol{q}'}{\partial t}) + B\widehat{\boldsymbol{\Omega}} \times (\boldsymbol{\Omega} \times \frac{\partial \boldsymbol{q}'}{\partial t}).$$

4. Since

$$\nabla \cdot \boldsymbol{q}' = -ikq_x e^{i\omega t - ikx}, \qquad \nabla(\nabla \cdot \boldsymbol{q}') = (-k^2 q_x, 0, 0)e^{i\omega t - ikx},$$

$$\frac{\partial^2 \boldsymbol{q}'}{\partial t^2} = (-\omega^2 q_x, -\omega^2 q_y, 0)e^{i\omega t - ikx},$$

one finds the linear system

$$\begin{pmatrix} -\omega^2 + i\omega\Omega B + c_2^2 k^2 & -i\omega\Omega(2 - B') \\ i\omega\Omega(2 - B') & -\omega^2 + i\omega\Omega B \end{pmatrix} \begin{pmatrix} q_x \\ q_y \end{pmatrix} = \begin{pmatrix} 0 \\ 0 \end{pmatrix}$$

Nontrivial solutions exist only if the determinant of the matrix is zero, which means

$$\left(-1 + \frac{i\Omega B}{\omega} + \frac{c_2^2 k^2}{\omega^2}\right)\left(-1 + \frac{i\Omega B}{\omega}\right) - \frac{\Omega^2}{\omega^2}(2 - B')^2 = 0.$$

Assuming $\Omega/\omega \ll 1$

$$\left(-1 + \frac{i\Omega B}{\omega} + \frac{c_2^2 k^2}{\omega^2}\right)\left(-1 + \frac{i\Omega B}{\omega}\right) \approx 0,$$

which is

$$k^2 = \frac{\omega^2}{c_2^2}\left(1 - \frac{i\Omega B}{\omega}\right),$$

$$k = \frac{\omega}{c_2}\left(1 - \frac{i\Omega B}{2\omega} \cdots\right) = k_R + ik_I,$$

where $k_R = \omega/c_2$ and $k_I = -\Omega B/(2c_2) < 0$. Therefore the waves is proportional to

$$e^{i\omega t - ikx} = e^{i\omega t - ik_R x} e^{k_I x} = e^{i\omega t - ik_R x} e^{-x/\lambda}.$$

Hence it decays with x as it propagates. The attenuation coefficient is $\lambda = 2c_2/(\Omega B)$. A measurement of λ is thus a measurement of the vortex line density $n_v = 2\Omega/\kappa$.

8.6 *Vortex tension:*

For simplicity of notation, here we drop the subscript s for superfluid. Let $\boldsymbol{\omega} = (\omega_x, \omega_y, \omega_z) = \omega\widehat{\boldsymbol{\omega}}$ be the superfluid vorticity, $\omega = |\boldsymbol{\omega}| = \sqrt{\omega_x^2 + \omega_y^2 + \omega_z^2}$ be its magnitude, and $\widehat{\boldsymbol{\omega}} = \boldsymbol{\omega}/\omega$ be the unit vector along the direction of $\boldsymbol{\omega}$.

Substitute $\boldsymbol{A} = \boldsymbol{\omega}$ and $\boldsymbol{B} = \widehat{\boldsymbol{\omega}}$ into the vector identity

$$\nabla(\boldsymbol{A} \cdot \boldsymbol{B}) = \boldsymbol{A} \times (\nabla \times \boldsymbol{B}) + \boldsymbol{B} \times (\nabla \times \boldsymbol{A}) + (\boldsymbol{A} \cdot \nabla)\boldsymbol{B} + (\boldsymbol{B} \cdot \nabla)\boldsymbol{A}.$$

Since $\nabla(\omega \cdot \widehat{\boldsymbol{\omega}}) = \nabla\omega$ because $\widehat{\boldsymbol{\omega}} \cdot \widehat{\boldsymbol{\omega}} = 1$, we obtain

$$\nabla\omega = \boldsymbol{\omega} \times (\nabla \times \widehat{\boldsymbol{\omega}}) + (\boldsymbol{\omega} \cdot \nabla)\widehat{\boldsymbol{\omega}} + \widehat{\boldsymbol{\omega}} \times (\nabla \times \boldsymbol{\omega}) + (\widehat{\boldsymbol{\omega}} \cdot \nabla)\boldsymbol{\omega},$$

which is

$$\boldsymbol{\omega} \times (\nabla \times \widehat{\boldsymbol{\omega}}) + (\boldsymbol{\omega} \cdot \nabla)\widehat{\boldsymbol{\omega}} = \nabla\omega - \widehat{\boldsymbol{\omega}} \times (\nabla \times \boldsymbol{\omega}) - (\widehat{\boldsymbol{\omega}} \cdot \nabla)\boldsymbol{\omega}.$$

We need to show that the right hand side of this equation is zero. We do it component by component. Consider the x-component. The first term is

$$[\nabla\omega]_x = \partial_x(\omega_x^2 + \omega_y^2 + \omega_z^2)^{1/2} = \widehat{\omega}_x \partial_x \omega_x + \widehat{\omega}_y \partial_x \omega_y + \widehat{\omega}_z \partial_x \omega_z.$$

Since

$$\nabla \times \boldsymbol{\omega} = (\partial_y\omega_z - \partial_z\omega_y, \, \partial_z\omega_x - \partial_x\omega_z, \, \partial_x\omega_y - \partial_y\omega_x),$$

then the second term is

$$[\widehat{\boldsymbol{\omega}} \times (\nabla \times \boldsymbol{\omega})]_x = \widehat{\omega}_y(\partial_x\omega_y - \partial_y\omega_x) - \widehat{\omega}_z(\partial_z\omega_x - \partial_x\omega_z),$$

and the third term is

$$[(\widehat{\boldsymbol{\omega}} \cdot \nabla)\boldsymbol{\omega}]_x = (\widehat{\omega}_x \partial_x + \widehat{\omega}_y \partial_y + \widehat{\omega}_z \partial_z)\omega_x.$$

Putting the three terms together, we find that the the x-component of the right hand side is zero. Similarly we can show that the y and z-components are zero, therefore $\boldsymbol{\omega} \times (\nabla \times \widehat{\boldsymbol{\omega}}) + (\boldsymbol{\omega} \cdot \nabla)\widehat{\boldsymbol{\omega}} = 0$. Hence we conclude that

$$\boldsymbol{\omega} \times (\nabla \times \widehat{\boldsymbol{\omega}}) = -(\boldsymbol{\omega} \cdot \nabla)\widehat{\boldsymbol{\omega}}.$$

8.7 *Vortex in a disk:*
The complex potential is

$$\Omega(z) = -\frac{i\kappa}{2\pi}\log(z - z_0) + \frac{i\kappa}{2\pi}\log(z - z_1),$$

where $z_0 = x_0 + iy_0 = Re^{i\omega t}$ and $z_1 = a^2/z_0^* = a^2 z_0/(z_0 z_0^*) = a^2 z_0/R^2$.

1. The boundary of the disk has equation $z = ae^{i\theta} = a(\cos\theta + i\sin\theta)$, thus, letting $z - z_0 = |z - z_0|e^{i\theta_0}$, we have

$$z - z_0 = a(\cos\theta + i\sin\theta) - x_0 - iy_0 = (a\cos\theta - x_0) + i(a\sin\theta - y_0),$$

where

$$|z - z_0| = \sqrt{(a\cos\theta - x_0)^2 + (a\sin\theta - y_0)^2},$$

$$\theta_0 = \tan^{-1}\left(\frac{a\sin\theta - y_0}{a\cos\theta - x_0}\right).$$

Similarly, letting $z - z_1 = |z - z_1|e^{i\theta_1}$, we have

$$z - z_1 = a(\cos\theta + i\sin\theta) - \frac{a^2}{R^2}(x_0 + iy_0) = (a\cos\theta - \frac{a^2}{R^2}x_0) + i(a\sin\theta - \frac{a^2}{R^2}y_0),$$

where

$$|z - z_1| = \sqrt{(a\cos\theta - \frac{a^2}{R^2}x_0)^2 + (a\sin\theta - \frac{a^2}{R^2}y_0)^2},$$

$$\theta_1 = \tan^{-1}\left(\frac{a\sin\theta - \frac{a^2}{R^2}y_0}{a\cos\theta - \frac{a^2}{R^2}x_0}\right).$$

Since

$$\Omega = \Phi + i\Psi = -\frac{i\kappa}{2\pi}\left(\ln|z - z_0| + i\theta_0 - \ln|z - z_1| + i\theta_1\right),$$

the streamfunction is

$$\Psi = -\frac{\kappa}{2\pi} \left(\ln |z - z_0| - \ln |z - z_1| \right) =$$

$$= -\frac{\kappa}{2\pi} \frac{1}{2} \ln \left(\frac{(a\cos\theta - x_0)^2 + (a\sin\theta - y_0)^2}{(a\cos\theta - \frac{a^2}{R^2} x_0)^2 + (a\sin\theta - \frac{a^2}{R^2} y_0)^2} \right)$$

which is, using $x_0^2 + y_0^2 = R^2$ and $\sin^2\theta + \cos^2\theta = 1$,

$$\Psi = -\frac{\kappa}{4\pi} \ln \left(\frac{a^2 + R^2 - 2a(x_0\cos\theta + y_0\sin\theta)}{a^2 + \frac{a^4}{R^2} - 2\frac{a^3}{R^2}(x_0\cos\theta + y_0\sin\theta)} \right) =$$

$$= -\frac{\kappa}{4\pi} \ln \left(\frac{a^2 + R^2 - 2a(x_0\cos\theta + y_0\sin\theta)}{(a^2/R^2)(R^2 + a^2 - 2a(x_0\cos\theta + y_0\sin\theta))} \right) =$$

$$= -\frac{\kappa}{4\pi} \ln \left(R^2/a^2 \right) = \text{constant}.$$

We conclude that the value of Ψ is constant along the circle $z = ae^{i\theta}$, which means that the circle is a streamline, hence the flow cannot cross it.

2. The velocity of the vortex at z_0 is

$$u_0 - iv_0 = \frac{d}{dz}\left(\frac{i\kappa}{2\pi} \log(z - z_1) \right)_{z=z_0} = \frac{i\kappa}{2\pi} \frac{1}{(z_0 - \frac{a^2}{R^2} z_0)} = \frac{i\kappa}{2\pi R} \frac{e^{-i\omega t}}{(1 - \frac{a^2}{R^2})}$$

$$= \frac{i\kappa R}{2\pi} \frac{\cos\omega t - i\sin\omega t}{(a^2 - R^2)}$$

Thus

$$u_0 = -V\sin\omega t, \qquad v_0 = V\cos\omega t, \qquad \text{with } V = \frac{\kappa R}{2\pi(a^2 - R^2)}.$$

This means that the vortex makes a circular orbit of radius R and angular velocity

$$\omega = \frac{V}{R} = \frac{\kappa}{2\pi(a^2 - R^2)}.$$

3. If the vortex is at the centre of the disk ($R = 0$), then it does not move ($u_0 = v_0 = 0$).

If the vortex is near the boundary of the disk, let $a - R = \delta$ and assume $\delta \ll a$. Then $a^2 - R^2 = (a + R)(a - R) = (2a - \delta)\delta = 2a\delta(1 - \frac{\delta}{2a})$ and

$$V = \frac{\kappa R}{2\pi(a^2 - R^2)} = \frac{\kappa a(1 - \frac{\delta}{a})}{4\pi a\delta(1 - \frac{\delta}{2a})} \approx \frac{\kappa}{4\pi\delta},$$

which means that the vortex and its image move like a vortex-antivortex pair of separation 2δ.

8.8 *Particle trapped in a vortex line*

Let $v_s = \kappa/(2\pi r)$ be the velocity field around a vortex line. We approximate the spherical region of radius a_p with a cylinder of radius a_p and height a_p and assume that the vortex core is hollow ($\rho_s = 0$ for $r < a_0$). The kinetic energy is

$$\Delta E = \int_0^{2\pi} d\theta \int_0^{a_p} dz \int_{a_0}^{a_p} dr\, r \frac{1}{2}\left(\rho_s v_s^2\right) = \frac{\rho_s \kappa^2 a_p}{4\pi} \ln\left(a_p/a_0\right).$$

Using $a_0 = 10^{-10}\,\mathrm{m}, a_p = 10^{-6}\,\mathrm{m}, \kappa = 9.97 \times 10^{-8}\,\mathrm{m^2/s}$ and $\rho_s = 145\,\mathrm{kg/m^3}$, we find $\Delta E \approx 10^{-18} >> k_B T = 2.8 \times 10^{-23}\,\mathrm{J}$ at $T = 2\,\mathrm{K}$.

Problems of Chap. 9

9.1 *Vortex-antivortex pair with friction:*

Let $z_1 = X + iY)$ and $z_2 = X - iY$. The total complex potential is

$$\Omega(z) = \Omega_1(z) + \Omega_2(z) = -\frac{ik}{2\pi} \log\left(z - z_1\right) + \frac{ik}{2\pi} \log\left(z - z_2\right).$$

The velocity of vortex 1 is (u_1, v_1) with

$$u_1 - iv_1 = \frac{d}{dz}\Omega_2(z_1) = \frac{ik}{2\pi}\frac{1}{(z_1 - z_2)} = \frac{ik}{2\pi}\frac{1}{2iY} = \frac{k}{4\pi Y}.$$

The velocity of vortex 2 is (u_2, v_2) with

$$u_2 - iv_2 = \frac{d}{dz}\Omega_1(z_2) = -\frac{ik}{2\pi}\frac{1}{(z_2 - z_1)} = -\frac{ik}{2\pi}\frac{1}{-2iY} = \frac{k}{4\pi Y}.$$

Call $U = k/(4\pi Y)$. We have found that the self induced velocities of the two vortices are $u_1 = U$, $v_1 = 0$ and $u_2 = U$ and $v_2 = 0$. Therefore at temperature $T = 0$ the vortex pair would travel to the right at this self-induced velocity.
At $T \neq 0$ the motion is determined by Scharz's equation

$$\frac{ds}{dt} = v_{si} + \alpha s' \times (v_n - v_{si}) - \alpha' s' \times [s' \times (v_n - v_{si})],$$

but here the normal fluid velocity is zero ($v_n = 0$) and we are left with

$$\frac{ds}{dt} = v_{si} - \alpha s' \times v_{si} + \alpha' s' \times [s' \times v_{si}].$$

For vortex 1 we have $s = (X, Y)$, $s' = \widehat{z}$, $v_{si} = U\widehat{x}$, $ds/dt = (\dot{X}, \dot{Y})$ where a dot is a time derivative. Schwarz equation is

$$(\dot{X}, \dot{Y}) = U\widehat{x} - \alpha\widehat{z} \times U\widehat{x} + \alpha'\widehat{z} \times [\widehat{z} \times U\widehat{x}] = U\widehat{x} - \alpha U\widehat{y} - \alpha' U\widehat{x},$$

which is

$$\frac{dX}{dt} = \frac{(1-\alpha')k}{4\pi Y}, \qquad \frac{dY}{dt} = -\frac{\alpha k}{4\pi Y}.$$

For vortex 2 we have $\boldsymbol{x} = (X, -Y)$, $\boldsymbol{s}' = -\widehat{\boldsymbol{z}}$, $\boldsymbol{v}_{si} = U\widehat{\boldsymbol{x}}$ and $d\boldsymbol{s}/dt = (\dot{X}, -\dot{Y})$, hence

$$(\dot{X}, -\dot{Y}) = U\widehat{\boldsymbol{x}} + \alpha\widehat{\boldsymbol{z}} \times U\widehat{\boldsymbol{x}} + \alpha'\widehat{\boldsymbol{z}} \times [\widehat{\boldsymbol{z}} \times U\widehat{\boldsymbol{x}}] = U\widehat{\boldsymbol{x}} + \alpha U\widehat{\boldsymbol{y}} - \alpha' U\widehat{\boldsymbol{x}},$$

which is the same equation again:

$$\frac{dX}{dt} = \frac{(1-\alpha')k}{4\pi Y}, \qquad \frac{-dY}{dt} = \frac{\alpha k}{4\pi Y}.$$

The solution of the equation for Y is

$$Y(t) = Y_0 \sqrt{1 - \frac{\alpha k t}{2\pi Y_0^2}},$$

where $Y_0 = Y(0)$. Solving for $X(t)$ we find

$$X(t) = \frac{(1-\alpha')Y_0}{2\alpha}\left(-2\sqrt{1 - \frac{\alpha k t}{2\pi Y_0^2}} + C\right),$$

where C is a constant of integration. Impose $C = 2$ so that $X = 0$ at $t = 0$ and find

$$X(t) = \frac{(1-\alpha')Y_0}{\alpha}\left(1 - \sqrt{1 - \frac{\alpha k t}{2\pi Y_0^2}}\right).$$

The vortex-antivortex pair annihilates at time $\tau = 2\pi Y_0^2/(\alpha k)$ when the separation $2Y(t)$ vanishes. The distance travelled is $X(\tau) = (1-\alpha')Y_0/\alpha$. If the temperature is lowered α decreases so τ and $X(\tau)$ increase.

9.2 Donnelly-Glaberson instability on a vortex line:

1. Since $A \ll 1$, then $\zeta \approx z$, $d/d\zeta \approx d/dz$, hence $\boldsymbol{s} \approx (A\cos\phi, A\sin\phi, \zeta)$, and $\boldsymbol{s}' = (-kA\sin\phi, kA\cos\phi, 1)$, $\boldsymbol{s}'' = (-k^2 A\cos\phi, -k^2 A\sin\phi, 0)$, and $\boldsymbol{v}_i = (\beta k^2 A\sin\phi, -\beta k^2 A\cos\phi, 0)$, where we have neglected terms which are quadratic in A.

   Since $\boldsymbol{v}_n - \boldsymbol{v}_i = (-\beta k^2 A\sin\phi, \beta k^2 A\cos\phi, U)$, then $\boldsymbol{s}' \times (\boldsymbol{v}_n - \boldsymbol{v}_i) = (kA(U - \beta k)\cos\phi, kA(U - \beta k)\sin\phi, 0)$ and $\boldsymbol{s}' \times (\boldsymbol{s}' \times (\boldsymbol{v}_n - \boldsymbol{v}_i)) = (-kA(U - \beta k)\sin\phi, kA(U - \beta k)\cos\phi, 0)$.

Substitute into the Schwarz equation using

$$\frac{ds}{dt} = (\frac{dA}{dt}\cos\phi + \omega A \sin\phi, \ \frac{dA}{dt}\sin\phi - \omega A \cos\phi, 0).$$

Then, setting equal to zero the coefficients of $\sin\phi$ and $\cos\phi$, we have:

$$\frac{dA}{dt} = \alpha k A(U - \beta k), \quad \omega = \beta k^2 + \alpha'(U - \beta k),$$

The solution of the first equation is $A(t) = A(0)e^{\sigma t}$ where the growth rate $\sigma = \alpha k(U - \beta k)A$ is positive (i.e. the perturbation grows) if $U > \beta k$.

Notice that the growth rate depends only on α, not on α', which corrects the frequency of the Kelvin wave at $T = 0$ which would be simply $\omega = \beta k^2$ (shorter waves rotate faster).

2. The maximum growth rate σ_0 is found by setting $d\sigma/dk = \alpha U - 2\alpha\beta k = 0$ which has solution $k_0 = U/(2\beta)$. Thus $\sigma_0 = \sigma(k_0) = \alpha U^2/(4\beta)$.

9.3 Donnelly-Glaberson instability on a vortex lattice:

1. We have $\boldsymbol{\lambda} = 2\boldsymbol{\Omega} + \boldsymbol{\omega} = (\omega_x, \omega_y, \omega_z + 2\Omega)$, hence

$$\lambda = |\boldsymbol{\lambda}| = \sqrt{\omega_x^2 + \omega_y^2 + (\omega_z + 2\Omega)^2} \approx \sqrt{4\Omega^2 + 4\Omega\omega_z + \cdots},$$

$$\frac{1}{\lambda} = \frac{1}{2\Omega}\frac{1}{(1 + \omega_z/\Omega + \cdots)} \approx \frac{1}{2\Omega}\left(1 - \frac{\omega_z}{2\Omega} + \cdots\right).$$

Therefore

$$\widehat{\boldsymbol{\lambda}} = \frac{1}{\lambda}\boldsymbol{\lambda} = \frac{1}{2\Omega}\left(1 - \frac{\omega_z}{2\Omega} + \cdots\right)(\omega_x, \omega_y, \omega_z + 2\Omega) \approx \frac{1}{2\Omega}(\omega_x, \omega_y, 2\Omega).$$

2. Notice that

$$(\boldsymbol{\lambda}\cdot\nabla)\widehat{\boldsymbol{\lambda}} = \frac{1}{2\Omega}(\omega_x\partial_x + \omega_y\partial_y + \omega_z\partial_z)(\omega_x, \omega_y, 2\Omega) \approx (\partial_z\omega_x, \partial_z\omega_y, 0),$$

$$\widehat{\boldsymbol{\lambda}} \times (\boldsymbol{\lambda}\cdot\nabla)\widehat{\boldsymbol{\lambda}} \approx (-\partial_z\omega_y, \partial_z\omega_x, 0).$$

Since $\boldsymbol{q} = \boldsymbol{v}_n - \boldsymbol{v} = (-v_x, -v_y, U - v_z)$ then

$$\boldsymbol{\lambda}\times\boldsymbol{q} \approx (U\omega_y + 2\Omega v_y, -2\Omega v_x - U\omega_x, 0), \quad \widehat{\boldsymbol{\lambda}}\times(\boldsymbol{\lambda}\times\boldsymbol{q}) = (2\Omega v_x + U\omega_x, 2\Omega v_y + U\omega_y, 0).$$

Since $\boldsymbol{\Omega} \times \boldsymbol{v} = (-\Omega v_y, \Omega v_x, 0)$ then, substituting into the superfluid equation we have

$$\begin{pmatrix} \partial_t v_x \\ \partial_t v_y, \\ \partial_t v_z \end{pmatrix} = -\begin{pmatrix} \partial_x \phi \\ \partial_y \phi \\ \partial_z \phi \end{pmatrix} - \begin{pmatrix} -2\Omega v_y \\ 2\Omega v_x \\ 0 \end{pmatrix} - \alpha \begin{pmatrix} 2\Omega v_x + U\omega_x \\ 2\Omega v_y + U\omega_y \\ 0 \end{pmatrix} - \alpha v_s \begin{pmatrix} -\partial_z \omega_y \\ \partial_z \omega_x \\ 0 \end{pmatrix} + v_s \begin{pmatrix} \partial_z \omega_x \\ \partial_z \omega_y \\ 0 \end{pmatrix}$$

3. Express ω_x, ω_y and ω_z in terms of v_x, v_y and v_z:

$$\boldsymbol{\omega} = \nabla \times \boldsymbol{v} = (\partial_y v_z - \partial_z v_y, \partial_z v_x - \partial_x v_z, \partial_x v_y - \partial_y v_x).$$

We find

$$\begin{pmatrix} \partial_t v_x \\ \partial_t v_y \\ \partial_t v_z \end{pmatrix} = \begin{pmatrix} \partial_x \phi \\ \partial_y \phi \\ \partial_z \phi \end{pmatrix} - \begin{pmatrix} -2\Omega v_y \\ 2\Omega v_x \\ 0 \end{pmatrix} - \alpha \begin{pmatrix} 2\Omega v_x + U(\partial_y v_z - \partial_z v_y) \\ 2\Omega v_y + U(\partial_z v_x - \partial_x v_z) \\ 0 \end{pmatrix} +$$

$$-\alpha v_s \begin{pmatrix} -\partial_{zz}^2 v_x + \partial_{xz}^2 v_z \\ \partial_{yz}^2 v_z - \partial_{zz}^2 v_y \\ 0 \end{pmatrix} + v_s \begin{pmatrix} \partial_{yz}^2 v_z - \partial_{zz}^2 v_y \\ \partial_{zz}^2 v_x - \partial_{xz}^2 v_z \\ 0 \end{pmatrix},$$

together with the continuity equation $\partial_x v_x + \partial_y v_y + \partial_z v_z = 0$.

4. Assume that $v_x, v_y, v_z, \phi \sim e^{ik_z z + i\omega t}$:

$$i\omega v_x = 2\Omega v_y - \alpha(2\Omega v_x - ik_z U v_y) - \alpha v_s k_z^2 v_x + v_s k_z^2 v_y,$$
$$i\omega v_y = -2\Omega v_x - \alpha(2\Omega v_y + ik_z U v_x) - \alpha v_s k_z^2 v_y - v_s k_z^2 v_x,$$
$$i\omega v_z = -ik_z \phi,$$
$$ik_z v_z = 0.$$

From the last two equations we have $v_z = \phi = 0$. The first two equations form a linear system:

$$(i\omega + 2\Omega\alpha + \alpha v_s k_z^2)v_x + (-2\Omega - i\alpha k_z U - v_s k_z^2)v_y = 0,$$
$$(2\Omega + i\alpha k_z U + v_s k_z^2)v_x + (i\omega + 2\Omega\alpha + \alpha v_s k_z^2)v_y = 0.$$

Nontrivial solutions exists only if the determinant is zero, which means

$$(i\omega + \alpha\eta_0)^2 + (\eta_0 + i\alpha k_z U)^2 = 0,$$

where $\eta_0 = 2\Omega + v_s k_z^2$. At all temperatures (but the region near the lambda point) $\alpha < 1$ so we can neglect terms proportional to α^2 obtaining

$$\omega = i\alpha\eta_0 \pm \eta\sqrt{1 + 2i\alpha k_z U/\eta_0}.$$

If $T = 0$ then $\alpha = 0$ and $\omega = \pm\eta_0 = \pm\sqrt{2\Omega + \nu_s k_z^2}$. If $T \neq 0$, assuming that $\alpha k_z U/\eta_0 \ll 1$, we find $\omega = \omega_R + i\omega_I$ where $\omega_R = \pm\eta_0$ and $\omega_I = \alpha(\eta_0 \pm k_z U)$. The perturbations are proportional to $e^{ik_z z + i\omega t} = e^{i\omega_R t} e^{-\omega_I t}$ hence if $\omega_I < 0$ they grow exponentially with t, hence the vortex lattice is unstable. Since $U > 0$ and $\Omega > 0$, ω_I can be negative only if we take the minus sign in fron of the square root: $\omega_I = \alpha(\eta_0 - k_x U)$. Thus we have an instability if $U > (2\Omega + \nu_s k_z^2)/k_z$.

9.4 Hamilton equations for a vortex ring:
We have $dR/dP = 1/(dP/dR) = 1/(2\pi\rho\kappa R)$ and

$$\frac{\partial E}{\partial P} = \frac{\partial E}{\partial R}\frac{dR}{dP} = \frac{1}{2\pi\rho\kappa R}(\frac{\rho\kappa^2}{2})(\ln(8R/a_0) - 3/2 + 1),$$

which is

$$\frac{\partial E}{\partial P} = \frac{\kappa}{4\pi R}(\ln(8R/a_0) - 1/2) = V.$$

Problems of Chap. 10

10.1 Navier-Stokes dissipation:
Using the vector identity

$$\nabla(\boldsymbol{A} \cdot \boldsymbol{B}) = (\boldsymbol{A} \cdot \nabla)\boldsymbol{B} + (\boldsymbol{B} \cdot \nabla)\boldsymbol{A} + \boldsymbol{A} \times (\nabla \times \boldsymbol{B}) + \boldsymbol{B} \times (\nabla \times \boldsymbol{A}),$$

with $\boldsymbol{A} = \boldsymbol{B} = \boldsymbol{u}$ and $\boldsymbol{\omega} = \nabla \times \boldsymbol{u}$, Eq. (10.4) becomes

$$\frac{\partial \boldsymbol{u}}{\partial t} + \boldsymbol{\omega} \times \boldsymbol{u} = -\nabla(\frac{p}{\rho} + \frac{q^2}{2}) + \nu\nabla^2\boldsymbol{u},$$

where $q^2 = \boldsymbol{u} \cdot \boldsymbol{u}$. Take the dot product with $\boldsymbol{u}$ and note that $\boldsymbol{u}$ and $\boldsymbol{\omega} \times \boldsymbol{u}$ are perpendicular to each other, so their dot product is zero. Therefore, since $\partial q^2/\partial t = 2\boldsymbol{u} \cdot \partial\boldsymbol{u}/\partial t$, we have

$$\frac{\partial}{\partial t}(\frac{q^2}{2}) = -\boldsymbol{u} \cdot \nabla\left(\frac{p}{\rho} + \frac{q^2}{2}\right) + \nu\boldsymbol{u} \cdot \nabla^2\boldsymbol{u}.$$

Use the vector identity $\nabla \cdot (f\boldsymbol{A}) = f(\nabla \cdot \boldsymbol{A}) + \boldsymbol{A} \cdot \nabla f$ with $\boldsymbol{A} = \boldsymbol{u}$ and $f = p/\rho + q^2/2$ and find

$$\frac{\partial}{\partial t}(\frac{q^2}{2}) = -\nabla \cdot \left(\boldsymbol{u}\left(\frac{p}{\rho} + \frac{q^2}{2}\right)\right) + \nu\boldsymbol{u} \cdot \nabla^2\boldsymbol{u}.$$

Notice that $\nabla^2 \boldsymbol{u} = \nabla(\nabla \cdot \boldsymbol{u}) - \nabla \times \nabla \times \boldsymbol{u} = -\nabla \times \boldsymbol{\omega}$ because $\nabla \cdot \boldsymbol{u} = 0$ and $\nabla \times \boldsymbol{u} = \boldsymbol{\omega}$. Hence we have

$$\frac{\partial}{\partial t}\left(\frac{q^2}{2}\right) = -\nabla \cdot \left(\boldsymbol{u}\left(\frac{p}{\rho} + \frac{q^2}{2}\right)\right) - \nu \boldsymbol{u} \cdot (\nabla \times \boldsymbol{\omega}).$$

Use the identity $\nabla \cdot (\boldsymbol{A} \times \boldsymbol{B}) = (\nabla \times \boldsymbol{A}) \cdot \boldsymbol{B} - (\nabla \times \boldsymbol{B}) \cdot \boldsymbol{A}$ with $\boldsymbol{A} = \boldsymbol{u}$ and $\boldsymbol{B} = \boldsymbol{\omega}$, obtaining $\boldsymbol{u} \cdot (\nabla \times \boldsymbol{\omega}) = \omega^2 - \nabla \cdot (\boldsymbol{u} \times \boldsymbol{\omega})$ where $\omega^2 = \boldsymbol{\omega} \cdot \boldsymbol{\omega}$. We have

$$\frac{\partial}{\partial t}\left(\frac{q^2}{2}\right) = -\nu\omega^2 - \nabla \cdot \left(\boldsymbol{u}\left(\frac{p}{\rho} + \frac{q^2}{2}\right) - \nu \boldsymbol{u} \times \boldsymbol{\omega}\right).$$

Now integrate over the volume V and apply the Divergence Theorem:

$$\frac{\partial}{\partial t}\int_V \left(\frac{q^2}{2}\right)dV = -\nu\int_V \omega^2 dV - \int_S \left(\boldsymbol{u}\left(\frac{p}{\rho} + \frac{q^2}{2}\right) - \nu \boldsymbol{u} \times \boldsymbol{\omega}\right)dS,$$

where S is the surface of the volume V. The surface term vanishes because $\boldsymbol{u} = \boldsymbol{0}$ on S. Dividing by V we conclude that

$$\frac{dE}{dt} = -\frac{\nu}{V}\int_V \omega^2 dV.$$

10.2 *Vorticity decay:*
Integrating Equation (10.12) from k_D to $k_\eta \gg k_D$ we have

$$E = \int_{k_D}^{k_\eta} C\epsilon^{2/3}k^{-5/3}dk \approx \frac{3C}{2}k_D^{-2/3}\epsilon^{2/3}.$$

Using $dE/dt = -\epsilon$ we find

$$\epsilon^{-4/3}\frac{d\epsilon}{dt} = -\frac{1}{C}k_D^{2/3},$$

hence for large times

$$\epsilon \approx \frac{(3C)^3 D^2}{4\pi^2}t^{-3},$$

and since $\epsilon = \nu\omega^2$

$$\omega \approx \frac{(3C)^{3/2} D}{2\pi \nu^{1/2}}t^{-3/2},$$

and finally

$$E(t) \approx \frac{(3C)^3 D^2}{8\pi^2}t^{-2}.$$

10.3 *Decay of counterflow turbulence:*
Consider Vinen's equation in the absence of drive ($V_{ns} = 0$):

$$\frac{dL}{dt} = -\frac{\chi_2 \kappa}{2\pi} L^2.$$

Integrate in time,

$$-\frac{1}{L} + \frac{1}{L_0} = -\frac{\chi_2 \kappa}{2\pi} t,$$

where $L_0 = L(0)$ is the initial vortex line density at $t = 0$. So or large times

$$L \approx \frac{2\pi}{\chi_2 \kappa t}.$$

10.4 *Energy of Vinen turbulence and its decay:*
Consider a straight vortex of length h in cylindrical coordinates (r, θ, z). The velocity field around the vortex is $u = \kappa/(2\pi r)$ where κ is the quantum of circulation. The kinetic energy in a cylindrical region of radius R and height h is

$$E = \int_V \frac{\rho u^2}{2} dV = \int_0^{2\pi} d\theta \int_0^h dz \int_{a_0}^R dr\, r \frac{\rho u^2}{2} = \frac{\rho \kappa^2 h}{4\pi} \ln\left(R/a_0\right),$$

where $d\theta\, dz\, r dr$ is the infinitesimal volume element dV in cylindrical coordinates. Here we have taken into account that the density goes to zero at the axis of the vortex by assuming the 'hollow core' approximation that the superfluid density ρ is constant far from the vortex, but $\rho = 0$ for $r < a_0$ where a_0 is the effective vortex core radius (of the order of few times the healing length as in Fig. 5.3a). The kinetic energy per unit length is therefore

$$\frac{E}{h} = \frac{\rho \kappa^2 h}{4\pi} \ln\left(R/a_0\right).$$

Now assume that the vortices in the tangle are randomly oriented and uniformly distributed. At a given position, the total velocity field is the sum of the contributions from all vortex lines. The far-field contributions arising from distances larger than the characteristic intervortex spacing $\ell \approx L^{-1/2}$ cancel each others out. Therefore, in evaluating the energy, we restrict the integration to distances $R < \ell$, multiplying the resulting energy per unit length E/h times the total length of vortex lines which is LD^3. We obtain

$$E = \frac{\rho \kappa^2 L D^3}{4\pi} \ln\left(\ell/a_0\right).$$

The rate of kinetic energy dissipation per unit mass is

$$\frac{dE'}{dt} = \nu \omega^2,$$

where v is the kinematic viscosity and ω the vorticity. We have found that

$$E = \frac{\rho \kappa^2 L D^3}{4\pi} \ln\left(\ell/a_0\right),$$

where $V = D^3$, κ is the quantum of circulation, a_0 is the vortex core radius and $\ell \approx L^{-1/2}$ the typical distance between vortices in the tangle. We have,

$$E' = \frac{\kappa^2 L}{4\pi} \ln\left(\ell/a_0\right).$$

Letting $\omega = \kappa L$ we have

$$\frac{dL}{dt} = -\frac{v}{c} L^2,$$

with

$$c = \frac{1}{4\pi} \ln\left(\ell/a_0\right).$$

Since c changes slowly with L, we approximate it to a constant and, solving the differential equation for L, find that

$$\frac{1}{L(t)} = \frac{v}{c} t + \frac{1}{L_0},$$

where $L_0 = L(0)$ is the vortex line density at $t = 0$ at the beginning of the decay. We conclude that at sufficiently large times the vortex line density decays as

$$L(t) \approx \frac{c}{v} t^{-1}.$$

Problems of Chap. 11

11.1 *Miscible-immiscible condition:*
From Eq. (11.16) using $V_2 = V - V_1$ we have,

$$0 = \frac{dE_{imm}}{dV_1} = \frac{1}{2} \frac{d}{dV_1}\left(g_{11}\frac{N_1^2}{V_1} + g_{22}\frac{N_2^2}{(V - V_1)}\right) = \frac{1}{2}\left(-g_{11}\frac{N_1^2}{V_1^2} + g_{22}\frac{N_2^2}{(V - V_1)^2}\right),$$

hence

$$g_{11}\frac{N_1^2}{V_1^2} = g_{22}\frac{N_2^2}{(V - V_1)^2}$$

which is

$$\frac{V - V_1}{V_1} = \sqrt{g_{22}/g_{11}}\,\frac{N_1}{N_2}.$$

Solving for $\mathcal{V}_1$ we recover

$$\frac{\mathcal{V}_1}{\mathcal{V}} = \frac{1}{\left(1 + \sqrt{g_{22}/g_{11}}\, N_2/N_1\right)},$$

which is Eq. (11.17). Finally,

$$\frac{\mathcal{V}_2}{\mathcal{V}} = 1 - \frac{\mathcal{V}_1}{\mathcal{V}} = \frac{\sqrt{g_{22}/g_{11}}\,(N_2/N_1)}{\left(1 + \sqrt{g_{22}/g_{11}}\,(N_2/N_1)\right)} = \frac{1}{\left(1 + \sqrt{g_{11}/g_{22}}\,(N_1/N_2)\right)},$$

which is Eq. (11.18).

11.2 *Thomas-Fermi solution 1:*
We write the time-independent solutions in the form $\psi_j(\mathbf{r}, t) = \phi_j(\mathbf{r}) \exp(-i\mu_j t/\hbar)$, where μ_j are the chemical potentials of condensate 1 and 2 respectively. Inserting into the coupled Gross-Pitaevskii equations (11.1) and (11.2), cancelling common factors of $\exp(-i\mu_j t/\hbar)$ and ϕ_j, introducing the densities $n_j(\mathbf{r}) = |\phi_j(\mathbf{r})|^2$ and dropping the Laplacian terms gives,

$$\mu_1 = g_{11}n_1(\mathbf{r}) + g_{12}n_2(\mathbf{r}) + V_1(\mathbf{r}), \tag{A.18}$$
$$\mu_2 = g_{22}n_2(\mathbf{r}) + g_{12}n_1(\mathbf{r}) + V_2(\mathbf{r}). \tag{A.19}$$

Solving as simultaneous equations we obtain the TF solution,

$$n_1(\mathbf{r}) = \frac{[\mu_1 g_{11} - \mu_2 g_{12}] - [g_{22} - g_{12}]V_1(\mathbf{r})}{g_{11}g_{22} - g_{12}^2}, \tag{A.20}$$

$$n_2(\mathbf{r}) = \frac{[\mu_2 g_{11} - \mu_1 g_{12}] - [g_{11} - g_{12}]V_2(\mathbf{r})}{g_{11}g_{22} - g_{12}^2}. \tag{A.21}$$

11.3 *Thomas-Fermi solution 2:*
Starting from the TF solution (11.22) and (11.23), letting $V_1(\mathbf{r} = V_2(\mathbf{r}) = m\omega^2 r^2/2$ and rearranging slightly,

$$n_1(r) = \frac{\mu_1 g_{11} - \mu_2 g_{12}}{g_{11}g_{22} - g_{12}^2}\left[1 - \frac{(g_{22} - g_{12})\omega^2}{2(\mu_1 g_{11} - \mu_2, g_{12})}r^2\right] \tag{A.22}$$

$$n_2(r) = \frac{\mu_2 g_{11} - \mu_1 g_{12}}{g_{11}g_{22} - g_{12}^2}\left[1 - \frac{(g_{11} - g_{12})\omega^2}{2(\mu_2 g_{11} - \mu_1 g_{12})}r^2\right], \tag{A.23}$$

which have each been written in the form $n(r) = n_0(1 - r^2/R^2)$. We can immediately identity the ratio of the TF radii as,

$$\frac{R_2}{R_1} = \sqrt{\frac{\mu_2 g_{11} - \mu_1 g_{12}}{\mu_1 g_{11} - \mu_2 g_{12}}}\sqrt{\frac{g_{22} - g_{12}}{g_{11} - g_{12}}}. \tag{A.24}$$

Setting $g_{12} = 0$ this immediately reduces down to $R_2/R_1 = \sqrt{\mu_2/\mu_1}$, which is what would be expected for two independent condensates.

Recall Equations (B.17) and (B.18). For equal chemical potentials (denoted μ), intra-species nonlinearity (denoted g) and equal harmonic external potential these become,

$$\mu = gn_1(r) + g_{12}n_2(r) + \frac{1}{2}m\omega^2 r^2, \tag{A.25}$$

$$\mu = gn_2(r) + g_{12}n_1(r) + \frac{1}{2}m\omega^2 r^2. \tag{A.26}$$

We immediately see that the density profiles $n_1(r)$ and $n_2(r)$ must be the same. We can write the density profile as both as,

$$\mu = gn_1(r) + g_{12}n_2(r) + \frac{1}{2}m\omega^2 r^2, \tag{A.27}$$

$$\mu = gn_2(r) + g_{12}n_1(r) + \frac{1}{2}m\omega^2 r^2. \tag{A.28}$$

$$n(r) = \frac{\mu(g - g_{12})}{g^2 - g_{12}^2}\left[1 - \frac{(g - g_{12})\omega^2}{2(\mu(g - g_{12})}r^2\right]. \tag{A.29}$$

Noting that $g^2 - g_{12}^2 = (g + g_{12})(g - g_{12})$ we can cancel the $g - g_{12}$ terms to obtain,

$$n(r) = \frac{\mu}{g + g_{12}}\left[1 - \frac{\omega^2 r^2}{2\mu)}\right], \tag{A.30}$$

from which we see that the effective nonlinear coefficient acting on each condensate is $g + g_{12}$.

11.4 *Domain wall solution:*
Equations 11.27 become (dropping the prime for simplicity),

$$i\frac{\partial \psi_1}{\partial t} = -\frac{1}{2}\frac{\partial^2 \psi_1}{\partial x^2} + |\psi_1|^2\psi_1 + 3|\psi_2|^2\psi_1 - \psi_1, \tag{A.31}$$

$$i\frac{\partial \psi_2}{\partial t} = -\frac{1}{2}\frac{\partial^2 \psi_2}{\partial x^2} + |\psi_2|^2\psi_2 + 3|\psi_1'|^2\psi_2' - \mu_2'\psi_2'. \tag{A.32}$$

Writing $\psi_j = \phi_j \exp(-i\mu_j t)$ we obtain their time-independent form,

$$\mu_1\phi_1 = -\frac{1}{2}\frac{\partial^2\phi_1}{\partial x^2} + |\phi_1|^2\phi_1 + 3|\phi_2|^2\phi_1 - \phi_1, \quad \mu_2\phi_2 = -\frac{1}{2}\frac{\partial^2\phi_2}{\partial x^2} + |\phi_2|^2\phi_2 + 3|\phi_1'|^2\phi_2' - \mu_2'\phi_2'. \tag{A.33}$$

We look for time-independent solutions of the form $\phi_j(x) = A[1 + (-1)^j \tanh(\frac{x}{\sigma})]$. We differentiate twice to obtain $\frac{\partial^2\phi_j}{\partial x^2} = (-1)^j\frac{2A}{\sigma^2}\text{sech}^2(x/\sigma)\tanh(x/\sigma)$. Inserting into the time-independent equations, using the relation $\text{sech}^2 x = 1 - \tanh^2 x$, and comparing terms in zeroth, first, second and third order in $\tanh(x/\sigma)$, we obtain $\mu = 4A^2$ and $\sigma = 1/(2A)$.

Problems of Chap. 12

12.1 *Manipulating the DDI:*
(a) Substitution of $\hat{\boldsymbol{e}}_1 = \hat{\boldsymbol{e}}_2 = (0, 0, 1)$ and $\boldsymbol{r} = r\hat{\boldsymbol{r}}$, with $\hat{\boldsymbol{r}} = (\sin\theta, 0, \cos\theta)$ into the unpolarised DDI gives

$$U_{\mathrm{dd}}(\boldsymbol{r}) = \frac{C_{\mathrm{dd}}}{4\pi} \left(\frac{r^2 - 3r^2\cos^2\theta}{r^5} \right),$$

$$= \frac{C_{\mathrm{dd}}}{4\pi} \left(\frac{1 - 3\cos^2\theta}{r^3} \right),$$

as required.
(b) Substitution of the given $\hat{\boldsymbol{e}}_1 = \hat{\boldsymbol{e}}_2 = (\sin\varphi\cos\Omega t, \sin\varphi\sin\Omega t, \cos\varphi)$ into the unpolarised DDI gives

$$U_{\mathrm{dd}}(\boldsymbol{r}) = \frac{C_{\mathrm{dd}}}{4\pi r^3} \left[1 - 3\left(\sin^2\theta\sin^2\varphi\cos^2\Omega t + \cos^2\theta\cos^2\varphi + 2\sin\theta\sin\varphi\cos\theta\cos\varphi\cos\Omega t \right) \right]$$

which after averaging the result over one period, i.e. calculating

$$\langle U_{\mathrm{dd}}(\boldsymbol{r}) \rangle = \frac{1}{T} \int_0^T U_{\mathrm{dd}}(\boldsymbol{r}),$$

we obtain the solution.

12.2 *Fourier transform of the DDI:*
Obtaining an expression of the Fourier transform of the DDI involves evaluating the following expression

$$\tilde{U}_{\mathrm{dd}}(\boldsymbol{k}) = \frac{C_{\mathrm{dd}}}{4\pi} \int \mathrm{d}^3 r \frac{1 - 3\cos^2\theta}{r^3} e^{i\boldsymbol{k}\cdot\boldsymbol{r}} = \frac{C_{\mathrm{dd}}}{4\pi} \int_0^\infty \frac{\mathrm{d}r}{r} \int \mathrm{d}\Omega (1 - 3\cos^2\theta) e^{i\boldsymbol{k}\cdot\boldsymbol{r}},$$

where the final expression has been transformed into spherical polar coordinates with $\mathrm{d}\Omega = \sin\theta \mathrm{d}\theta \mathrm{d}\phi$. The integral over the solid angle $\mathrm{d}\Omega$ can be carried out using a few substitutions. Firstly, we state an expansion of a plane wave in terms of spherical harmonics $Y_l^m(\hat{\boldsymbol{r}})$ and spherical Bessel functions $j_l(kr)$, given as

$$e^{i\boldsymbol{k}\cdot\boldsymbol{r}} = 4\pi \sum_{l=0}^\infty i^l j_l(kr) \sum_{m=-l}^l Y_l^m(\hat{\boldsymbol{k}})^* Y_l^m(\hat{\boldsymbol{r}}),$$

and the relation $(1 - 3\cos^2\theta) = -\sqrt{16\pi/5} Y_2^0(\hat{\boldsymbol{r}})$, where $\hat{\boldsymbol{r}} = (\sin\theta\cos\phi, \sin\theta\sin\phi, \cos\theta)$. Using the orthonormality relations for the spherical harmonic functions the only remaining terms over the integral are defined by $\int \mathrm{d}\Omega\, Y_l^m(\hat{\boldsymbol{r}}) Y_2^0(\hat{\boldsymbol{r}}) = \delta_{l2}\delta_{m0}$. Thus, the solid angle integral above becomes

$$\int \mathrm{d}\Omega (1 - 3\cos^2\theta) e^{i\boldsymbol{k}\cdot\boldsymbol{r}} = -4\pi j_2(kr) \left[-\sqrt{\frac{16\pi}{5}} Y_2^0(\hat{\boldsymbol{k}})^* \right] = -4\pi j_2(kr)(1 - 3\cos^2\theta_k),$$

where θ_k is the angle between $\boldsymbol{k}$ and the direction of the dipoles, i.e. if the dipoles are polarised along z this would read $\cos\theta_k = k_z/k$. The remaining integration over r can be evaluated with the relation

$$\int_0^\infty dr\,\frac{j_2(kr)}{r} = \frac{1}{3}\,.$$

Finally, we obtain

$$\tilde{U}_{\mathrm{dd}}(\boldsymbol{k}) = \frac{C_{\mathrm{dd}}}{3}(3\cos^2\theta_k - 1)\,.$$

12.3 *Dipolar Point Vortex Model:*
Firstly, note that

$$\frac{\partial F(z|n)}{\partial z} = \frac{1}{\sqrt{1 - n\sin^2 z}}\,,$$

$$\frac{\partial \Pi(m;z|n)}{\partial z} = \frac{1}{(1 - m\sin^2 z)\sqrt{1 - n\sin^2 z}}\,.$$

Starting from the analytic phase profile,

$$S(x,y) = q\Lambda\left\{(\lambda^2 - 1)F\left[\phi\,|\,1 - \lambda^4\right] + \lambda^2\Pi\left[1 - \lambda^2\,;\phi\,|\,1 - \lambda^4\right]\right\},\quad \text{where } \phi = \arctan\left(\frac{y}{x}\right).$$

We first compute the gradient using the chain rule,

$$\nabla S = \frac{dS}{d\phi}\nabla\phi$$

$$= \left\{(\lambda^2 - 1)\frac{1}{\sqrt{1 - (1 - \lambda^4)\sin^2\phi}} + \frac{\lambda^2}{(1 - (1 - \lambda^2)\sin^2\phi)\sqrt{1 - (1 - \lambda^4)\sin^2\phi}}\right\}\nabla\arctan\left(\frac{y}{x}\right).$$

The gradient of $\phi = \arctan(y/x)$ is,

$$\nabla\phi = \frac{1}{x^2 + y^2}\begin{pmatrix} -y \\ x \end{pmatrix}.$$

Substituting back,

$$\nabla S = \frac{1}{x^2 + y^2}\cdot\frac{1}{\sqrt{1 - (1 - \lambda^4)\sin^2\phi}}\cdot\left[(\lambda^2 - 1) + \frac{\lambda^2}{1 - (1 - \lambda^2)\sin^2\phi}\right]\begin{pmatrix} -y \\ x \end{pmatrix}.$$

We now simplify using the identity

$$\sin^2\left[\arctan\left(\frac{y}{x}\right)\right] = \frac{y^2}{x^2 + y^2}\,.$$

This gives,

$$\nabla S = \frac{1}{x^2 + y^2} \cdot \frac{1}{\sqrt{1 - (1 - \lambda^4)\frac{y^2}{x^2+y^2}}} \cdot \left[(\lambda^2 - 1) + \frac{\lambda^2}{1 - (1 - \lambda^2)\frac{y^2}{x^2+y^2}} \right] \begin{pmatrix} -y \\ x \end{pmatrix}.$$

Simplifying the square root and denominator terms,

$$\sqrt{1 - \frac{(1 - \lambda^4)y^2}{x^2 + y^2}} = \sqrt{\frac{x^2 + \lambda^4 y^2}{x^2 + y^2}}, \quad \text{and} \quad 1 - (1 - \lambda^2)\frac{y^2}{x^2 + y^2} = \frac{x^2 + \lambda^2 y^2}{x^2 + y^2}.$$

Substituting gives,

$$\nabla S = \frac{1}{x^2 + y^2} \cdot \frac{\sqrt{x^2 + y^2}}{\sqrt{x^2 + \lambda^4 y^2}} \cdot \left[(\lambda^2 - 1) + \frac{\lambda^2(x^2 + y^2)}{x^2 + \lambda^2 y^2} \right] \begin{pmatrix} -y \\ x \end{pmatrix}.$$

Canceling terms,

$$\nabla S = \frac{1}{\sqrt{x^2 + y^2}\sqrt{x^2 + \lambda^4 y^2}} \cdot \left[(\lambda^2 - 1) + \frac{\lambda^2(x^2 + y^2)}{x^2 + \lambda^2 y^2} \right] \begin{pmatrix} -y \\ x \end{pmatrix}.$$

We combine into a single fraction,

$$\left[(\lambda^2 - 1) + \frac{\lambda^2(x^2 + y^2)}{x^2 + \lambda^2 y^2} \right] = \frac{(\lambda^2 - 1)(x^2 + \lambda^2 y^2) + \lambda^2(x^2 + y^2)}{x^2 + \lambda^2 y^2}.$$

Expanding and simplifying,

$$\text{Numerator:} \ (\lambda^2 - 1)(x^2 + \lambda^2 y^2) + \lambda^2(x^2 + y^2) = (2\lambda^2 - 1)x^2 + \lambda^4 y^2.$$

We then arrive at the final expression,

$$\nabla S = \frac{(2\lambda^2 - 1)x^2 + \lambda^4 y^2}{\sqrt{x^2 + y^2}(x^2 + \lambda^2 y^2)\sqrt{x^2 + \lambda^4 y^2}} \begin{pmatrix} -y \\ x \end{pmatrix}.$$

This can be used in the equation provided in the problem to give the velocity contribution due to the vortex phase:

$$\mathbf{v}_j^{(P)} = \frac{\hbar \Lambda}{m} \sum_{\substack{k=1 \\ k \neq j}}^{N_v} \frac{q_k}{r_{jk}} \frac{\lambda^4 y_{jk}^2 + x_{jk}^2(2\lambda^2 - 1)}{(x_{jk}^2 + \lambda^2 y_{jk}^2)\sqrt{x_{jk}^2 + \lambda^4 y_{jk}^2}} \begin{pmatrix} -y_{jk} \\ x_{jk} \end{pmatrix}.$$

To find the long-range interaction contribution, we begin with

$$U_{\mathrm{DD}}(x, y) = \frac{\hbar^2}{m} \xi_v \varepsilon_{\mathrm{dd}} \frac{1}{r^3} \left(1 - 3 \sin^2 \alpha \, \frac{x^2}{r^2} \right), \qquad r^2 = x^2 + y^2.$$

The first step is to compute the gradient ∇U_{DD}. Write

$$U_{\mathrm{DD}} = C \left(r^{-3} - 3 \sin^2 \alpha \, x^2 r^{-5} \right), \qquad C = \frac{\hbar^2}{m} \xi_v \varepsilon_{\mathrm{dd}}.$$

It is useful to note

$$\frac{\partial r^{-n}}{\partial x} = -nx \, r^{-n-2}, \qquad \frac{\partial r^{-n}}{\partial y} = -ny \, r^{-n-2}.$$

Thus

$$\frac{\partial}{\partial x} \left(r^{-3} \right) = -3xr^{-5}, \qquad \frac{\partial}{\partial y} \left(r^{-3} \right) = -3yr^{-5},$$

and

$$\frac{\partial}{\partial x}(x^2 r^{-5}) = 2xr^{-5} - 5x^3 r^{-7}, \qquad \frac{\partial}{\partial y}(x^2 r^{-5}) = -5x^2 y \, r^{-7}.$$

Now assemble,

$$\partial_x U_{\mathrm{DD}} = C \left[-3xr^{-5} - 3 \sin^2 \alpha \, (2xr^{-5} - 5x^3 r^{-7}) \right],$$

$$\partial_y U_{\mathrm{DD}} = C \left[-3yr^{-5} + 15 \sin^2 \alpha \, x^2 y \, r^{-7} \right].$$

Thus

$$\nabla U_{\mathrm{DD}} = C \begin{pmatrix} -3xr^{-5} + 15 \sin^2 \alpha \, x^3 r^{-7} - 6 \sin^2 \alpha \, xr^{-5} \\ -3yr^{-5} + 15 \sin^2 \alpha \, x^2 y \, r^{-7} \end{pmatrix}.$$

Next, we use this to compute

$$\hat{\mathbf{z}} \times \nabla U_{\mathrm{DD}} = \begin{pmatrix} -\partial_y U_{\mathrm{DD}} \\ \partial_x U_{\mathrm{DD}} \end{pmatrix}.$$

This gives

$$\hat{\mathbf{z}} \times \nabla U_{\mathrm{DD}} = C \begin{pmatrix} 3yr^{-5} - 15 \sin^2 \alpha \, x^2 y \, r^{-7} \\ -3xr^{-5} + 15 \sin^2 \alpha \, x^3 r^{-7} - 6 \sin^2 \alpha \, xr^{-5} \end{pmatrix}.$$

Factoring out $\begin{pmatrix} -y \\ x \end{pmatrix}$,

$$\hat{\mathbf{z}} \times \nabla U_{\mathrm{DD}} = \frac{3C}{r^7} \left[\left(5x^2 \sin^2 \alpha - r^2 \right) \begin{pmatrix} -y \\ x \end{pmatrix} - 2 \sin^2 \alpha \, r^2 \begin{pmatrix} 0 \\ x \end{pmatrix} \right].$$

As a note, velocity terms like $A\begin{pmatrix} -y \\ x \end{pmatrix}$ describe circular motion, so this form of the equation suggests the dipolar interaction can lead to non-circular orbits, sheared by $B\begin{pmatrix} 0 \\ x \end{pmatrix}$.

Finally, inserting this into the equation from the problem gives

$$\mathbf{v}_j^{(\mathrm{DD})} = \frac{3\hbar^2 \xi_v \varepsilon_{\mathrm{dd}}}{m} \sum_{\substack{k=1 \\ k \neq j}}^{N_v} \frac{q_k}{r_{jk}^7} \left[\left(5x_{jk}^2 \sin^2 \alpha - r_{jk}^2 \right) \begin{pmatrix} -y_{jk} \\ x_{jk} \end{pmatrix} - 2 \sin^2 \alpha \, r_{jk}^2 \begin{pmatrix} 0 \\ x_{jk} \end{pmatrix} \right].$$

Problems of Chap. 13

13.1 *Self-bound solutions in 3D free-space:*
(a) We are asked to evaluate the per-particle energy functional,

$$E[\phi] = 4\pi \int_0^\infty dr \, r^2 \left[\frac{1}{2}\phi^* \nabla_r^2 \phi - \frac{3}{2}|\phi|^4 + |\phi|^5 \right], \tag{A.34}$$

for the Gaussian Ansatz,

$$\phi(r) = \sqrt{\frac{N}{l^3 \pi^{3/2}}} \, e^{-r^2/(2l^2)}. \tag{A.35}$$

To compute the required integrals, we will repeatedly use the identity,

$$\int_0^\infty r^n e^{-r^2} \, dr = \frac{1}{2}\Gamma\left(\frac{n+1}{2}\right), \qquad n \in \mathbb{Z}. \tag{A.36}$$

Before evaluating the three terms separately, note the useful powers,

$$|\phi(r)|^2 = \frac{N}{l^3 \pi^{3/2}} \, e^{-r^2/l^2}, \tag{A.37}$$

$$|\phi(r)|^4 = \frac{N^2}{l^6 \pi^3} \, e^{-2r^2/l^2}, \tag{A.38}$$

$$|\phi(r)|^5 = \left(\frac{N}{l^3 \pi^{3/2}}\right)^{5/2} e^{-5r^2/(2l^2)}. \tag{A.39}$$

Approaching this term-by-term, the contact contribution is,

$$E_{\mathrm{c}} = -6\pi \int_0^\infty dr \, r^2 |\phi|^4. \tag{A.40}$$

Substituting $|\phi|^4$,

$$E_c = -6\pi \int_0^\infty r^2 \left(\frac{N^2}{l^6 \pi^3} e^{-2r^2/l^2} \right) dr \tag{A.41}$$

$$= -6 \frac{N^2}{l^6 \pi^2} \int_0^\infty r^2 e^{-2r^2/l^2} \, dr. \tag{A.42}$$

Make the substitution,

$$x = \sqrt{2}\frac{r}{l} \quad \Rightarrow \quad r = \frac{lx}{\sqrt{2}}, \qquad dr = \frac{l}{\sqrt{2}} dx.$$

Then

$$\int_0^\infty r^2 e^{-2r^2/l^2} \, dr = \int_0^\infty \left(\frac{l^2 x^2}{2} \right) e^{-x^2} \left(\frac{l}{\sqrt{2}} dx \right) \tag{A.43}$$

$$= \frac{l^3}{2\sqrt{2}} \int_0^\infty x^2 e^{-x^2} \, dx. \tag{A.44}$$

Using the identity with $n = 2$,

$$\int_0^\infty x^2 e^{-x^2} \, dx = \frac{1}{2} \Gamma\left(\frac{3}{2} \right) = \frac{1}{2} \cdot \frac{\sqrt{\pi}}{2} = \frac{\sqrt{\pi}}{4},$$

we obtain,

$$E_c = -6 \frac{N^2}{l^6 \pi^2} \cdot \frac{l^3}{2\sqrt{2}} \cdot \frac{\sqrt{\pi}}{4} \tag{A.45}$$

$$= -\frac{3N^2}{4\sqrt{2}\,\pi^{3/2}\, l^3}. \tag{A.46}$$

Next, the LHY contribution is,

$$E_{\mathrm{LHY}} = 4\pi \int_0^\infty r^2 |\phi|^5 \, dr. \tag{A.47}$$

Insert $|\phi|^5$:

$$E_{\mathrm{LHY}} = 4\pi \left(\frac{N}{l^3 \pi^{3/2}} \right)^{5/2} \int_0^\infty r^2 e^{-5r^2/(2l^2)} \, dr. \tag{A.48}$$

Let,

$$x = \sqrt{\frac{5}{2}}\frac{r}{l} \quad \Rightarrow \quad r = \frac{lx}{\sqrt{5/2}}, \qquad dr = \frac{l}{\sqrt{5/2}} dx.$$

Then,

$$\int_0^\infty r^2 e^{-5r^2/(2l^2)} \mathrm{d}r = \left(\frac{l^2}{5/2}\right) \cdot \left(\frac{l}{\sqrt{5/2}}\right) \int_0^\infty x^2 e^{-x^2} \mathrm{d}x \tag{A.49}$$

$$= \frac{l^3}{(5/2)^{3/2}} \cdot \frac{\sqrt{\pi}}{4}. \tag{A.50}$$

Thus,

$$E_{\mathrm{LHY}} = 4\pi \left(\frac{N^{5/2}}{l^{15/2} \pi^{15/4}}\right) \frac{l^3}{(5/2)^{3/2}} \cdot \frac{\sqrt{\pi}}{4} \tag{A.51}$$

$$= \frac{N^{5/2}}{l^{9/2}} \frac{1}{\pi^{9/4}} \frac{1}{(5/2)^{3/2}}. \tag{A.52}$$

Simplify the constant,

$$(5/2)^{3/2} = \frac{5^{3/2}}{2^{3/2}} = \frac{5\sqrt{5}}{2\sqrt{2}}.$$

Therefore,

$$E_{\mathrm{LHY}} = \frac{2\sqrt{2}}{5\sqrt{5}} \frac{N^{5/2}}{\pi^{9/4} l^{9/2}}. \tag{A.53}$$

Finally, the kinetic term. The radial Laplacian in 3D spherical symmetry is,

$$\nabla_r^2 \phi = \frac{1}{r^2} \frac{\mathrm{d}}{\mathrm{d}r} \left(r^2 \frac{\mathrm{d}\phi}{\mathrm{d}r}\right).$$

Compute the derivative,

$$\frac{\mathrm{d}\phi}{\mathrm{d}r} = \sqrt{\frac{N}{l^3 \pi^{3/2}}} \left(-\frac{r}{l^2}\right) e^{-r^2/(2l^2)}, \tag{A.54}$$

$$r^2 \frac{\mathrm{d}\phi}{\mathrm{d}r} = -\sqrt{\frac{N}{l^3 \pi^{3/2}}} \frac{r^3}{l^2} e^{-r^2/(2l^2)}. \tag{A.55}$$

Then,

$$\nabla_r^2 \phi = \sqrt{\frac{N}{l^3 \pi^{3/2}}} e^{-r^2/(2l^2)} \left(-\frac{3}{l^2} + \frac{r^2}{l^4}\right). \tag{A.56}$$

Thus,

$$E_{\text{kin}} = 2\pi \int_0^\infty r^2 \phi^* (\nabla_r^2 \phi)\, dr \tag{A.57}$$

$$= 2\pi \frac{N}{l^3 \pi^{3/2}} \int_0^\infty r^2 e^{-r^2/l^2} \left(-\frac{3}{l^2} + \frac{r^2}{l^4} \right) dr. \tag{A.58}$$

Define,

$$I_n = \int_0^\infty r^n e^{-r^2/l^2}\, dr.$$

With substitution $x = r/l$, one finds,

$$I_2 = \frac{\sqrt{\pi}}{4} l^3, \qquad I_4 = \frac{3\sqrt{\pi}}{8} l^5.$$

Hence,

$$E_{\text{kin}} = 2\pi \frac{N}{l^3 \pi^{3/2}} \left[-\frac{3}{l^2} I_2 + \frac{1}{l^4} I_4 \right] \tag{A.59}$$

$$= 2\pi \frac{N}{l^3 \pi^{3/2}} \left[-\frac{3}{l^2} \frac{\sqrt{\pi}}{4} l^3 + \frac{1}{l^4} \frac{3\sqrt{\pi}}{8} l^5 \right] \tag{A.60}$$

$$= 2\pi \frac{N}{l^3 \pi^{3/2}} \left[-\frac{3\sqrt{\pi}}{4} l + \frac{3\sqrt{\pi}}{8} l \right] \tag{A.61}$$

$$= 2\pi \frac{N}{l^3 \pi^{3/2}} \left(-\frac{3\sqrt{\pi}}{8} l \right). \tag{A.62}$$

This simplifies to,

$$E_{\text{kin}} = -\frac{3N}{4l^2}. \tag{A.63}$$

Putting all of this together gives,

$$E(N, l) = -\frac{3N}{4l^2} - \frac{3N^2}{4\sqrt{2}\, \pi^{3/2} l^3} + \frac{2\sqrt{2}}{5\sqrt{5}} \frac{N^{5/2}}{\pi^{9/4} l^{9/2}}$$

This is the energy of a Gaussian quantum droplet of size l and particle number N. (b) The figure shows the solution to the above equation with $N = 100$ with and without LHY. In the absence of LHY, the energy is always negative, reaching a minimum towards $l \to 0$, i.e. the state is collapsing. Beyond mean-field effects instead stabilise the solution, with a numerically found minimum at finite width $l \approx 1.9624$ and $E \approx -52.8268$ (Fig. A.3).

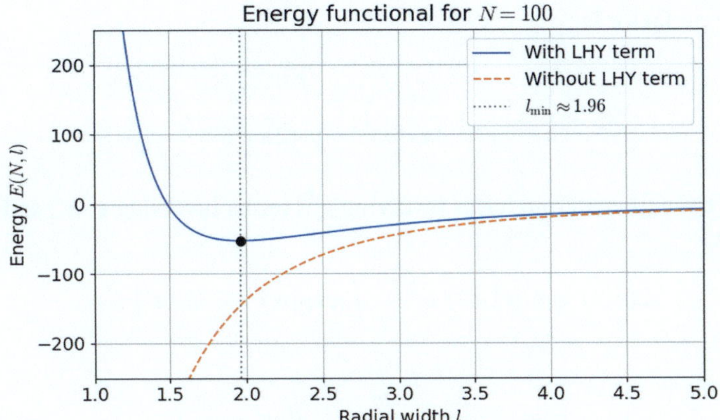

Fig. A.3 Energy minimum widths for a Gaussian ball in free space, with and without beyond-mean-field effects

13.2 *Interference pattern of dipolar droplets in TOF:*
(a) We are given the initial Gaussian wavepacket

$$\psi(\mathbf{r}, 0) = \sqrt{\frac{N}{(2\pi)^{3/2}\sigma^3}} \, \exp\left[-\frac{r^2}{4\sigma^2} + i\phi_0\right],$$

and the 3D free particle propagator (with $t_0 = 0$)

$$K(\mathbf{r}, t; \mathbf{r}', 0) = \left(\frac{m}{2\pi i \hbar t}\right)^{3/2} \exp\left[\frac{im(\mathbf{r} - \mathbf{r}')^2}{2\hbar t}\right].$$

The time evolved wavefunction is

$$\psi(\mathbf{r}, t) = \int K(\mathbf{r}, t; \mathbf{r}', 0)\, \psi(\mathbf{r}', 0)\, \mathrm{d}^3 r'.$$

Substituting both expressions gives

$$\psi(\mathbf{r}, t) = \left(\frac{m}{2\pi i \hbar t}\right)^{3/2} \sqrt{\frac{N}{(2\pi)^{3/2}\sigma^3}} \, e^{i\phi_0} \int \exp\left[-\frac{r'^2}{4\sigma^2} + \frac{im(\mathbf{r} - \mathbf{r}')^2}{2\hbar t}\right] \mathrm{d}^3 r'. \tag{A.64}$$

Expand the propagator term:

$$(\mathbf{r} - \mathbf{r}')^2 = r^2 + r'^2 - 2\mathbf{r} \cdot \mathbf{r}'.$$

Insert into the equation for ψ,

$$\psi(\mathbf{r}, t) = C(t)\, e^{i\phi_0} \int \exp\left[-\frac{r'^2}{4\sigma^2} + \frac{im}{2\hbar t}(r^2 + r'^2 - 2\mathbf{r} \cdot \mathbf{r}')\right] \mathrm{d}^3 r', \tag{A.65}$$

where the prefactor is

$$C(t) = \left(\frac{m}{2\pi i \hbar t}\right)^{3/2} \sqrt{\frac{N}{(2\pi)^{3/2}\sigma^3}}.$$

Group the exponent into terms involving r'^2, terms involving $\mathbf{r} \cdot \mathbf{r}'$, and the constant term,

$$\psi(\mathbf{r}, t) = C(t) \, e^{i\phi_0} \, e^{\frac{imr^2}{2\hbar t}} \int \exp\left[\alpha \, r'^2 + \boldsymbol{\beta} \cdot \mathbf{r}'\right] \mathrm{d}^3 r',$$

with

$$\alpha = -\frac{1}{4\sigma^2} + \frac{im}{2\hbar t}, \qquad \boldsymbol{\beta} = -\frac{im}{\hbar t}\,\mathbf{r}.$$

In order to proceed it is convenient to complete the square in the exponential. We write the integral as,

$$\int \exp\left[\alpha r'^2 + \boldsymbol{\beta} \cdot \mathbf{r}'\right] \mathrm{d}^3 r'.$$

Complete the square,

$$\alpha r'^2 + \boldsymbol{\beta} \cdot \mathbf{r}' = \alpha\left(\mathbf{r}' + \frac{\boldsymbol{\beta}}{2\alpha}\right)^2 - \frac{\boldsymbol{\beta}^2}{4\alpha}.$$

Thus,

$$\int e^{\alpha r'^2 + \boldsymbol{\beta} \cdot \mathbf{r}'} \mathrm{d}^3 r' = \exp\left[-\frac{\boldsymbol{\beta}^2}{4\alpha}\right] \int \exp\left[\alpha\left(\mathbf{r}' + \frac{\boldsymbol{\beta}}{2\alpha}\right)^2\right] \mathrm{d}^3 r'.$$

Shift the integration variable. The remaining Gaussian integral is standard,

$$\int e^{\alpha r'^2} \mathrm{d}^3 r' = \left(\frac{\pi}{-\alpha}\right)^{3/2} \qquad (\Re\, \alpha < 0),$$

which holds since, $\Re\, \alpha = -1/(4\sigma^2) < 0$.

Hence,

$$\int e^{\alpha r'^2 + \boldsymbol{\beta} \cdot \mathbf{r}'} \mathrm{d}^3 r' = \left(\frac{\pi}{-\alpha}\right)^{3/2} \exp\left[-\frac{\boldsymbol{\beta}^2}{4\alpha}\right].$$

Evaluating $\boldsymbol{\beta}^2$ and α explicitly gives,

$$\boldsymbol{\beta}^2 = \left(\frac{m}{\hbar t}\right)^2 r^2,$$

and

$$\frac{\beta^2}{4\alpha} = \frac{\left(\frac{m}{\hbar t}\right)^2 r^2}{4\left(-\frac{1}{4\sigma^2} + \frac{im}{2\hbar t}\right)}.$$

Factor the denominator,

$$4\alpha = -\frac{1}{\sigma^2} + \frac{2im}{\hbar t}.$$

Thus,

$$-\frac{\beta^2}{4\alpha} = -\frac{\left(\frac{m}{\hbar t}\right)^2 r^2}{-\frac{1}{\sigma^2} + \frac{2im}{\hbar t}}.$$

After algebra one obtains,

$$-\frac{\beta^2}{4\alpha} = -\frac{r^2}{4\sigma^2}\frac{1}{1 + i\frac{\hbar t}{2m\sigma^2}}.$$

Define the complex, time-dependent width,

$$\sigma(t) = \sigma\sqrt{1 + i\frac{\hbar t}{2m\sigma^2}}.$$

Then

$$-\frac{r^2}{4\sigma^2}\frac{1}{1 + i\frac{\hbar t}{2m\sigma^2}} = -\frac{r^2}{4\sigma(t)^2}.$$

Assembling the final expressions, the wavefunction is now,

$$\psi(\mathbf{r}, t) = C(t)\, e^{i\phi_0}\, e^{\frac{imr^2}{2\hbar t}} \left(\frac{\pi}{-\alpha}\right)^{3/2} \exp\left[-\frac{r^2}{4\sigma(t)^2}\right].$$

One may show by direct simplification that,

$$C(t)\left(\frac{\pi}{-\alpha}\right)^{3/2} = \sqrt{\frac{N}{(2\pi)^{3/2}\sigma(t)^3}}.$$

Thus the final answer is a Gaussian with a time-dependent width,

$$\psi(\mathbf{r}, t) = \sqrt{\frac{N}{(2\pi)^{3/2}\sigma(t)^3}} \exp\left[-\frac{r^2}{4\sigma(t)^2} + i\frac{mr^2}{2\hbar t} + i\phi_0\right],$$

with

$$\sigma(t) = \sigma\sqrt{1 + i\frac{\hbar t}{2m\sigma^2}}.$$

Taking the modulus shows the real spreading width,

$$|\sigma(t)|^2 = \sigma^2 \left[1 + \left(\frac{\hbar t}{2 m \sigma^2} \right)^2 \right].$$

This is the full, exact time-dependent solution of a 3D Gaussian wavepacket in free space under the free-particle propagator.

(b) Since the three droplets are identical and equally spaced from the origin, the wavefunction amplitude at the origin will be the same for each droplet—differing only by a phase factor. Therefore, we may write

$$\Psi(\mathbf{0}, t) = \sum_{j=1}^{3} F(\mathbf{0}, t) e^{i\phi_{0,j}} = F(\mathbf{0}, t) \sum_{j=1}^{3} e^{i\phi_{0,j}},$$

for some unknown function $F(\mathbf{0}, t) > 0$, whose exact form is not needed for our purposes.

Thus, the amplitude at the origin is determined entirely by the sum of the initial phase factors.

Case 1: All droplets in phase, $\phi_{0,j} = 0$:

$$\sum_{j=1}^{3} e^{i\phi_{0,j}} = 1 + 1 + 1 = 3,$$

so the central amplitude remains finite at all times. This is a hallmark of phase coherence and superfluidity, a key signatures of the supersolid state.

Case 2: Droplets with a 2π phase winding around the triangle: for an equilateral triangle centered at the origin, define the droplet positions as

$$\mathbf{r}_j = R \begin{pmatrix} \cos\phi_{0,j} \\ \sin\phi_{0,j} \\ 0 \end{pmatrix},$$

where $\phi_{0,1} = \varphi$, $\phi_{0,2} = 2\pi/3 + \varphi$, and $\phi_{0,3} = 4\pi/3 + \varphi$ for arbitrary offset φ. Then,

$$\sum_{j=1}^{3} e^{i\phi_{0,j}} = e^{i\varphi} + e^{i(2\pi/3+\varphi)} + e^{i(4\pi/3+\varphi)}$$

$$= e^{i\varphi} \left(1 + e^{i2\pi/3} + e^{i4\pi/3} \right) = 0.$$

Thus, the three contributions destructively interfere at the origin when a 2π winding is present. This implies that the presence or absence of a central density peak after time-of-flight expansion is a direct signature of phase winding, providing a robust method to distinguish supersolids with and without vortices.

Index

A
Annihilation, 175, 252, 273
Arc length, 147
Atomic parameters, 36, 215

B
Beyond-mean-field effects, 259
Binormal, 147
Biot-Savart law, 146
Bogoliubovâ£"de Gennes method, 60
Boiling, 106
Bose-Einstein Condensate (BEC), 4, 28
  atomic, 5
  collapse, 46, 63, 93, 95, 97
  energy, 37, 38, 43, 45
  ideal gas, 3, 19, 22
  macroscopic wavefunction, 35
  mass, 36
  momentum, 37
  one-dimensional, 46
  two-dimensional, 48
Bose-Einstein condensation, 3, 19, 22
Bosons, 3, 17
Breather, 98, 169
Bright soliton, 87, 262
  collapse, 95
  collisions, 88, 90, 95
  dipolar BEC, 254
  energy, 87, 94
  formation, 97
  in 3D, 93
  in an optical lattice, 97
  integrals of motion, 99, 101
  interaction with a step, 90
  observation, 97
  particle model, 91
  solution, 87
  supersolid, 277
  trains, 97
  trapped system, 91
  variational solution, 94

C
Cauchy-Riemann relations, 142
Chemical potential, 15, 37, 38, 43
Circulation, 124
Classical
  turbulence, 189
Classical particles, 12
  distinguishability, 12
  statistics, 14
Coflow, 189
Collapse, 46, 63, 93, 95, 97, 217, 220, 260, 262, 264
Collective modes, 65
Complex potential, 142
Condensate fraction, 23, 27, 108, 114
Counterflow, 189
  instability, 228
  thermal, 112
  turbulence, 187
Critical number of particles, 22
Critical temperature, 5
Critical velocity, 115, 163
Curvature, 147

Pressure, 39
  degeneracy pressure, 6, 29
  ideal gas, 25
  quantum pressure, 39
Principle of equal a priori probabilities, 14
Pseudo-vorticity, 136
Pure superflow, 189

Q

Quantum droplet, 261, 262, 267
  and collapse, 262
  Bose-Bose system, 261
  dipolar gas, 262
Quantum fluctuations, 259
  Lee-Huang-Yang correction, 259
Quantum of circulation, 124
Quantum particles, 16
  indistinguishability, 17
  statistics, 17
Quantum reflection, 90
Quasi-one-dimensional BEC, 46
Quasi-two-dimensional BEC, 48
Quench, 200, 275

R

Rayleigh-Taylor instability, 231
Reconnection, 168, 171, 173, 202
Reimann zeta function, 21
Reynolds number, 190, 201
Rosensweig instability, 264
Rotation, 123
  in a bucket, 130, 131
  in a harmonic trap, 134
Roton, 63, 116, 246, 264
  instability, 266
Rydberg atoms, 238

S

Sbright oliton
  collapse, 97
Scalar potential, 142
Scaling solutions, 65
Scattering length, 34
Schwarz equation, 148
Second sound, 110
Snake instability, 84, 254
Solid body rotation, 125
Soliton, 73, 74
  bright-dark, 224
  bright solitons, 87
  characteristics, 74
  dark solitons, 75
  dipolar BEC, 253

energy, 74
  Hasimoto soliton, 169
  higher-order, 98
  integrals of motion, 74
  Jones-Roberts soliton, 175, 206, 207
  momentum, 74
  norm, 74
  oscillations, 78, 81, 254
  particle model, 76, 78, 91
  snake instability, 254
  supersolid, 277
  trapped systems, 254
  two-component condensates, 223
  two-component solitons, 223
Solitonic vortex, 180
Sound, 61, 63
  absorption, 83
  emission, 175
  first sound, 110
  second sound, 110
  speed of sound, 64
Sound emission, 79
Specific heat, 107
Split-step Fourier method, 281
State
  classical, 12
  excited state, 14
  ground state, 14
  quantum state, 17
Stokes drag, 152
Stokes theorem, 127
Stream function, 142
Superconductivity, 2
Superfluid component, 107
Superfluid fraction, 268
Superfluidity, 2, 106, 113, 115
Superfluid Reynolds number, 201
Supersolid, 264
  dipolar, 266
  one-dimensional, 267
  synchronisation, 276
  two-dimensional, 269
  vortex lattice, 273
  vortices, 270
Surface tension, 223

T

Tangent, 147
Tangle, 189
Temperature
  critical, 23
  Fermi temperature, 29
  negative, 209
Thermal gas, 25, 28

If you have any concerns about our products,
you can contact us on
ProductSafety@springernature.com

In case Publisher is established outside the EU,
the EU authorized representative is:
Springer Nature Customer Service Center GmbH
Europaplatz 3, 69115 Heidelberg, Germany

Printed by Libri Plureos GmbH
in Hamburg, Germany